Fifth International Conference on Physics, Mathematics and Statistics

Journal of Physics: Conference Series Volume 2313

Online
21-22 May 2022

Editor:

Wanyang Dai

ISBN: 978-1-7138-6035-8
ISSN: 1742-6588

Printed from e-media with permission by:

Curran Associates, Inc.
57 Morehouse Lane
Red Hook, NY 12571

Some format issues inherent in the e-media version may also appear in this print version.

Institute of Physics
Dirac House, Temple Back
Bristol BS1 6BE UK

Phone: 44 1 17 929 7481
Fax: 44 1 17 920 0979

techtracking@iop.org

Additional copies of this publication are available from:

Curran Associates, Inc.
57 Morehouse Lane
Red Hook, NY 12571 USA
Phone: 845-758-0400
Fax: 845-758-2633
Email: curran@proceedings.com
Web: www.proceedings.com

TABLE OF CONTENTS

Journal of Physics: Conference Series **2313** (2022) 011001 doi:10.1088/1742-6596/2313/1/011001

Preface

Wanyang Dai

Editor, Proceedings of the Fifth International Conference on Physics, Mathematics and Statistics

Professor, Department of Mathematics, Nanjing University, China

Email: nan5lu8@nju.edu.cn

Since its inception in 2018, ICPMS (International Conference on Physics, Mathematics and Statistics) were successfully held for the last five years, Shanghai (2018), Hangzhou (2019), and online (2020, 2021, 2022), attracted delegates from more than 10 countries and regions including China, Thailand, Japan, the United States, South Korea. As an annual international conference, ICPMS aims at being a fast and efficient platform for researchers and scholars worldwide to discuss recent developments in the area of Physics, Mathematics and Statistics.

The Fifth International Conference on Physics, Mathematics and Statistics (ICPMS2022) was successfully held on May 21-22, 2022. Under the effect of COVID-19, the conference was chosen to be online, which gave everyone access to attend. This year the online meeting was composed of three sessions: Keynote Speeches, Oral Presentations and Poster Presentations, attracted about 50 delegates from all over the world. Details about the presentations can be found in the Conference Overview part.

This conference proceedings included 33 accepted articles selected from 71 submissions, all the papers have been through rigorous review and process to meet the requirements of International publication standard. We would like to express our gratitude to the reviewers of these manuscripts, who provided constructive criticism and stimulated comments and suggestions to the authors. We are extremely grateful as organizers, technical program committee and editors and extend our most sincere thanks to all the authors for their excellent contribution and work. Our sincere gratitude also goes to the IOP Publishing editors and managers for their helpful cooperation during the preparation of the proceedings.

On behalf of the Organizing Committees of ICPMS2022.

Journal of Physics: Conference Series **2313** (2022) 011001 doi:10.1088/1742-6596/2313/1/011001

Technical Program Committee

Conference Chair
Wanyang Dai, Nanjing University, China

Conference Co-chair
Yingxu Wang, University of Calgary, Canada

TPC Chair
Ping Xiang, Central South University, China

TPC member
Huijin Xu, Shanghai Jiao Tong University, China
Boris I. Gramatikov, Johns Hopkins University, USA
Irena Lasiecka, The University of Memphis, USA
Chingiz Hajiyev, Istanbul Technical University, Turkey
Florentin Smarandache, University of New Mexico, USA
S. Ponnusamy, Indian Institute of Technology, India
Xueping Chen, Jiangsu University of Technology, China
Gonçalo dos Reis, University of Edinburgh, UK
Tadeusz Kosztolowicz, Jan Kochanowski University in Kielce, Poland
Jin Peng, Huanggang Normal University, China
Quanxin Zhu, Hunan Normal University, China
Hongwei Wu, University of Hertfordshire, UK
Yang Yu, University of Technology Sydney (UTS), Australia
Jin Li, University of Cambridge, UK
M Shadi Mohamed, Heriot-Watt University, UK
Andry Sedelnikov, Samara National Research University, Russia
Huamin Li, University at Buffalo, The State University of New York, USA

Journal of Physics: Conference Series **2313** (2022) 011001 doi:10.1088/1742-6596/2313/1/011001

Conference Overview

The Fifth International Conference on Physics, Mathematics and Statistics (ICPMS2022) was successfully held on May 21-22, 2022. Under the effect of COVID-19, the conference was chosen to be online, which gave everyone access to attend. This year the online meeting was composed of three sessions: Keynote Speeches, Oral Presentations and Poster Presentations, attracted about 50 delegates from all over the world.

Conference Schedule

Date	Time	Program	Online Platform
May 21	10:00-17:30	Online Registration	VooV Meeting (腾讯会议) **ID: 957 6123 9484**
May 22	09:00-09:10	Opening Ceremony	
	09:10-12:00	Keynote Speech	
	14:00-16:00	Oral Presentation	
	16:00-16:30	Poster Presentation	

At the opening ceremony, Prof Wanyang Dai from Nanjing University, extended a warm welcome to all online participants and looked back on the development of ICPMS series over the past four years.

Journal of Physics: Conference Series **2313** (2022) 011001 doi:10.1088/1742-6596/2313/1/011001

Keynote Speeches

It was our great honor to have the following impressive keynote speeches during the conference.

Prof Lixing Zhou from Tsinghua University, China
Speech Title: Advances in Validation of the Second-Order Moment Combustion Model by Direct Numerical Simulation

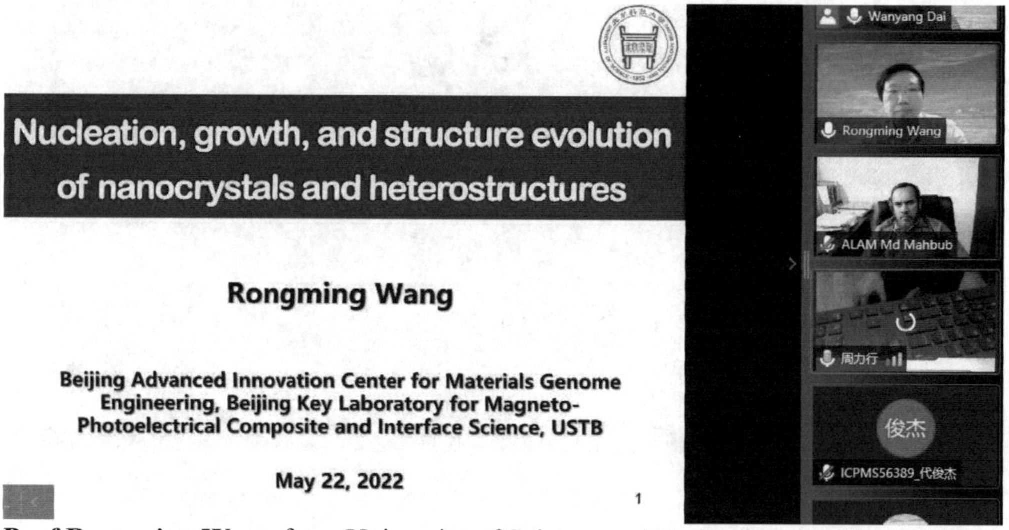

Prof Rongming Wang from University of Science and Technology Beijing, China
Speech Title: Nucleation, growth, and structure evolution of nanocrystals and heterostructures

Journal of Physics: Conference Series **2313** (2022) 011001 doi:10.1088/1742-6596/2313/1/011001

Mathematical modeling the influence of nanomaterials' morphology in bioassays

Professor Paulo De Morais

Catholic University of Brasília

&

University of Brasilia

ICPMS22 May 2022

Prof Paulo César DE MORAIS from University of Brasilia, Brazil
Speech Title: Mathematical modelling the influence of nanomaterials' morphology in bioassays

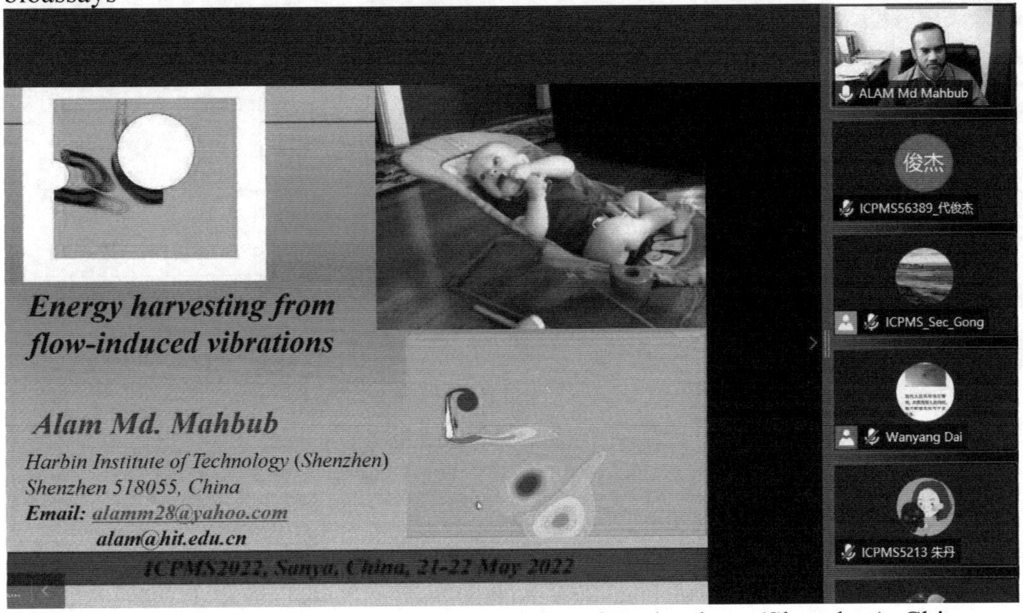

Prof Alam Md. Mahbub from Harbin Institute of Technology (Shenzhen), China
Speech Title: Energy from flow-induced vibration

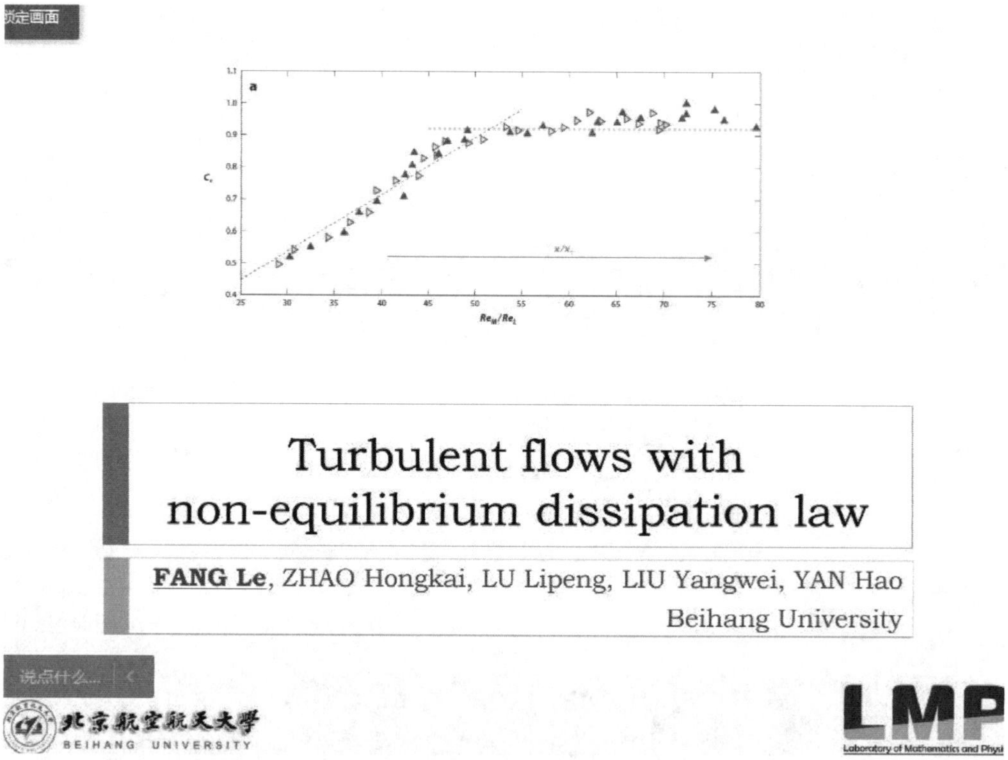

Prof Le FANG from Beihang University, China
Speech Title: Turbulent flows with non-equilibrium dissipation law

Oral Presentations

Following the keynote speakers, 7 scholars made the oral presentations respectively in the afternoon. Information of speakers and their speeches can be found as below.

Speech Title	Speaker	Affiliation
Mechanisms of Ultrafast Processing with Temporally Shaped Femtosecond Laser Ablation of Metal	Jingya Sun	Beijing Institute of Technology
Lax-Wendroff method for incompressible flow	Md Mizanur Rahman	Hangzhou Dianzi University
Efficient third-harmonic generation control with dielectric nanostructures of designated rotational symmetry	Bingyi Liu	Beijing Institute of Technology
Construction and properties analysis of Z-A-Z graphene	Hanxi Wang	Hefei University of Technology

Journal of Physics: Conference Series **2313** (2022) 011001 doi:10.1088/1742-6596/2313/1/011001

Study on thrust performance of small water rocket launch	Yuqi Yang	National University of Defense Technology
Identification and analysis of elements in liquid by using laser-induced breakdown spectroscopy	Ali Khumaeni	Diponegoro University
Haze spectra exhibited by human cells	Carlos Torres-Torres	Instituto Politecnico Nacional

Poster Presentations

Sixteen poster presentations came after oral presentations. Below is a brief list of presenters and their poster titles.

Presenter	Affiliation	Presentation Title
Haodi Ji	Beihang University	A reliability estimation method for spring group balanced system at different degradation temperatures
Q Zhang	Xi'an University of Technology	Analysis of discharge characteristics of pulsed plasma power supply in the microsecond scale
Xiangyu Zhang	Zhejiang Light Industrial Products Inspection and Research Institute	Radiation assessment of household appliances and stone materials from residences in Hangzhou
T Z Tian	Beihang University	Reliability evaluation of balanced systems with common bus performance sharing considering balance degree base on battery storage systems
Junchi Zhou	Meteorological Disaster Protection Center of Jiangsu	A study of lightning electromagnetic effect on the metal roof constructions
Yuan-Shuo Liu	University of Jinan	Theoretical simulation on Fraunhofer diffraction of arbitrarily shaped aperture
Lei Zhang	University of Science and Technology of China	Simulation system of lighting environment for optical imaging test
Zhihuan Wang	Tsinghua University	An analytical method for the spherical stress wave equation in linear hardening materials
Xiaomei Hu	Anhui University	Upper critical field beyond Pauli limit in the exfoliated $RbCr_3As_3$ nanowires

Journal of Physics: Conference Series **2313** (2022) 011001 doi:10.1088/1742-6596/2313/1/011001

Yunfeng Yang	Xi'an University of Science and Technology	Synchronization time in two coupled cluster networks based on discontinuous map
Ronghui Yuan	Zhejiang Light Industrial Products Inspection and Research Institute	Anticorrosive behavior assessment of electrodeposited zinc salicylate coating on carbon steel
W Yang	China Institute of Nuclear Industry Strategy	Uncertainty analysis and visualization for a coupled thermohydraulics-neutronics simulation by deterministic sampling
Xiuxiu Gao	CORESING SEMICONDUCTOR TECHNOLOGY CO. LTD	Simulation of gradient floating field limiting rings for 4H-SiC power devices
Z J Chen	South China University of Technology	A novel voltage protection method for multi-cell Lithium-ion battery protection IC
Y H Liu	China Academy of Space Technology	Numerical analysis and optimization of radio-frequency ion trap potential
Hanxi Wang	Hefei University of Technology	Construction and properties analysis of Z-A-Z graphene nanoribbons transistors

Closing Ceremony
At the closing ceremony, Prof Wanyang Dai made closing remarks and announced that the next conference will be held at Sanya during May 20-22, 2023. Thanks for the support of all attendees, ICPMS2022 concluded with a fruitful success.

Journal of Physics: Conference Series **2313** (2022) 011002 doi:10.1088/1742-6596/2313/1/011002

Peer Review Statement

All papers published in this volume have been reviewed through processes administered by the Editors. Reviews were conducted by expert referees to the professional and scientific standards expected of a proceedings journal published by IOP Publishing Publishing.

- **Type of peer review:** Double Anonymous
- **Conference submission management system:** Morressier
- **Number of submissions received:** 71
- **Number of submissions sent for review:** 48
- **Number of submissions accepted:** 33
- **Acceptance Rate (Submissions Accepted / Submissions Received × 100):** 46.5
- **Average number of reviews per paper:** 3
- **Total number of reviewers involved:** 207
- **Contact person for queries:**
  Name: Lily Gong
  Email: secretary@pmsconf.org
  Affiliation: Wuhan Bencheng Culture Spreads Co., Ltd

Journal of Physics: Conference Series **2313** (2022) 012001 doi:10.1088/1742-6596/2313/1/012001

Impact of Pitot tube diameter on the centre line mean flow velocity in multi-scale grid-generated turbulence

Hiroto Yamaguchi[1], Hiroki Suzuki[1, 2], Kento Tanaka[1] and Toshinori Kouchi[1]

[1] Graduate School of Natural Science and Technology, Okayama University, 3-1-1 Tsushima-naka, Kita-ku, Okayama-shi, Okayama 700-8530, Japan

[2] Corresponding author's e-mail: h.suzuki@okayama-u.ac.jp

Abstract. In this study, the impact of the leading-edge diameter of a Pitot tube on the mean flow of the decaying turbulence generated by a multi-scale grid was investigated. Standard Pitot tubes with leading edge diameters of 4 and 6 mm were used in this study, and a multi-scale grid was installed at the entrance of the test section to generate the decaying turbulence. The characteristic velocity was observed to be 2.8 m/s. In the downstream region of the test section, the mean velocities measured with the two Pitot tubes were in sufficiently good agreement, as well as the acceleration parameters calculated in this region. However, in the upstream region, the mean flow velocities obtained with the two Pitot tubes were slightly different. The thin Pitot tube showed a higher mean flow velocity in this region, corresponding to the jet-like structure that has been described in previous research. In contrast, this higher mean velocity in the upstream region was not observed in the mean flow velocity distribution obtained with the thick Pitot tube. This difference is discussed here from the perspective of the turbulent component and measurement uncertainty.

1. Introduction

Turbulent flow phenomena in the field of fluid mechanics are widely observed in fluids engineering devices [1]. Moreover, these flows are often of an incompressible nature. Therefore, understanding incompressible turbulence phenomena is essential for engineering purposes. Of particular interest is the case of flows with a high Reynolds number, which have a greater likelihood of universality at small scales. Large-eddy simulations are often used to evaluate this small-scale universality, and in recent years, the use of fluid analysis solvers, such as Open FOAM, has also become commonplace. Although these solvers are sometimes validated with canonical flows (e.g., [2]), the most commonly used numerical analysis method is a comparison with experimental results obtained under similar conditions. Therefore, in this field of study, experimental measurements are necessary to validate the accuracy of numerical analyses. In the analysis of turbulent flow fields, the mean velocity, which is one of the most fundamental measurements of the flow field, must be measured with extremely high accuracy.

Over the past 15 years, decaying turbulence generated by multi-scale turbulent grids has been a particular area of focus [3–9]. This problem has been approached using wind tunnel experiments and numerical simulations. One of the features of this decaying turbulence is that its dissipation characteristics are qualitatively different from those of other types of flows. Under high Reynolds number conditions, the dissipation constant is constant, whereas in multi-scale decaying turbulence, this dissipation constant is small and still depends on the Reynolds number. When the

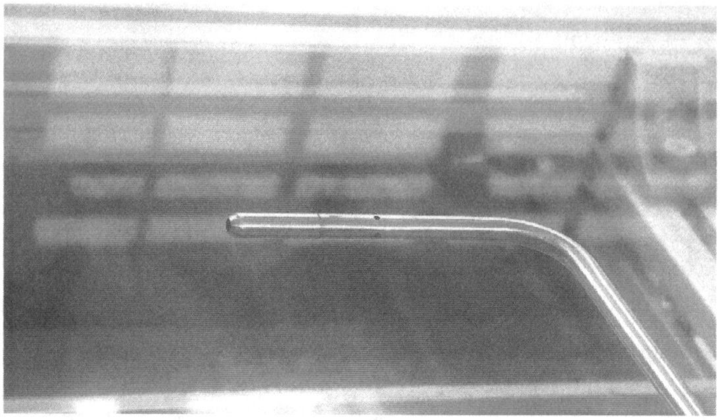

Figure 1. Schematic of the fine Pitot tube set in the present test section. Here the flow is left to right.

mean flow velocity is not constant in the flow direction, the acceleration of the mean flow may affect the decay characteristics of the turbulence. The impact of weak fluid acceleration due to wind tunnel blockage on turbulent decay has been studied (e.g., [10,11]). To accurately measure the acceleration of the mean flow, the velocity profiles of the mean flow should be measured with sufficient accuracy.

Hot-wire anemometers and Pitot tubes are commonly used to measure velocities in flow fields [12]. The hot-wire anemometer measures both the mean velocity and the velocity variation in the flow field, whereas the Pitot tube measures only the mean velocity. In previous studies, the effect of the leading edge of the sensor probe on a turbulent field was investigated in the case of a hot-wire anemometer (e.g., [13]). The present study focuses on the fact that the edge of a Pitot tube is larger than that of a hot wire probe. This suggests that the sensitivity of the measured flow field to the size of the Pitot tube edge should be examined to accurately investigate the characteristics of the turbulence produced by the multi-scale turbulence grids. Previous studies have investigated the characteristics of multi-scale grid-generated turbulence along the centre line [7,9]. Therefore, the purpose of this study is to examine the impact of different Pitot tube leading edge diameters on the observed mean velocity of multi-scale grid-generated turbulence along the centre line.

2. Experimental methods

Present experiments were conducted in a wind tunnel located in the Fluid Mechanics Laboratory of Okayama University. The test section had cross-sectional dimensions of 400 mm × 400 mm. Here, the characteristic length D was the length of one side in the cross section of this test section. At the entrance of the test section, a multi-scale turbulence grid was installed. A coordinate system was defined with its origin located at the centre of this turbulence grid plane and with the streamwise, transverse, and spanwise directions corresponding to the x-, y-, and z-directions, respectively.

A square multi-scale turbulence grid was used to generate turbulence. Here, the shape of the turbulence grid was selected on the basis of periodicity. This turbulence grid used here is similar to the one used in previous studies [7,9]. To increase the rigidity of the present turbulence grid, a conventional turbulence grid with the same grid spacing was used as the finest stage. The thickness of this turbulence grid in the flow direction was kept uniform in the plane, as in previous studies [3,9].

The impact of the leading edge diameter of the Pitot tube on the mean flow of the turbulence generated by the multi-scale turbulence-generating grid was examined. Two standard Pitot tubes [12] with diameters of 6 and 4 mm were used in this study; they are hereafter referred to as the thick and thin Pitot tubes (Figure 1), respectively. The cross-sectional area at the leading edge of the thick Pitot

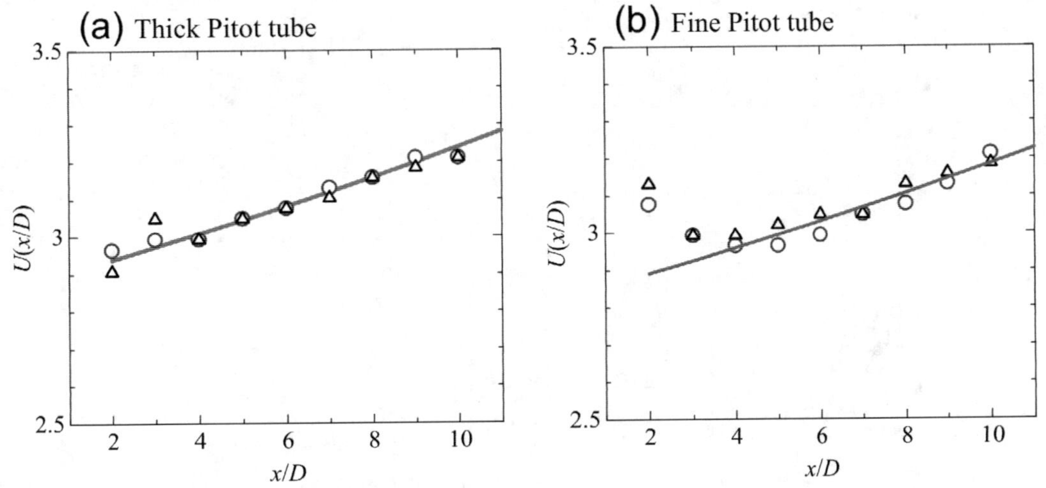

Figure 2. Streamwise mean flow velocity profiles on the centre line obtained by (a) the thick Pitot tube and (b) the thin Pitot tube. The grey lines are linear fits to the profiles.

tube is larger than that of the thin Pitot tube. The same standard differential pressure gauge was used for the measurement with both Pitot tubes. In this study, the measurement results of the thin Pitot tube were considered to be more accurate than those of the thick Pitot tube. In the next section, we compare the results obtained with the two Pitot tubes.

The streamwise velocity was set to approximately 3 m/s. In the wind tunnel used here, the streamwise velocity is set via the rotation speed of the motor. Therefore the setting of the streamwise velocity produces some uncertainty. As detailed in the results, the value of the characteristic velocity was measured to be approximately 2.85 m/s. The bulk Reynolds number Re computed from this characteristic velocity and the characteristic length, which is the width of the cross-section of the test section, is then approximately 76,000.

3. Experimental results

Figure 2(a) shows the mean velocity distribution U on the centre line obtained with the thick Pitot tube in the direction of the flow velocity. Here, the results of a second measurement are also shown for reference. As shown in this figure, the two sets of results are in good agreement with each other. These results demonstrate that as the normalized distance along the tunnel x/D increases, the mean flow velocity in the flow direction increases. Therefore, the mean velocity does not remain constant along the centre line of the turbulence generated by the multi-scale grid. This result is qualitatively consistent with previously obtained results [9]. The mean flow velocity was also found to increase linearly with x/D in the case of the thick Pitot tube. This linear increase in the mean flow velocity in the direction of the flow velocity suggests that the production term in the turbulent kinetic energy equation is not quite zero. This production term consists of the longitudinal first derivatives of the velocity fluctuation intensity and the mean velocity. On the centre line, the Reynolds stress is zero because of the symmetry of the flow field. Therefore, the generation term due to Reynolds stress in the turbulent kinetic energy equation is considered to be sufficiently small.

Figure 2(b) shows the mean velocity distribution in the direction of the flow obtained with the thin Pitot tube. As with Figure 2(a), two sets of measurements are shown here. As demonstrated in this figure, in the downstream region, the mean flow velocity in the flow direction obtained with the thin Pitot tube increases with increasing x/D, as with the thick Pitot tube. This result is in qualitative agreement

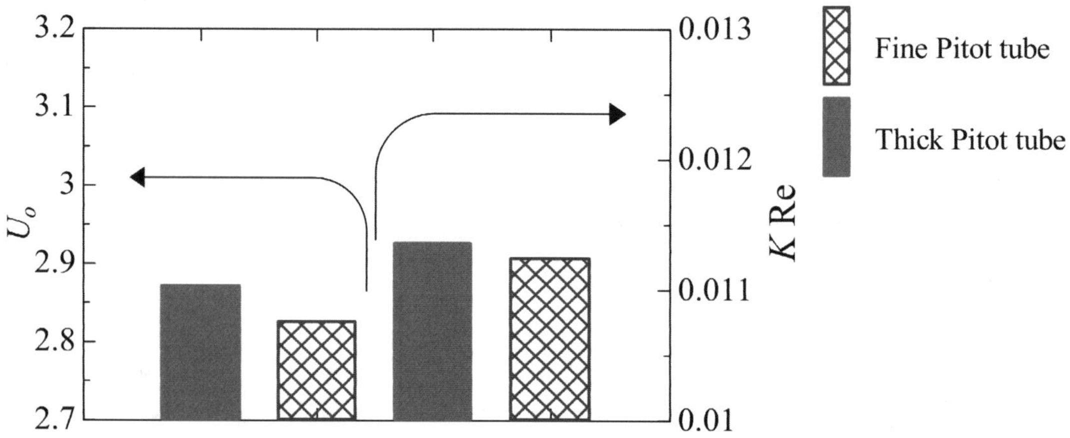

Figure 3. Values of obtained characteristic velocity and acceleration quantified by *K*Re.

with the results of the previous studies (e.g., [9]). As shown in the figure, the uncertainty of the measurements obtained by the Pitot tubes in the downstream region is estimated to be about ±1%. Since we can consider this magnitude of uncertainty to be comparable to the general uncertainty of Pitot tube measurements, the study can consider that the values in the downstream region are sufficiently accurate. However, in the upstream region, the trend obtained with the thin Pitot tube deviates from that obtained with the thick Pitot tube; in this region of low x/D, the mean flow velocity obtained with the thin Pitot tube increases with decreasing x/D. Here, in this study, we refer to the region behind the grid as the upstream region, as opposed to the sufficiently downstream region. This behaviour in the upstream region has also been observed in the previous studies (e.g., [9]). The increased velocity in the upstream region may indicate the presence of the jet-like structure seen in previous studies (e.g., [8-9]).

In previous studies, the acceleration parameter has often been used to quantify the fluid acceleration in the streamwise direction [10,14]. The acceleration parameter K is used to measure the turbulence generated by the grid in addition to the turbulent boundary layer and is defined as

$$K = \frac{\nu}{U(x)^2} \frac{\mathrm{d}U}{\mathrm{d}x}, \tag{1}$$

where ν is the kinematic viscosity. This equation can be further rewritten using the representative velocity and length scales. Here, the mean velocity in the freestream direction at $x/D = 0$ is taken to be the characteristic velocity U_0, and the length of one side of the cross section of the test section is taken as the characteristic length D. This definition of the acceleration parameter can be considered as a first-order ordinary differential equation, and it can be solved analytically. With this characteristic velocity, the velocity distribution represented by the acceleration parameter can be written as follows

$$U(x) = \frac{U_0}{1 - (x/D)\, K\, (U_0 D/\nu)}. \tag{2}$$

In the present study, the value of the acceleration parameter was obtained from the distribution of the mean velocity.

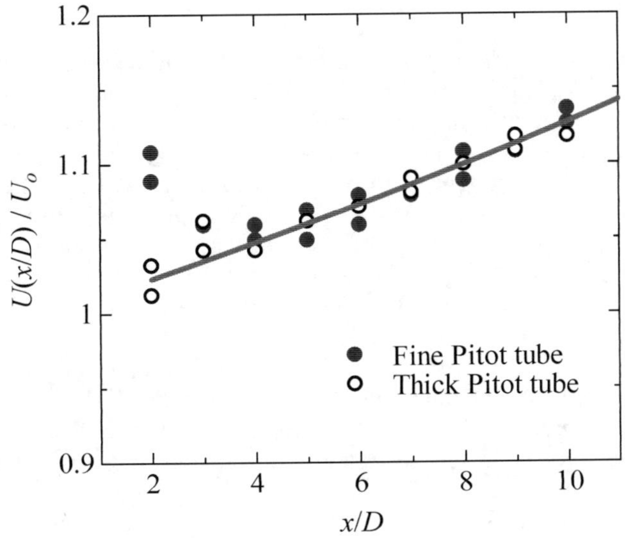

Figure 4. Profiles of normalised mean flow velocity in the streamwise direction. The grey solid line represents the derived velocity function fit to the measured velocity data. Here, the set values of the acceleration are given as the mean for the two conditions.

The function given in Eq. (2), which describes the distribution of the mean velocity derived from the definition of the acceleration parameter, reproduces the measurement results with sufficient accuracy in the downstream region, as shown in Figure 2. This was observed in the results obtained for both the thick and thin Pitot tubes. Fitting Eq. (2) to the measurement results allows the characteristic velocity U_0 and the acceleration, quantified as KRe, to be obtained from the measurement data. The characteristic velocity and acceleration obtained in this way are shown in Figure 3 for the thick and thin Pitot tubes. Here, in Figure 3, the mean velocity values in the mainstream direction and the acceleration rate in the upstream region are compared between fine and thick Pitot tubes. As shown in this figure, the characteristic velocities and accelerations obtained with the thick and thin pitot tubes are generally in good agreement. The small difference can be attributed to the uncertainty of the rotational speed of the wind tunnel motor, which sets the freestream velocity. In the case of the fine Pitot tube, it was considered that the velocity measured in the upstream region is significantly higher than the fitted velocity distribution function, as shown in Figure 2(b), whereas this deviation from the fitted velocity distribution was not observed in the case of the thick Pitot tube.

Figure 4 shows the distribution of the mean velocity in the streamwise direction normalised by the characteristic velocity. This normalisation by the characteristic velocity specific to each Pitot tube allows the two velocity distributions to be compared quantitatively. Here, the difference in the characteristic velocities between the two sets of experimental conditions is considered to be due to the uncertainty of the streamwise velocity setting in the main wind tunnel. As shown in Figure 4, the velocity distribution obtained with the thick Pitot tube corresponds to that obtained with the thin Pitot tube in the downstream region. This agreement suggests that the streamwise velocity values observed in the downstream region are likely insensitive to the diameter of the Pitot tube. However, in the upstream region, the results obtained by the thick and thin Pitot tubes were significantly different. The higher upstream velocities obtained with the thin Pitot tube are larger than the considered measurement uncertainty. Thus, it was concluded that this increased flow velocity in the upstream region is similar to that observed previously [9].

4. Discussion and conclusion

4.1. Discussion

The results obtained in this study are discussed in terms of the influence of turbulence on the results measured by the Pitot tubes [12]. If the free flow contains fluctuating components, the observed velocity obtained from the Pitot tube may be slightly larger than the actual velocity. In particular, in grid-generated turbulence, the intensity of turbulent fluctuations is larger in the upstream region than in the downstream region. This difference can be estimated as the order of the mean square of the relative velocity fluctuations. In a previous study where grid-generated turbulence generated by a multi-scale grid was experimentally measured, the relative root mean square of the velocity fluctuations in the direction of flow velocity in the upstream region was reported to be approximately 10% [9]. From this, the increase in the measured velocity due to the turbulent component can be estimated as approximately 1%. However, in the present study, the normalized mean velocity in the flow direction in the upstream region ($x/D = 2$) obtained with the thin Pitot tube was approximately 10% higher than that obtained with the thick Pitot tube. The magnitude of this increase is significantly larger than that of the measured velocity due to the turbulent component. This result indicates that the increase in the measured velocity in the upstream region of the figure is not due to the turbulent component.

The acceleration obtained in this study was dependent on the measurement conditions. In this study, this dependence is considered to be due to the uncertainty of the obtained acceleration. Previous studies [10,12] have applied a model in which the acceleration is used to estimate the effect of mean flow acceleration on the turbulent energy and dissipation rate. Although the definition of the acceleration used in this model is slightly different from that given by the acceleration parameters used in this and other previous studies, the results of the present study show that the accelerations obtained with the thick and thin Pitot tubes are almost the same. The effect of the acceleration uncertainty on the turbulent kinetic energy and its dissipation rate in the present study may be obtained by applying the Taylor series expansion to the form of the acceleration uncertainty in the model of the previous study. Because the obtained acceleration uncertainty is small, the effect of the uncertainty on the turbulent kinetic energy and its dissipation rate is considered insignificant.

4.2. Conclusion

The purpose of this study was to clarify the sensitivity of the measurement of the mean velocity of turbulence generated by a multi-scale grid to the diameter of the Pitot tube. If the Pitot tube is not sufficiently thin, the mean velocity of the grid turbulence generated by the multi-scale grid may be affected. In this study, measurements obtained using Pitot tubes with diameters of 6 and 4 mm in a wind tunnel with cross-sectional dimensions of 400 mm × 400 mm were compared. The flow velocity, which is considered to be the characteristic velocity, was measured to be approximately 2.8 m/s.

The distribution of the mean flow velocity in the streamwise direction was obtained along the centre line. To examine this measurement result from the point of view of the magnitude of the measurement uncertainty, two trials were conducted under the same measurement conditions for each Pitot tube. In the downstream region, the normalised streamwise mean velocities measured with the two Pitot tubes showed good agreement. The acceleration parameter was calculated for the downstream velocity distribution in the mainstream direction, and the acceleration parameters computed in the thick and thin Pitot tube cases were also in good agreement. However, there was a slight difference between the two Pitot tube measurements in the upstream area: here, the streamwise mean velocity obtained by the thin Pitot tube increased with decreasing x/D, whereas that obtained by the thick Pitot tube decreased. The magnitude of this increase was found to be sufficiently larger than that of the turbulent component, which increases in the upstream region. Here, the acceleration parameter and characteristic velocity were obtained using the formula for the velocity distribution derived from the definition of the acceleration parameter.

The present results indicate that the mean flow velocity profile is dependent on the thickness of the Pitot tube, suggesting that it is best to use a thin Pitot tube to measure the mean flow velocity. As shown in Figure 4, the mean flow velocity increases at a constant rate in the downstream region. Because the acceleration of the mean flow may affect the decay characteristics of grid-generated turbulence, this study considers that this impact should be investigated quantitatively. In particular, one of the characteristics of a decaying turbulence produced by a multi-scale turbulent grid is the dissipation of turbulent kinetic energy. Therefore, the effect of this mean flow acceleration on the dissipation of turbulent kinetic energy should be quantitatively investigated, which may be done using the model applied in previous studies.

Acknowledgments

This work was partly supported by the Japanese Ministry of Education, Culture, Sports, Science and Technology through Grants-in-Aid (Nos. 20H02069 and 21K03859). This work was supported (in part) by Grant for Environmental Research Projects from The Sumitomo Foundation.

References

[1] Pope S B 2000 *Turbulent Flows*. Cambridge University Press
[2] Komen E, Shams A, Camilo L, and Koren B 2014 Quasi-DNS capabilities of OpenFOAM for different mesh types *Comput. Fluids* 96 87-104
[3] Seoud R E, Vassilicos J C 2007 Dissipation and decay of fractal-generated turbulence *Phys Fluids* 19 105108
[4] Nagata K, Suzuki H, Sakai Y, Hayase T, Kubo T 2008 Direct numerical simulation of turbulent mixing in grid-generated turbulence *Phys Scr* T132 014054
[5] Suzuki H, Nagata K, Sakai Y, Ukai R 2010 High-Schmidt-number scalar transfer in regular and fractal grid turbulence *Phys Scr* T142 014069
[6] Suzuki H, Nagata K, Sakai Y, Hasegawa Y 2013 An attempt to improve accuracy of higher-order statistics and spectra in direct numerical simulation of incompressible wall turbulence by using the compact schemes for viscous terms *Phys Scr* 2013 T155 014062
[7] Suzuki H, Nagata K, Sakai Y, Hayase T, Hasegawa Y, Ushijima T 2013 Direct numerical simulation of fractal-generated turbulence *Fluid Dyn. Res.* 45 061409
[8] Nagata K, Sakai Y, Inaba T, Suzuki H, Terashima O, Suzuki H 2013 Turbulence structure and turbulence kinetic energy transport in multiscale/fractal-generated turbulence *Phys. Fluids* 25 065102
[9] Valente P C, Vassilicos J C 2014 The non-equilibrium region of grid-generated decaying turbulence *J. Fluid Mech.* 744 5-37
[10] Suzuki H, Mochizuki S, Hasegawa Y 2016 Validation scheme for small effect of wind tunnel blockage on decaying grid-generated turbulence *J. Fluid Sci. Tech.* 11 JFST0012
[11] Suzuki H, Mochizuki S, Hasegawa Y 2018 Numerical-based theoretical analysis on effects of weak fluid acceleration of free-stream due to wind-tunnel blockage on grid-generated turbulence *Flow Meas. Instrum.* 62 1-8
[12] Tropea C, Yarin A L, Foss J F 2007 *Springer Handbook of Experimental Fluid Mechanics*. Berlin: Springer.
[13] Tsuji T, Nagano Y 1999 Recent advances in Hot-wire measurements *Nagare* 18 191-198 (in Japanese)
[14] De Graaff D B, Eaton J K 2000 Reynolds-number scaling of the flat-plate turbulent boundary layer *J. Fluid Mech.* 422 319-346

Journal of Physics: Conference Series **2313** (2022) 012002 doi:10.1088/1742-6596/2313/1/012002

Lax-Wendroff method for incompressible flow

Md Mizanur Rahman[1,2], **K. Hasan**[1], **Zhiqian Sang**[1] **and Zing Ni**[1]

[1]Hangzhou Dianzi University, School of Mechanical Engineering, 310018 Hangzhou, China

[2]Email: mizanur_rahman@hdu.edu.cn

Abstract. The Lax-Wendroff scheme is extended to incompressible fluid flow problems within the framework of artificial compressibility method (ACM) utilizing a "cell-centered finite-volume" Δ-approximation on a non-orthogonal non-staggered grid. An ACM induces a transformation between conservative and primitive variables; the physical relevance of ACM is to bring about "matrix preconditionings", provoking density perturbations. The coupled algorithm is pressure-based; it benefits from enhanced accuracy and greater flexibility using the "monotone upstream-centered schemes for conservation laws (MUSCL)" approach. Numerical experiments in reference to "buoyancy-driven" flows with strong sources illustrate that the entire strategy augments overall damping capability and robustness adhering to the factored pseudo-time integration method. Conclusively, the associated limiter function has little influence on the high resolution for selected test cases.

Nomenclature

C = artificial sound speed
CFL = Courant number
f = source term
F, G = x and y directional flux vectors
Gr = Grashoff number
M = Mach number
Nu = Nusselt number
p = static pressure
Pr = Prandtl number
S = cell-face area
Δt = pseudo-time step
U, V = non-dimensional velocity components
β = compressibility parameter
θ = non-dimensional temperature
λ = eigenvalue
μ = dynamic viscosity
ν = kinematic viscosity
ξ = monitor function
$\forall$ = cell volume

Subscript
l = local condition
r = reference condition

1. Introduction

The first-order upwind (FOU) expression to the inviscid term in 1-D (one-dimensional) equation $\partial\phi/\partial t + u\,\partial\phi/\partial x = 0$ with $F = u\phi$ can be provided with:

$$\frac{\partial \phi}{\partial t} + \frac{1}{\Delta x}\left(\frac{F_{i+1}+F_i}{2} - \frac{F_i+F_{i-1}}{2}\right) - \frac{u}{2\Delta x}\left(\phi_{i+1}-\phi_i\right) + \frac{u}{2\Delta x}\left(\phi_i-\phi_{i-1}\right) \tag{1}$$

where Δx is the grid-length, t signifies the time, u is the velocity and ϕ implies a scalar. The corresponding expression for the Lax-Wendroff (LW) scheme becomes [1]:

$$\frac{\partial \phi}{\partial t} + \frac{1}{\Delta x}\left(\frac{F_{i+1} + F_i}{2} - \frac{F_i + F_{i-1}}{2}\right) - \frac{u^2 \Delta t}{2\Delta x^2}(\phi_{i+1} - \phi_i) + \frac{u^2 \Delta t}{2\Delta x^2}(\phi_i - \phi_{i-1}) \tag{2}$$

Both FOU and LW methods have their merits and demerits; the FOU is only first-order accurate and very diffusive but conserves extrema while the LW scheme cannot conserve extrema but second-order accurate. The objective herein is to combine these two schemes based on the diffusion/dissipation scaling, capturing desired aspects of each. Introducing the slope limiter function Ψ, the inviscid flux on the cell face $(i + 1/2)$ can be cast as

$$F^{FOU}_{i+1/2} = \frac{F_{i+1} + F_i}{2} - \frac{(1 - \Psi_{i+1/2})}{2}|u_{i+1/2}|(\phi_{i+1} - \phi_i)$$

$$F^{LW}_{i+1/2} = \frac{F_{i+1} + F_i}{2} - \frac{\Psi_{i+1/2}}{2}\left(\frac{u^2 \Delta t}{\Delta x}\right)_{i+1/2}(\phi_{i+1} - \phi_i)$$

Clearly, a simple interpolation between F^{FOU} and F^{LW} yields a modified TVD (total variation diminishing) LW at the cell-face $(i + 1/2)$ as:

$$F^{LW}_{i+1/2}|_{TVD} = \frac{1}{2}\left\{(F_i + F_{i+1}) - |u_{i+1/2}|\left[1 + \Psi_{i+1/2}\left(\frac{\Delta t}{\Delta x}|u_{i+1/2}| - 1\right)\right](\phi_{i+1} - \phi_i)\right\} \tag{3}$$

where the limiter function "$\Psi_{i+1/2} = \Psi\left[\min(r^-, r^+)_{i+1/2}\right]$"; the slope ratios are: "$r^-_{i+1/2} = (\phi_i - \phi_{i-1})/(\phi_{i+1} - \phi_i)$ and $r^+_{i+1/2} = (\phi_{i+2} - \phi_{i+1})/(\phi_{i+1} - \phi_i)$". Obviously, the linear combination of dissipation terms attempts at developing a higher-order accurate method in a way that it functions as an FOU scheme if non-physical oscillations appear due to discontinuities in its derivatives or in the solution while maintaining 2nd-order accuracy in rest of the domain. Unfortunately, the freedom to match the dissipation with a unique Courant number $\overline{CFL} = u\Delta t/\Delta x$ could be broken down when considering a system of equations [2].

The current research attempts at extending the TVD LW method (characterized by its simplicity in coding and its ability in computing complicated flows with strong shock accompanied by contact discontinuities) to incompressible flows within the framework of artificial compressibility method (ACM) [3–5]. This is accomplished by appending a compatible non-linear cell-face interpolation method [6, 7] to the TVD LW approach which convokes precisely an accurate amount of dissipation required at each cell-face to suppress spurious grid-scale oscillations when calculating the mass flux. A monitor function ξ based on the reference Mach number M_r is included with the TVD limiter Ψ, adjusting the coefficient of dissipation on the cell-face. Moreover, the mass flux is manipulated as a cell-face dissipation scaling and the constant time-step is replaced by a variable pseudo-time step. "Buoyancy-driven" cavity flows could be a proper numerical experiment to justify the accuracy and efficacy of proposed method. A detailed spatial discretization together with other relevant aspects has been presented.

2. Governing Equations

The Einstein notation has been used to represent continuity, momentum and scalar ϕ equations, dominating an incompressible flow:

$$\frac{\partial \rho}{\partial t} + \frac{\partial \rho u_i}{\partial x_i} = 0 \tag{4}$$

$$\frac{\partial \rho u_i}{\partial t} + \frac{\partial \rho u_i u_j}{\partial x_j} + \frac{\partial p}{\partial x_i} = \frac{\partial}{\partial x_j}\left(\Gamma \frac{\partial u_i}{\partial x_j}\right) + f_i \tag{5}$$

$$\frac{\partial \rho \phi}{\partial t} + \frac{\partial \rho u_j \phi}{\partial x_j} = \frac{\partial}{\partial x_j}\left(\Gamma \frac{\partial \phi}{\partial x_j}\right) \tag{6}$$

where ρ the density, Γ the viscosity or conductivity of fluid, p the pressure, u_i the velocity vector, f_i the source term and x_i the Cartesian coordinates.

2.1. *Discretization with Δ-formulation*

A "cell-centered finite-volume" method is used to solve the governing equations, comprising 2-D Navier-Stokes and ϕ equations; an integral form is obtained as follows:

$$\frac{d}{dt}\int_{\forall} W \, d\forall + \int_{S} \vec{F}(W) \cdot d\vec{S} = \int_{\forall} f \, d\forall \tag{7}$$

with a boundary S for an arbitrary regime $\forall$; the source term and conservative variables are indicated by f and $W = (\rho, \rho u_i, \rho\phi)^T$, respectively. Integrating over a computational cell i gives:

$$\forall_i \frac{dW_i}{dt} = \sum_{faces} -S\hat{F} + \forall_i f_i \tag{8}$$

where the "sum" stands for computational cell faces. Each cell face retains a unit normal vector $\vec{n}$, defined by

$$\vec{n} = n_x\vec{i} + n_y\vec{j} = \frac{S_x}{S}\vec{i} + \frac{S_y}{S}\vec{j} \tag{9}$$

The cell-face flux correlation gives rise to the following expression:

$$\hat{F} = n_x F + n_y G \tag{10}$$

where in the respective x and y directions, convective and viscous fluxes are indicated by F and G; Eqs. (1)-(3) involve both fluxes. A 2nd-order central differencing with a "thin-layer approximation" is applied to discretize viscous fluxes on a curvilinear grid. A "rotational matrix", accounting for "directional dominance on upwinding" schemes [8] is utilized to calculate the inviscid flux on a cell-face $(i + 1/2)$. "Contravariant" momentum-flux components are converted to "primitive variables" within the framework of a Cartesian-coordinate system using the "rotational matrix". "Primitive variables" $\tilde{W} = (u_i, p, \phi)^T$ on the cell-face have been evaluated using Van Leer's [9] "MUSCL" approach.

2.2. *Lax-Wendroff method*

Convective fluxes on control volume faces are determined using the TVD Lax-Wendroff (LW) scheme; flow equations on the cell-face $(i + 1/2)$ with the modified TVD LW method read:

$$\hat{F}^{LW}_{i+1/2}|_{TVD} = \frac{1}{2}\left\{F_i + F_{i+1} - |U_{i+1/2}|\left[1 + \Psi(r)_{i+1/2}\left(\overline{CFL}_{i+1/2} - 1\right)\right](W_{i+1} - W_i)\right\} \tag{11}$$

where $U = un_x + vn_y$ represents a "contravariant" velocity component and $\overline{CFL} = \min(1, |U|\,\Delta t/\Delta n)$ with a distance between cell centers, given by $\Delta n_{i+1/2} = (\forall_i + \forall_{i+1})/2S_{i+1/2}$. An LW scheme customarily accommodates compressible flows; nonetheless, it has been reformulated to evaluate convective fluxes for an incompressible flow:

$$\begin{aligned} \hat{F}^{LW}_{i+1/2}|_{TVD} &= \frac{1}{2}\left[\dot{m}\left(\tilde{W}^R + \tilde{W}^L\right) - \xi\,|\dot{m}|\left(\tilde{W}^R - \tilde{W}^L\right)\right]_{i+1/2} \\ \xi &= 1 + \min(1, M_r)\,\Psi(r)\,\left(\overline{CFL} - 1\right) \end{aligned} \tag{12}$$

where left (L) and right (R) primitive state variables $\tilde{W} = (u_i, p, \phi)^T$ can be estimated using the "MUSCL" scheme. The mass flux $\dot{m}$ is evaluated with a compatible cell-face interpolation scheme [6, 7]:

$$\dot{m}_{i+1/2} = \left(\overline{\rho U}\right)_{i+1/2} - \frac{\Delta n_{i+1/2}}{4}\left(\frac{1}{C_{i+1}} - \frac{1}{C_i}\right)(f_{i+1} - f_i) - \left(\frac{1}{2C}\right)_{i+1/2}\left(p^R - p^L\right) \tag{13}$$

where an averaging between adjacent grid nodes is denoted by the overbar, $f = f_x n_x + f_y n_y$, state variables (p^L, p^R) on L and R sides of the cell-face are calculated with a "fully upwinded second-order (FUS)" scheme and C is the "artificial sound speed" defined afterward. The formulation eliminates non-physical oscillations on a collocated grid. The reference Mach number $M_r = U_r/C^*$ with U_r is the reference velocity and $C^* = \sqrt{\gamma p^*/\rho}$ is the real sound speed where γ signifies the specific heat capacity ratio (typically $\gamma = 1.4$ for air) and p^* implies the absolute pressure. The limiter function is given as "$\Psi(r)_{i+1/2} = \Psi\left[\min(r^-, r^+, 1)_{i+1/2}\right]$ with $r^-_{i+1/2} = (\tilde{W}_i - \tilde{W}_{i-1})/(\tilde{W}_{i+1} - \tilde{W}_i)$ and $r^+_{i+1/2} = (\tilde{W}_{i+2} - \tilde{W}_{i+1})/(\tilde{W}_{i+1} - \tilde{W}_i)$". A developed differentiable/smoother limiter function

$$\Psi(r) = \frac{2\min(r;1)}{\min(r;1)^2 + 1}; \quad \lim_{r\to\infty} \Psi(r) = 1$$

is applied to the modified LW method, equivalent to an improved version of the minmod limiter [10]. It is anticipated that Eq. (12) can be used for all Mach number flows since the monitor function ξ preserves the accuracy, stability and monotonicity of the scheme. In this situation, the cell-face density ρ in Eq. (12) can be estimated utilizing the "MUSCL" scheme accompanied by a "TVD limiter function" as devised to further support numerical stability; an FOU scheme can be used to evaluate state variables for hypersonic flows. It is worth pointing out that a variable pseudo-time step Δt_i (and therefore, the restriction on using constant time-step is avoided) involved in the LW method is approximated from Reference [11], configured with convection, diffusive, and artificial sound speeds.

2.3. Compressibility and associated features

For convenience, a 1-D analysis of the technique is applied. The mass continuity can be reformed with a non-conservative identity as:

$$\frac{\partial \rho}{\partial t} + u\frac{\partial \rho}{\partial x} + \rho\frac{\partial u}{\partial x} = 0 \tag{14}$$

For pseudo-compressibility formulation, the density is considered solely as $\rho = \rho(p)$. Therefore, the non-conservative "continuity equation" has been reformulated with a "chain rule" as:

$$\frac{\partial \rho}{\partial p}\left(\frac{\partial p}{\partial t} + u\frac{\partial p}{\partial x}\right) + \rho\frac{\partial u}{\partial x} = 0 \tag{15}$$

where the well-known "artificial sound speed" is recovered as

$$C^2 = \left(\frac{\partial \rho}{\partial p}\right)^{-1} = \beta^2 \max\left(\sum u_i^2; \frac{1}{2}U_r^2\right) \tag{16}$$

where the compressibility parameter $\beta = 3$, indicating that the local Mach number $M_l = |u_i|/C \approx 0.3$; it is a critical M_l after which the flow transits from incompressible to compressible identity [12] with $C^* < C$ (real sound speed is less than "artificial sound speed"). Using Eq. (16) in combination with momentum and scalar (ϕ) equations, time-dependent inviscid flow equations in primitive forms comply with:

$$\begin{aligned} \frac{\partial \phi}{\partial t} + u\frac{\partial \phi}{\partial x} &= 0 \\ \frac{\partial u}{\partial t} + u\frac{\partial u}{\partial x} + \frac{1}{\rho}\frac{\partial p}{\partial x} &= 0 \\ \frac{\partial p}{\partial t} + u\frac{\partial p}{\partial x} + \rho C^2\frac{\partial u}{\partial x} &= 0 \end{aligned} \tag{17}$$

Equation (17) can be rationalized as

$$\frac{\partial \tilde{W}}{\partial t} + A\frac{\partial \tilde{W}}{\partial x} = 0 \tag{18}$$

with "primitive variables" $\tilde{W} = (\phi, u, p)^T$; the expression for "Jacobian matrix" A becomes:

$$A = \begin{pmatrix} u & 0 & 0 \\ 0 & u & \frac{1}{\rho} \\ 0 & \rho C^2 & u \end{pmatrix} \tag{19}$$

Characteristic speeds or eigenvalues λ of A can naturally be achieved as

$$\lambda = u, u - C, u + C \tag{20}$$

"Primitive variables" related R and L "eigen-vector matrices" of A can be devised as:

$$A = R\lambda R^{-1}, \quad R = \begin{pmatrix} 1 & 0 & 0 \\ 0 & -\frac{1}{\rho C} & \frac{1}{\rho C} \\ 0 & 1 & 1 \end{pmatrix}, \quad R^{-1} = \begin{pmatrix} 1 & 0 & 0 \\ 0 & -\frac{\rho C}{2} & \frac{1}{2} \\ 0 & \frac{\rho C}{2} & \frac{1}{2} \end{pmatrix} \tag{21}$$

"Characteristic variables" are obtainable from "$R^{-1}\Delta\tilde{W} = R^{-1}(\Delta\phi, \Delta u, \Delta p)^T$" and are given by

$$\begin{aligned} \tilde{w}_1 &= \Delta\phi \\ \tilde{w}_2 &= -\frac{1}{2}(\rho C \Delta u - \Delta p) \\ \tilde{w}_3 &= \frac{1}{2}(\rho C \Delta u + \Delta p) \end{aligned} \tag{22}$$

A differential change in density has been introduced to provide a conversion between primitive (non-conservative) and conservative variables as follows:

$$\Delta\rho = \frac{\partial \rho}{\partial p}\Delta p = \frac{\Delta p}{C^2} \tag{23}$$

The following relation is used to determine conservative explicit residuals ΔW:

$$\Delta W = (\Delta\rho,\ \Delta\rho u_i,\ \Delta\rho\phi)^T = \frac{\Delta t_i}{\forall_i} R_i \tag{24}$$

where the RHS of Eq. (8) assesses the residual R_i. The eigen-structure related matrix-inversion technique establishes a correlation between conservative and primitive variables as:

$$\begin{pmatrix} \Delta p \\ \Delta u_i \\ \Delta\phi \end{pmatrix} = \frac{1}{\rho} \begin{pmatrix} \rho C^2 & 0 & 0 \\ -u_i & 1 & 0 \\ -\phi & 0 & 1 \end{pmatrix} \begin{pmatrix} \Delta\rho \\ \Delta\rho u_i \\ \Delta\rho\phi \end{pmatrix} \tag{25}$$

Equation (25) comes up with an explicit representation of crucial fluidic relations. This transformation is performed before the residual smoothing strategy in the implicit stage. "Primitive variables" $\tilde{W} = (p,\ u_i,\ \phi)^T$ are updated after the implicit phase from the following relation:

$$\tilde{W}^{n+1} = \tilde{W}^n + (\Delta p,\ \Delta u_i,\ \Delta\phi)^T \tag{26}$$

where $\tilde{W}^{n+1}$ indicate current values at time level $(n+1)\Delta t$.

2.4. Solution algorithm

References [3–5] may provide details of boundary conditions. A "diagonally dominant (DD) ADI-factorization" [13] approach has been applied to integrate discretized equations in pseudo-time. "Approximate factorization and splitting of Jacobians" associated with viscous and inviscid fluxes have been enhanced by the DDADI method. The implicit stage utilizes backward and forward sweeps in each coordinate direction. Each sweep employs a first-order upwind differencing, preserving a diagonally dominant factorization. Nevertheless, the accuracy of solution remains apparently unaffected; Eq. (24) exemplifies explicit flux calculations, necessarily showing the order of solution accuracy. Boundary conditions accommodate explicit treatments. After factorization, the implicit phase yields:

$$\left[1 + \frac{\Delta t_i}{\forall_i}(\partial_i^- S_{i+1/2} A_i^+ - \partial_i^+ S_{i-1/2} A_i^-)\right]$$
$$\left[1 + \frac{\Delta t_i}{\forall_i}(\partial_j^- S_{j+1/2} B_j^+ - \partial_j^+ S_{j-1/2} B_j^-)\right] \left[1 - \Delta t_i \left.\frac{\partial f}{\partial \tilde{W}}\right|_i\right] \Delta\tilde{W}_i = \Delta\tilde{W}_i^* \tag{27}$$

where $\partial_{i,j}^-$ and $\partial_{i,j}^+$ signify the 1st-order "backward and forward spatial difference operators" respectively, in the i and j directions. Equation (19) defines the flux "Jacobian matrix" A for 1-D case; a 2-D extension is straightforward and an identical expression for the flux "Jacobian matrix" B can be made in the j direction. Primitive explicit residuals "$\Delta\tilde{W}_i^* = (\Delta p,\ \Delta u,\ \Delta v,\ \Delta\phi)^T$" are represented by Eq. (25). A detailed solution procedure can be found elsewhere [3–5,13].

Algorithmic features can be sequenced as follows:

1. Initialize "primitive variables" (e.g., velocity, pressure and scalar).
2. Calculate cell-face fluxes from Eq. (12).
3. Determine residuals of "primitive variables" (i.e., explicit changes) using Eq. (25).
4. Employ Eq. (27) to solve the flow equations.
5. Update "primitive variables" using Eq. (26) to obtain physical fields.
6. Keep continuing steps 2–5 until the solution converges.

Journal of Physics: Conference Series **2313** (2022) 012002 doi:10.1088/1742-6596/2313/1/012002

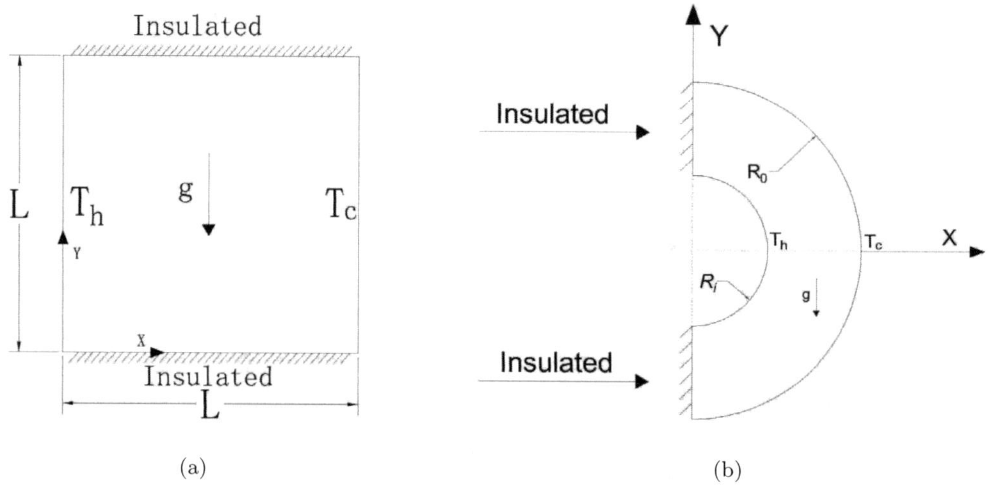

Figure 1. Geometries of "buoyancy-driven" flow: (a) square cavity and (b) annulus.

3. Numerical Tests

With insulated (adiabatic) bottom and top walls and a characteristic length of L, Fig. 1 shows the respective physical models for natural convection in the square cavity and half-concentric annulus. Temperatures T_c and T_h respectively, refer to vertical cold and hot walls. Reference scales are used to non-dimensionalize dimensional parameters; "restrictions on a particular system of units" can be avoided using dimensionless variables which reduce computer round-off errors. "Non-dimensional governing equations" for a 2-D case are available in Reference [6]. The test problem is dominated by two dimensionless parameters arising from the normalization: Prandtl number $Pr = \nu/\alpha$ and Grashof number $Gr = g\zeta(T_h - T_c)L^3/\nu^2$ where g, α and ζ are the gravitational acceleration, thermal diffusivity and thermal-expansion coefficient, respectively. Results are plotted in the moods of "$V = vL/\nu$ and $\theta = (T - T_c)/(T_h - T_c)$ versus $X = x/L$" with $Pr = 1.0$. For this test-case, "$u_r = \sqrt{g\zeta(T_h - T_c)L} = \nu\sqrt{Gr}/L$" is a dimensional reference velocity and hence, a dimensionless reference velocity can be obtained as $U_r = u_r L/\nu = \sqrt{Gr}$ [8]. The fluid is air; chosen reference values are: $p_r^* = 101325\ N/m^2$ (absolute pressure), $\rho_r = 1.18\ kg/m^3$, $\mu_r = 18.46 \times 10^{-6}$ N-s/m^2, $\gamma = 1.4$ and $L = 1\ m$. The reference sound speed can be calculated as $C_r^* = \sqrt{\gamma p_r^*/\rho_r}$.

Highly "diffusive central and convective near-wall flow" regimes are attached to this test-case. However, the analytic solution to this problem is unavailable and therefore, literature data [14] and computations from "CVFEM (control-volume based finite-element method) with stream-function vorticity formulation" [15] are accounted for estimating the present computational accuracy. The pseudo-mass source with "RMS (root-mean-square)" identification

$$\Delta\rho^* = \sqrt{\frac{\sum_{i=1}^{NP} |(\Delta p/C^2)_i|^2}{NP}} \leq \eta_t$$

has been utilized to assess convergence acceleration; in the above relation, NP implies the number of computational cells and η_t indicates the user-defined tolerance limit. Numerical problems employ a non-uniform denser grid resolution to capture sharp gradients in the wall-vicinity. Relying on grid dependence tests, numerical errors for all-field variables have been estimated as less than 2% due to the grid spacing. The temperature gradient $\partial\theta/\partial X$ with the "local Nusselt number $Nu = -(\partial\theta/\partial X)_{X=0}$" is approximated using a three-point one-sided finite-difference formula.

No-slip boundary conditions are applied to all walls. Isothermal temperature boundary conditions are: $\theta_h = 1.0$ and $\theta_c = 0.0$, respectively on the left and right vertical walls. The source term $f = (0,\, 0,\, Gr\,\theta,\, 0)^T$ and an FUS scheme is recruited to compute convective fluxes on the cell-face.

3.1. "Buoyancy-driven" flow in a square-cavity

A 60×60 non-uniform grid for $Gr = 10^7$ with denser grid spacings in the close proximity of wall than those of the core, is applied to the computation. Literature data [14] and "CVFEM" [15] solutions further support the present numerical formulation. The simulation commences with initial guesses $U = V = 0.1U_r$, $p = 0$ (pressure-difference) and $\theta = (\theta_h + \theta_c)/3$; the reference Mach number $M_r \approx 1.4 \times 10^{-4}$. Figures $2(a;b)$ exhibit the respective plots of V and θ profiles along the horizontal mid-plane. Plots also include literature data by Massarotti et al. [14] together with "CVFEM" computational data. Apparently, an encouraging qualitative agreement has been acquired by comparisons. Figure 2(c) displays local Nu-distributions along the hot vertical wall; comparisons also include results of "CVFEM". As can be seen, present results sustain a pretty good congruence with "CVFEM" data. Due to methodological differences, a slight variation in profiles is observed; a finer grid-refinement in the wall-vicinity may resolve this aspect to an extent. Figure 2(d) elucidates the artificial-density convergence with different CFL (e.g., variable pseudo-time step Δt_i) numbers. Obviously, a higher CFL number expedites a faster convergence; the smooth convergence indicates that the current method effectively suppresses non-physical oscillations.

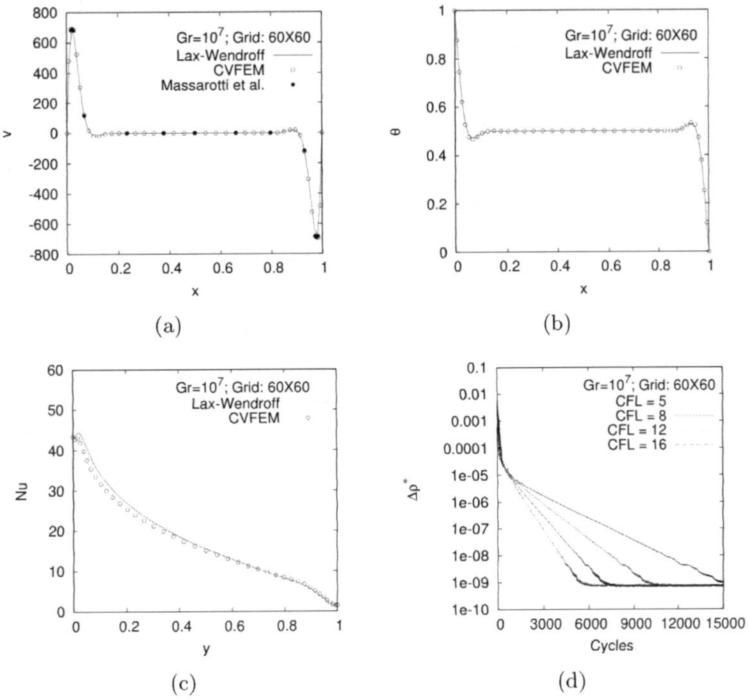

Figure 2. "Buoyancy-driven" cavity flow: (a) V; (b) θ; (c) local Nu and (d) convergence of pseudo-mass imbalance.

3.2. "Buoyancy-driven" flow in an annular-cavity

The non-uniform grid resolution for $Gr = 10^6$ is composed of 30 circumferential and 40 radial line segments; this grid distribution has been considered to be adequate to accurately describe transport characteristics. Respective radii of outer and inner surfaces are R_o and R_i with $(R_o/R_i) = 3$. The center of circles retains the system origin; the test-case has a characteristic length of $L = (R_o - R_i)$ with insulated straight bottom and top walls. The inner-curved surface is hot with a temperature of T_h and the cold outer one is T_c. Similar to the buoyancy-driven flow in a square-cavity, this flow contains both highly convective near-wall and diffusive central flows with identical governing equations, boundary and reference conditions. The reference Mach number $M_r \approx 4.5 \times 10^{-5}$. Profiles of V and θ are depicted in Figs. 3(a) and 3(b), respectively on the horizontal mid-plane for $Gr = 10^6$, where X is exactly measured from the inner surface. Comparisons also include "CVFEM" results. Present predictions maintain remarkable correspondence with "CVFEM" data; it seems likely that the agreement is quite satisfactory. Evidently, local Nu-distributions shown in Fig. 3(c) match "CVFEM" computations well. However, differences are visible at the lower section of inner wall; discrepancies probably come up with the fact that the "CVFEM" approximates a weighted nodal temperature. This inconsistency may be turned down to some extent using a higher-order formulation and a finer grid resolution with the proposed algorithm. Figure 3(d) represents the artificial mass convergence, showing that a faster convergence is facilitated with increasing CFL number.

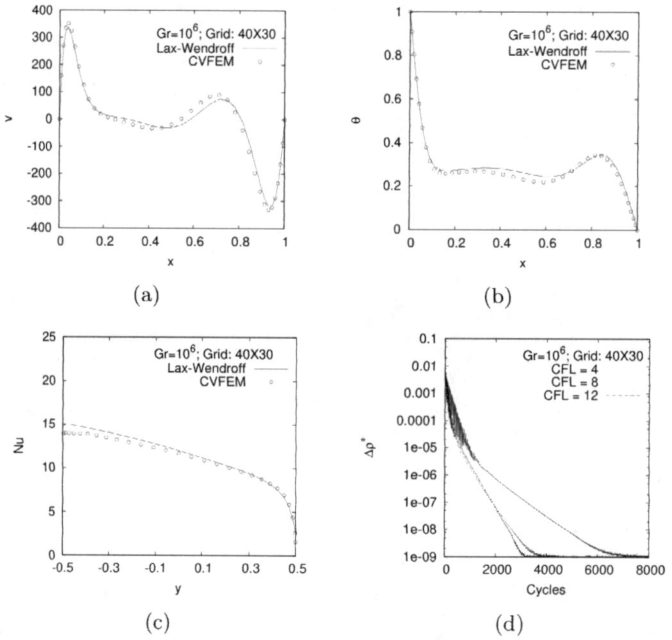

Figure 3. "Buoyancy-driven" flow in annulus: (a) V; (b) θ; (c) local Nu and (d) convergence of pseudo-mass imbalance.

4. Conclusions

An extension of the Lax-Wendroff scheme to an incompressible flow is proposed on a curvilinear collocated grid within the framework of pressure-based ACM using a "cell-centered finite-volume" Δ-formulation. The strength of newly developed limiter function is controlled by a reference Mach number and the constant time-step is replaced by a variable pseudo-time step.

Solutions to "buoyancy-driven" flows with strong source terms demonstrate the high reliability and competency of present coupled algorithm, indicating that physically meaningful oscillation-free numerical results are obtained, favorably comparable with literature and "CVFEM" data. The coupling of equations becomes strongly non-linear; however, the proposed algorithm could be utilized to calculate inviscid fluxes over the entire spectrum of subsonic to hypersonic flow by modifying certain attributes.

Acknowledgment

This work is supported by Hangzhou Dianzi University research supporting fund (Grant No. GK218803299037) of Zhejiang Province, P.R. China.

References

[1] Lax P, Wendroff B 1960: Systems of conservation laws. *Computations on Pure and Applied Mathematics*, 13(2):217-237.

[2] Davis SF 1984: TVD finite difference schemes and artificial viscosity. *ICASE Report 84-20, NASA CR-172373*, NASA Langley Research Center.

[3] Rahman MM, Siikonen T 2008: An artificial compressibility method for viscous incompressible and low Mach number flows. *Int. J. Numer. Meth. Engg.*, 75:1320-1340.

[4] Rahman MM, Siikonen T 2001: An artificial compressibility method for incompressible flows. *Numer. Heat Transfer Part B*, 40:391-409.

[5] Rahman MM, Rautaheimo P, Siikonen T 1997: Numerical study of turbulent heat transfer from a confined impinging jet using a pseudo-compressibility method, *Second Int. Symposium on Turbulence, Heat and Mass transfer*, Delft, The Netherlands, Hanjalic K and Peeters TWJ, (eds), Delft University Press: Delft, 511-520.

[6] Rahman MM, Miettinen A, Siikonen T 1996: Modified SIMPLE formulation on a collocated grid with an assessment of the simplified Quick scheme. *Numer. Heat Transfer Part B*, 30:291-314.

[7] Rahman MM, Siikonen T 2002: A dual-dissipation scheme for pressure-velocity coupling. *Numer. Heat Transfer Part B*, 42, 231-242.

[8] Rahman MM 2020: Avoiding under-relaxations in SIMPLE algorithm. *Numer. Heat Transfer Part B*, 52(5). doi.org/10.1080/10407790.2020.1787043.

[9] Anderson WK, Thomas JL, Van Leer B 1983: Comparison of finite volume flux vector splittings for the Euler equations. *AIAA J.*, 24:649-664.

[10] Roe PL 1985: Some contributions to the modeling of discontinuous flows. *Lecture Notes On Applied Mathematics*, 22:162-193.

[11] Rahman MM, Siikonen T 2000: An improved SIMPLE method on a collocated grid. *Numer. Heat Transfer Part B*, 38:177-201.

[12] Roller S, Munz CD 2000: A low Mach number scheme based on multi-scale asymptotics. *Computing and Visualization in Science*, 3:8591.

[13] Lombard C, Bardina J, Venkatapathy E, Oliger J 1983: Multi-dimensional formulation of CSCM - an upwind flux difference eigenvector split method for the compressible Navier-Stokes equations. *Sixth AIAA Comput. Fluid Dynamics Conference* , AIAA Paper 83-1895-CP, 649-664.

[14] Massarrotti N, Arpino F, Lewis RW, Nithiarasu P 2006: Explicit and semi-implicit CBS procedures for incompressible viscous flows. *Int. J. Numer. Meth. Engg.*, 66:16181640.

[15] Rahman MM, Lampinen MJ 1995: Numerical study of natural convection from a vertical surface due to combined buoyancies. *Numer. Heat Transfer A*, 28:409-429.

Study on aerodynamic characteristics of flying wing knapsack inlet duct

Shang Ma, Yang Tao, Junqiang Wu[1], Neng Xiong and Xingya Da

High Speed Aerodynamic Institute, China Aerodynamics Research and Development Center, Mianyang Sichuan, P, R. China

[1]Corresponding author's e-mail: Wujunqiang@cardc.cn

Abstract. Based on the extended shape of the existing flying wing standard model with small aspect ratio, this paper studies the flow field characteristics around the model at different angles of attack and the flow field characteristics of the knapsack dual release inlet when the Mach number is equal to 1.8. The pressure coefficient distribution, velocity distribution and eddy characteristics of typical states are illustrated. The results show that the lip of the knapsack inlet has a significant effect on the flow characteristics of the back and abdomen of the model. After passing through the inlet lip and the S-shaped bend in the front section, the internal pressure distribution of the inlet is relatively uniform, there are obvious vortices in the inlet, the core flow near the tail nozzle flows smoothly, and the strong vortices are only near the wall.

1. Introduction

As an associated component connecting the aircraft outflow field and the engine, the design performance of the inlet is directly related to the flight performance of the aircraft. In order to minimize the influence of inlet flow distortion on the engine and make the inlet compatible with the engine, it is necessary to start from the inlet layout and inlet configuration design to ensure that the inlet have good flow field quality [1]. Most of the existing high subsonic stealth UAVs adopt back inlet and tailless aerodynamic layout, and the inlet system configuration is embedded serpentine inlet with inlet and fuselage shape preservation. Structurally, the knapsack inlet has little interference with the weapon mounting and the layout space under the fuselage, and the structure is more compact, which is beneficial to the weight reduction of the aircraft. In terms of stealth, arranging the air inlet above the fuselage or wing can make full use of the forward shielding effect of the satellite communication radome. At the same time, the air inlet tilts back and the inner pipe bends greatly to realize the full line of sight shielding of the front end face of the engine, so that the forward and lateral radar scattering cross sections of this kind of aircraft are very small [2-4]. However, the performance of embedded and serpentine inlet is affected by body boundary layer, air flow separation and inlet lip shock wave, which will inevitably lead to large-area flow separation and vortex, reduce the flow field quality, and affect the matching performance of inlet and engine. In order to match the performance of aircraft outflow, inlet and engine, it is necessary to carry out the integrated design of aircraft and inlet to meet the relevant performance requirements [5, 6].

In this paper, the characteristics of the flow field around the small aspect ratio flying wing model of the knapsack double inlet at different angles of attack and the flow field in the knapsack double inlet are studied when the Mach number is equal to 1.8. The results show that the lip of the knapsack inlet has a significant effect on the flow characteristics of the back and abdomen of the model. The core flow near the tail nozzle of the inlet is stable, and there is strong eddy only near the wall.

2. Methodology

2.1. Governing equation

The governing equation of the flow field is N-S equation in conservative form. After dimensionless, it can be written in the following form in the curvilinear coordinate system.

$$V\frac{\partial \overline{Q}}{\partial t}+\frac{\partial \overline{E}}{\partial \xi}+\frac{\partial \overline{F}}{\partial \eta}+\frac{\partial \overline{G}}{\partial \zeta}=\frac{1}{R_e}\left(\frac{\partial \overline{E_v}}{\partial \xi}+\frac{\partial \overline{F_v}}{\partial \eta}+\frac{\partial \overline{G_v}}{\partial \zeta}\right) \tag{1}$$

Where V is the reciprocal of Jacobian, $\overline{Q}$ is the conserved variable, $\overline{E}$, $\overline{F}$, $\overline{G}$ is the inviscid convective flux, and $\overline{E_v}$, $\overline{F_v}$, $\overline{G_v}$ is the viscous diffusion flux. The calculation adopts the in house calculation program, and has carried out a number of research work. [7-9]

2.2. Turbulence model

Spalart-Allmaras（SA） turbulence model [10] is used for calculation. Its basic form is:

$$\frac{\partial \tilde{v}}{\partial t}+u_j\frac{\partial \tilde{v}}{\partial x_j}=C_{b1}\left[1-f_{t2}\right]\Omega\tilde{v}+\frac{M_\infty}{\mathrm{Re}}\left\{C_{b1}\left[\left(1-f_{t2}\right)f_{v2}+f_{t2}\right]\frac{1}{\kappa^2}-C_{w1}f_w\right\}\left(\frac{\tilde{v}}{d}\right)^2$$
$$-\frac{M_\infty}{\mathrm{Re}}\frac{C_{b2}}{\sigma}\tilde{v}\frac{\partial^2\tilde{v}}{\partial x_j^2}+\frac{M_\infty}{\mathrm{Re}}\frac{1}{\sigma}\frac{\partial}{\partial x_j}\left\{\left[v+\left(1+C_{b2}\right)\tilde{v}\right]\frac{\partial \tilde{v}}{\partial x_j}\right\} \tag{2}$$

where:

$$f_{t2}=C_{t3}\exp\left(-C_{t4}\chi^2\right),\quad f_w=g\left(\frac{1+C_{w3}^6}{g^6+C_{w3}^6}\right)^{\frac{1}{6}},\quad g=r+C_{w2}\left(r^6-r\right),\quad r=\frac{\tilde{v}}{\tilde{S}\left(\frac{\mathrm{Re}}{M_\infty}\right)\kappa^2d^2} \tag{3}$$

d is the distance to the object surface.

$$f_{v1}=\frac{\chi^3}{\chi^3+c_{v1}^3},\quad f_{v2}=1-\frac{\chi}{1+\chi f_{v1}},\quad \tilde{S}\equiv S+\frac{\tilde{v}}{\kappa^2d^2}f_{v2},\quad S=|\omega|=\Omega,\quad \frac{\mu_t}{\rho}=\tilde{v}f_{v1},\quad \chi\equiv\frac{\tilde{v}}{v} \tag{4}$$

the constants in the turbulence model used in the calculation is:

$$C_{b1}=0.1355\quad \sigma=\frac{2}{3}\quad C_{b2}=0.622\quad \kappa=0.41\quad C_{w2}=0.3$$
$$C_{w3}=2.0\qquad C_{v1}=7.1\quad C_{t3}=1.2\quad C_{t4}=0.5\qquad C_{w1}=\frac{C_{b1}}{\kappa^2}+\frac{\left(1+C_{b2}\right)}{\sigma} \tag{5}$$

2.3. Research model and calculation conditions

The research layout is based on the extended shape of the existing flying wing standard model with small aspect ratio. It is a diamond layout as a whole, which has a wingspan of 0.537m and a total length of 0.701m, with a leading edge sweep angle of 65 degrees, and a trailing edge sweep angle of 25 degrees. It adopts a dual engine layout, backpack left and right inlet and two-dimensional nozzles. Figure 1 shows the outline diagram of the model and the inlet profile diagram.

The calculation state is that the angle of attack range is -10°~18°, Mach number is 1.8, Reynolds number is 1.0×10^7 (consistent with the 1.2m wind tunnel force test), and the calculation of turbulent is 0.1%. The average aerodynamic chord length and Reynolds number reference length of the model is

0.467m. The reference area is $0.188m^2$. The scale is 23 million grid points and the surface grid is less than 1. The far field is greater than 20 times the average scale of the model. The grid height of the first layer is 0.0018mm, with Y+ $\approx$ 1. The free-stream condition is used in the far field and the no-slip condition is used on the model surface. Figure 2 shows the global shape of the grid, the near wall grid and the grid inside the inlet, Figure 3 shows the magnitude of Y+ of the first layer grid after calculation convergence. It can be seen that most of values are of the order of 1.

Figure 1. Research model.

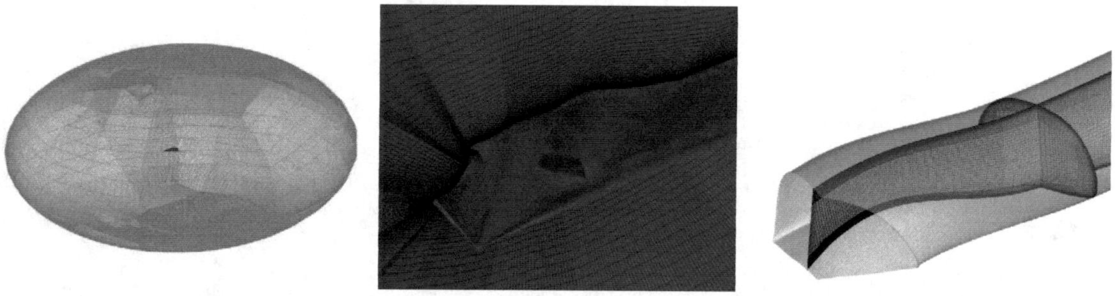

Figure 2. External flow field grid and inlet grid.

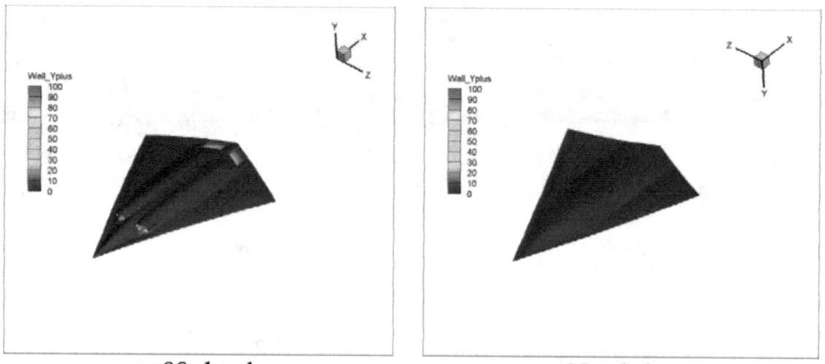

α=0°, back α=0°, abdomen

Figure 3. Y+ after calculation.

3. Results and discussion

3.1. Comparison between CFD and wind tunnel test results

From the comparison between CFD and the wind tunnel test results in Figure 4, it can be seen that the calculation is consistent with the test results in the range of medium and small angles of attack, which verifies the effectiveness of the CFD results.

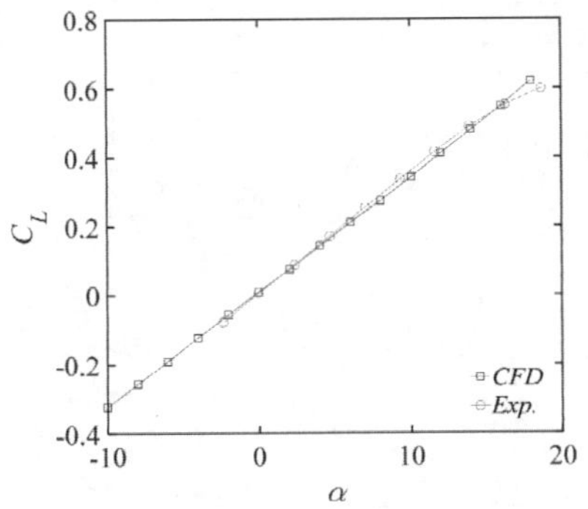

Figure 4. Comparison between CFD and Exp.

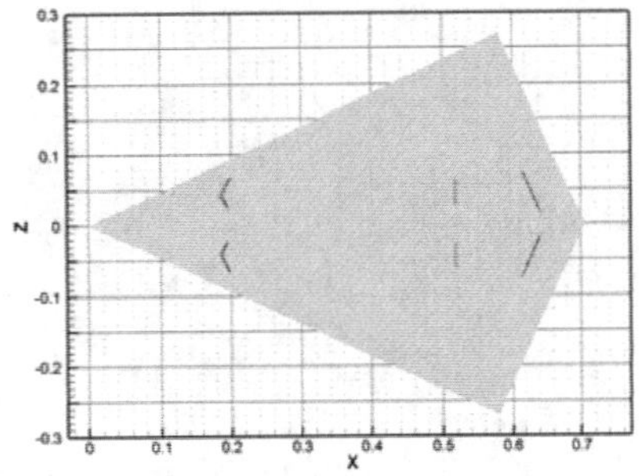

Figure 5. Model features scale.

3.2. *Characteristic analysis of model outflow field*

Figure 5 shows the scale of model features, Figure 6 shows pressure coefficients contour on the back and abdomen of the model at Mach 1.8 and angles of attack ranged from -6° to 12°. The back pressure coefficient of the model decreases with the increase of the angle of attack, and the abdominal pressure coefficient increases with the increase of the angle of attack, which conforms to the general aerodynamic rules. Figure 7 gives the pressure coefficients and velocity contour on the longitudinal symmetry section (z=0) and left inlet section (z=-0.045). The shock wave is generated at the lip of the model knapsack inlet and the leading edge, resulting in a high-pressure region near the corresponding area.

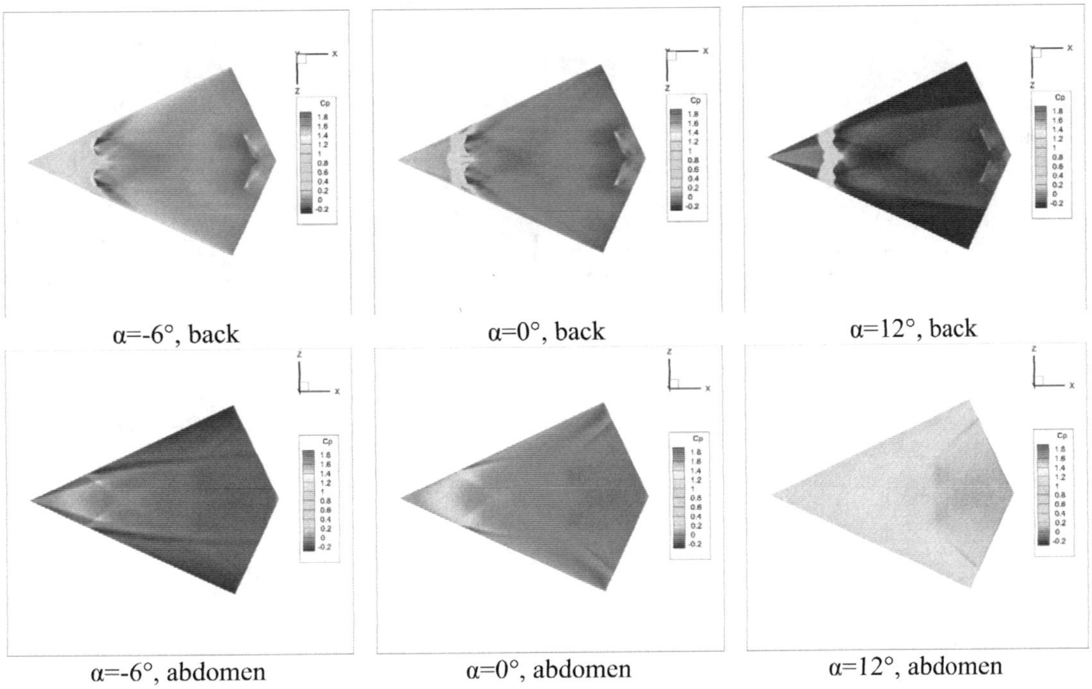

Figure 6. M=1.8, α=-6°~12°, pressure coefficients contour on back and abdomen.

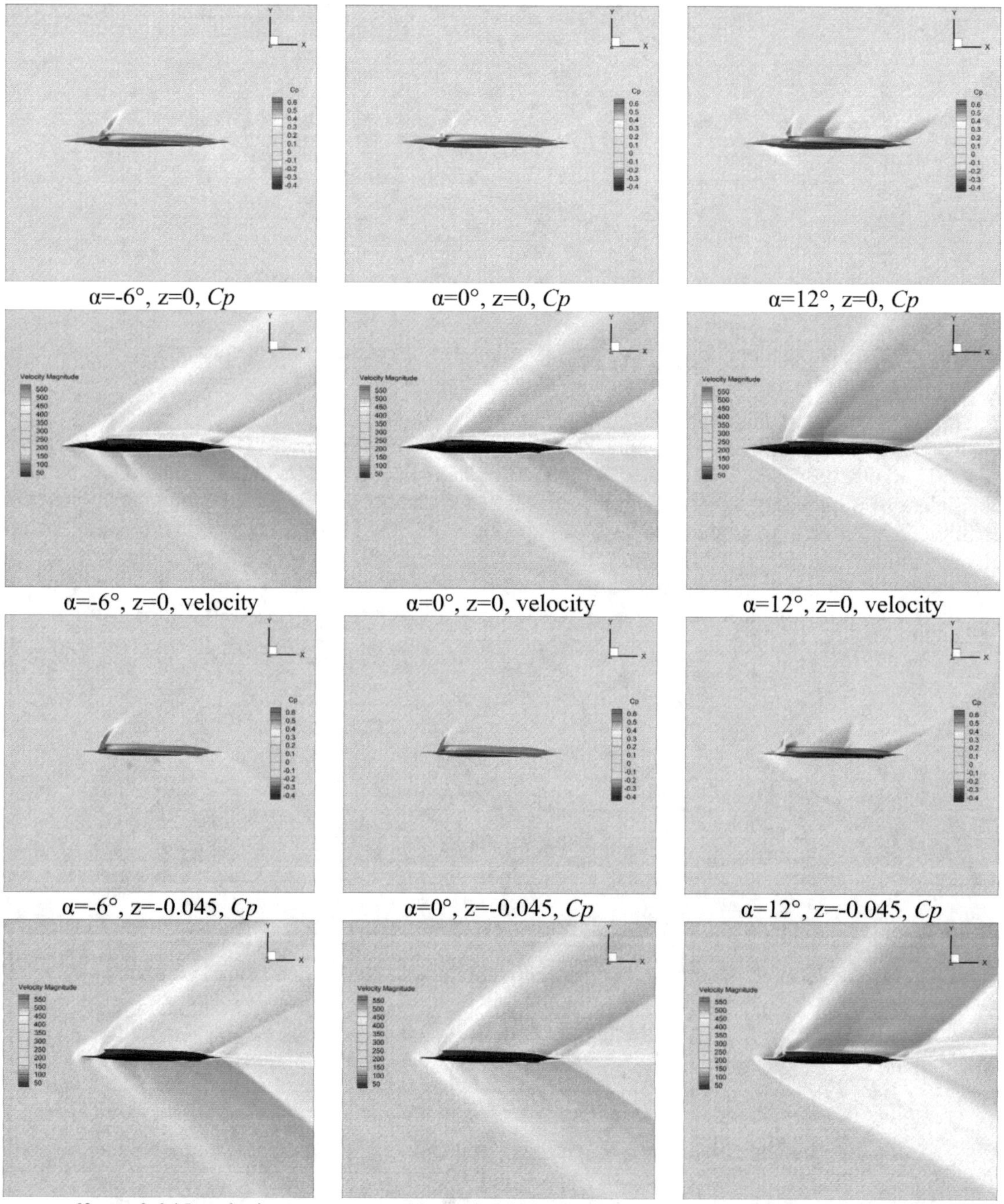

α=-6°, z=0, *Cp* | α=0°, z=0, *Cp* | α=12°, z=0, *Cp*

α=-6°, z=0, velocity | α=0°, z=0, velocity | α=12°, z=0, velocity

α=-6°, z=-0.045, *Cp* | α=0°, z=-0.045, *Cp* | α=12°, z=-0.045, *Cp*

α=-6°, z=-0.045, velocity | α=0°, z=-0.045, velocity | α=12°, z=-0.045, velocity

Figure 7. Pressure coefficients and velocity contour on z=0 and z=-0.045 section.

Figure 8 shows the pressure coefficients contour on Q [11] isosurface (Q=1000000) at three different angles of attack. There is an obvious leading edge vortex at the leading edge of the flying wing, so the negative angle of attack is in the abdomen of the model, and the positive angle of attack produces a low-pressure area at the back of the model (Figure 6). In addition, two distinct separated vortices are also generated at the inlet lip, resulting in a low-pressure region behind the inlet lip. And It should be noted that at a negative angle of attack, the high-pressure air flow generated in the back inlet will bypass the leading edge and produce a high pressure area in the abdomen of the model.

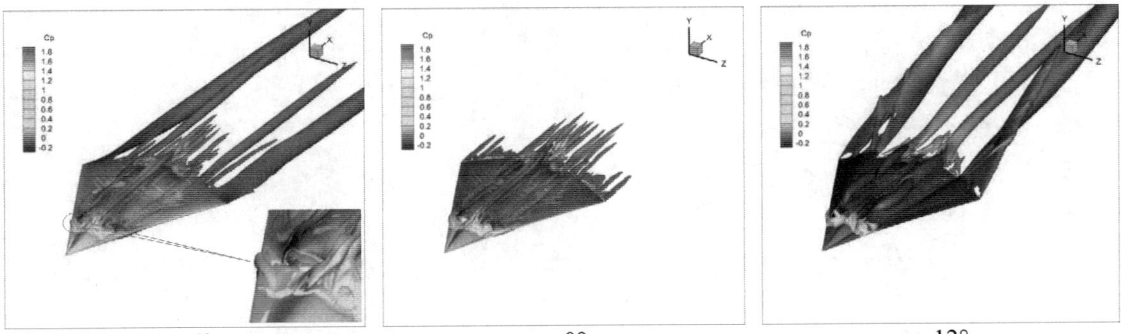

α=-6° α=0° α=12°

Figure 8. Pressure coefficients contour (Q=1000000).

3.3. *Characteristic analysis of inlet flow field*

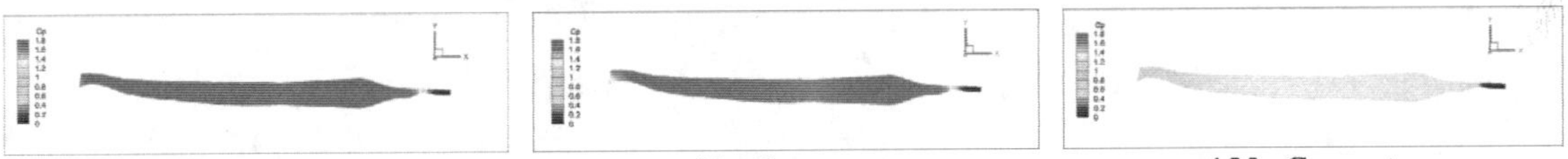

α=-6°, *Cp* contour α=0°, *Cp* contour α=12°, *Cp* contour

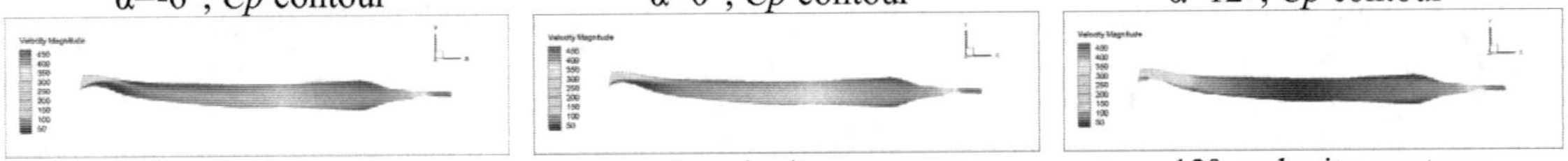

α=-6°, velocity contour α=0°, velocity contour α=12°, velocity contour

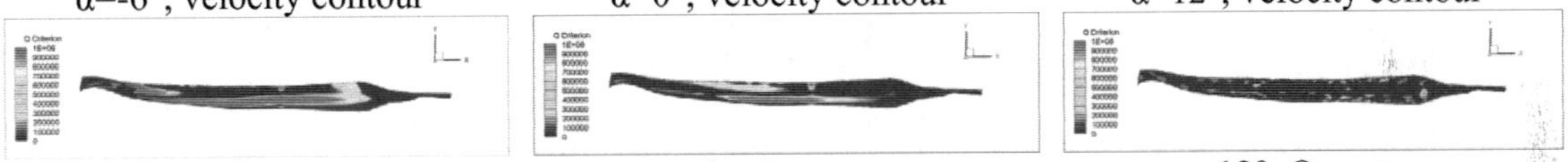

α=-6°, Q contour α=0°, Q contour α=12°, Q contour

Figure 9. Pressure coefficients, velocity and Q contour on the left inlet section.

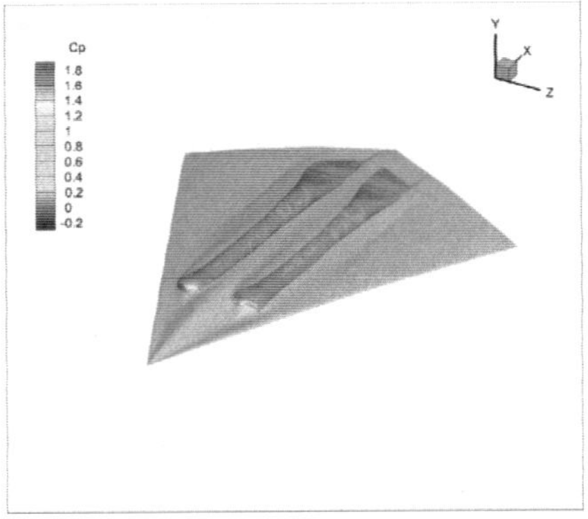

Figure 10. Inlet *Cp* contour. (α=0°, Q=1000000)

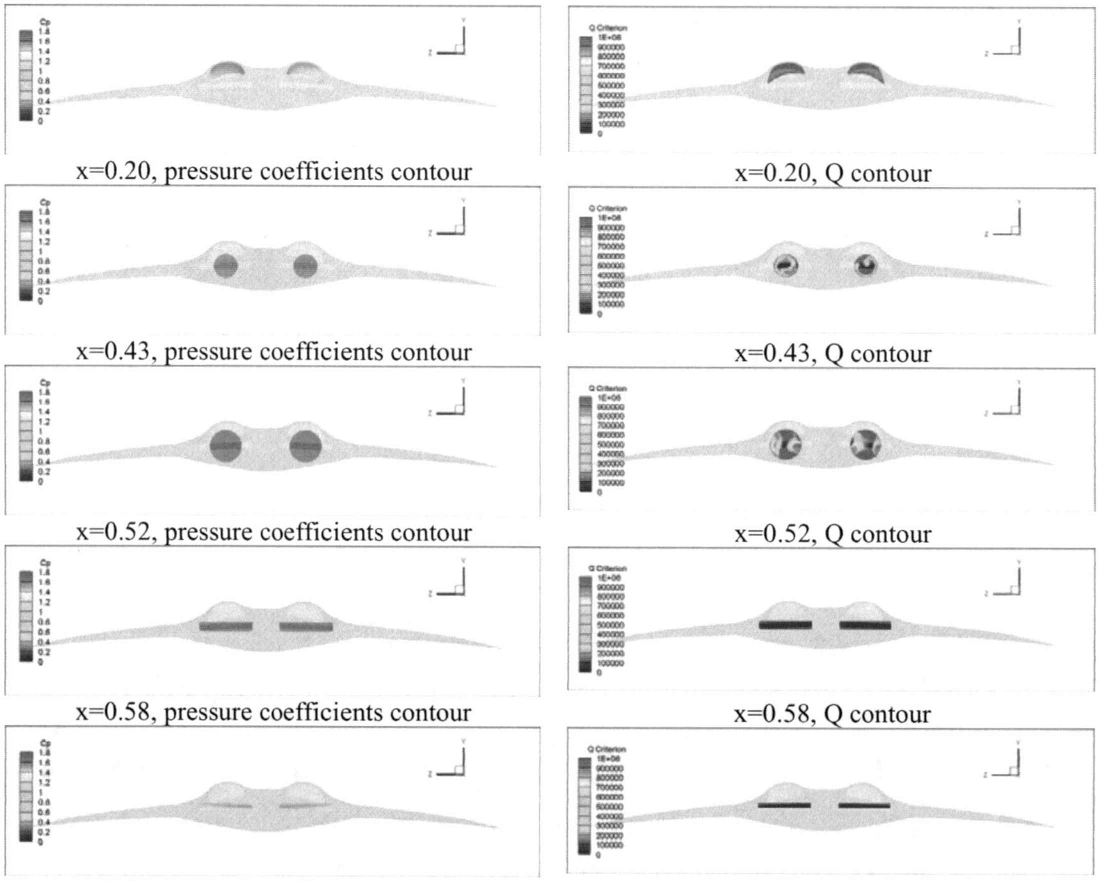

x=0.20, pressure coefficients contour; x=0.20, Q contour
x=0.43, pressure coefficients contour; x=0.43, Q contour
x=0.52, pressure coefficients contour; x=0.52, Q contour
x=0.58, pressure coefficients contour; x=0.58, Q contour
x=0.60, pressure coefficients contour; x=0.60, Q contour

Figure 11. Pressure coefficients and Q contour on different cross section of inlet (α=0°).

Figure 9 gives the pressure coefficients, velocity and Q contour on the left inlet section (z=-0.045) at angles of attack -6°, 0°, 12° respectively. Figure 10 shows the pressure coefficients contour on Q isosurface (Q=1000000) of the knapsack inlet at 0° angle of attack. Figure 11 shows the pressure coefficients and Q contour on typical cross section of the inlet at 0° angle of attack. The internal flow field of the inlet at -6° angles of attack and 0° angle of attack is similar. After passing through the inlet lip and the front S-bend, the internal pressure distribution of the inlet is relatively uniform. The supersonic air flow decelerates rapidly to subsonic flow through the contraction ant expansion profile at the front S-bend. After passing through the weak contraction profile at the middle and rear of the inlet, the air flow velocity increases slightly, the corresponding pressure coefficient decreases slightly, then expands gradually and the flow velocity further decreases. Then it shrinks rapidly at the tail of the inlet and expands into supersonic flow, which is ejected through the tail nozzle. At the 12° angle of attack, because the inlet is on the leeward side, the pressure at the lip is low, and the internal pressure coefficient of the inlet is also low. The variation of the internal velocity and the distribution of the pressure coefficient are consistent with the 0° and -6° angles of attack. At the three different angles of attack, the internal vortices in the inlet have similarities. In general, as the gradual increase of the angle of attack from negative to positive, the internal vortices in the inlet gradually break up, especially the separation vortices generated at the front S-bend break, and the strong vortices gradually close to the inlet wall. In addition, at different angles of attack, the core flow near the tail nozzle is stable, and the strong eddies are only near the wall.

4. Conclusion

In this paper, using CFD to study the characteristics of internal and external flow field of flying wing with knapsack dual inlet configuration. The conclusion is obtained through research.

(1) There is an obvious high pressure zone in the lip area of the knapsack inlet. At negative angle of attack, the high-pressure air flow generated in the back inlet will bypass the leading edge and produce a high pressure area in the abdomen of the model.

(2) After passing through the inlet lip and the front S-bend, the vortex inside the inlet gradually break up as the increase of angle of attack. Near the tail nozzle, the core flow is stable, and the strong eddies are only near the wall.

References

[1] Seddon J, Goldsmith E L 1985 Intake Aerodynamics [M]. Reston: AIAA Education Series.
[2] Gridley M C, Walker S H 1996 Inlet and Nozzle Technology for 21st Century Fighter Aircraft [R]. ASME-1996-GT-244.
[3] Wallis V C, Robert B 2003 Advanced Propulsion System Design and Integration for a Turbojet Powered Unmanned Aerial Vehicle [R] . AIAA 2003-415.
[4] Hamstra J W, Miller D N, Truax P P, et al. 2000 Active Inlet Flow Control Technology Demonstration [J]. Aeronautical Journal, 104: 473-479.
[5] Sun S, Guo R W 2006 Serpentine Inlet Performance Enhancement Using Vortex Generator Based Flow Control [J]. Chinese Journal of Aeronautics, 19(1): 10-17.
[6] Williams T L, Hunt B L, Smeltzer D B, et al. 1981 Top-Mounted Inlet System Feeaibility for Transonic-Supersonic Fighter Aircraft Applications [R]. AGARD-CP-301.
[7] Tao Yang, Li Yonghong, Zhang Zhao, Zhao Zhongliang, Liu Zhiyong 2016 Transonic wing stall of a blended flying wing common research model based on DDES method [J]. Chinese Journal of Aeronautics, 29(6): 1506-1516.
[8] Tao Yang, Xiong Neng, Wang Xiaobing, et al. 2021 Experimental and computational investigation of hybrid formation flight for aerodynamic gain at transonic speed [J], Chinese Journal of Aeronautics, 34(1): 32-43.
[9] Tao Yang, Liu Zhiyong, Xiong Neng, et al. 2018 Optimization of positional parameters of close-formation flight for blended-wing-body configuration[J]. Heliyon, 4(12).
[10] P. Spalart and S. Allmaras 1992 A One-equation Turbulence Model for Aerodynamic Flows [R]. AIAA-92-0439.
[11] Hunt J C R, Wray A A, Moin P. Eddies 1988 Stream, and Convergence Zones in Turbulent Flows [R]. Report CTR-S88, Centre for Turbulence Research.

A quick surface heat flux estimation method for typical structures of high-speed aircrafts

Long Wu[1], Feng Wang[1, 2], Qi Wang[1], Yiyu Han[1] and Zhengzhou Li[1]

[1]Aerospace Technology Institute, China Aerodynamics Research and Development Center, Mianyang 621000, China

[2]Corresponding author's e-mail: FengW@cardc.com

Abstract. Surface heat flux estimation of high-speed aircrafts is an inverse problem and high system ill-posedness usually leading to instability and low accuracy. In this study, an efficient method is proposed to solve it stably. The heat flux identification equation is established based on the superposition principle. Tikhonov regularization method with the parameter determined by generalized cross validation is applied to reduce the sensitivity to measurement error. A calibration method is presented to correct the pulse sensitivity coefficient matrix so that the estimation method can be used for complex three-dimensional structures. Numerical studies demonstrate the efficiency and accuracy.

1. Introduction

The temperature of air rises when an aircraft is traveling at a fast speed because of compression and friction. The air heats the aircraft structure in turn, which is aerodynamic heating. In order to verify the accuracy of thermal environment prediction and investigate the thermal insulation performance of the structure, it is necessary to monitor the temperature response of the structure. Ideally, heat flux or temperature sensors should be installed on the surface. However, considering service conditions of sensors, aerodynamic performance of aircrafts, strength of structure and some other factors, the sensor has to be installed in the structure which is at a certain depth from the surface. Therefore, the surface heat flux must be inversely estimated from the measured temperature inside, which is known as the inverse heat conduction problem (IHCP). The surface heat flux can only be estimated because measurement error is inevitable and the error affects the heat flux result significantly [1].

The earliest papers on IHCP were published around 1960s. Stolz [2] calculated heat transfer rates during quenching. Mirsepassi [3-4] studied IHCP for semi-infinite geometries both numerically and graphically. A variety of approaches for IHCP have been developed since then. For linear cases, there are convolution integral theorem and methods based on Laplace transforms. Beck [5] did a lot of work on the Duhamel's theorem. Sparrow and Grysa [6-7] presented the methods based on Laplace transforms. For nonlinear problems, numerical procedures such as finite differences and finite elements were employed. Blackwell [8] introduced an iterative technique based on finite differences to solve one-dimensional problems and a finite element method was applied by Bass [9]. IHCP is an inverse problem and is mathematically "ill-posed". The heat flux result is extremely sensitive to measurement errors of the interior temperature. Various methods have been developed to reduce the sensitivity which are called regularization methods. The most classic one was presented by Tikhonov [10], which reduces the ill-posedness by controlling the 2-norm of solution. Hansen [11] expounded, compared and summarized different kinds of regularization methods. With further research, many new approaches on IHCP have been developed for different application areas. Ku [12] presented a collocation method with space-time radial polynomials to solve two-dimensional IHCP. Sun [13] documented a meshless method for the three-dimensional IHCP based on the method of fundamental solution (MFS). Duda [14] formulated a new method of solving multidimensional heat conduction problems and used it to solve a non-linear two-dimensional IHCP. Gu [15] proposed a meshless singular boundary method for three-dimensional inverse heat conduction problems in general anisotropic media. Wang [16] developed a boundary element method (BEM) based on potential theory for inverse analysis of heat conduction in two-dimensional thin-walled structures. Fernandes [17] proposed an analytical transfer function method to solve IHCP. Najafi presented a real time solution [18] for IHCP in a two-dimensional plate with multiple heat fluxes at the surface, and a filter-based solution [19] is presented for IHCP in multi-layer mediums. Mohasseb [20] developed a novel method based on SBFEM (Scaled boundary finite element method) and a hybrid optimization algorithm. Ran [21] presented a modified space marching method using future temperature measurements for transient nonlinear IHCP.

This article presents a quick surface heat flux estimation method for typical structures of high-speed aircrafts. The heat flux identification equation is established in section 2.1. , and the regularization method is illustrated in section 2.2. A calibration method for three-dimensional structure is presented in section 2.3. Numerical examples are studied in section 3. , which shows the correctness and effectiveness of the method.

2. Problem formulation

2.1. Heat flux identification equation

Typical structures of high-speed aircrafts like aircraft skin, engine wall panel and compression surface usually have much larger size in the length and width direction than thickness direction. Therefore, one-dimensional simplified model can be adopted. In order to avoid iteration and improve the identification speed, it is assumed that the density, thermal conductivity and specific heat capacity are constants. Thus, we can derive the heat flux identification equation for a one-dimensional linear heat conduction problem based on the superposition principle.

Consider the problem: there is a one-dimensional model, the density ρ, thermal conductivity k and specific heat capacity c of the material are constants, heat flux $q(t)$ is applied at an input position and temperature $T(t)$ is recorded at the measuring point. The total length of the model is L .

Time is discretized into $M+1$ points: $t_0,t_1,\cdots,t_i,\cdots,t_M$, $i=0,1,\cdots,M$ and $t_0=0$ is the initial time point. $q(t)$ is discretized into $q_0,q_1,\cdots,q_i,\cdots,q_M$. $T(t)$ is discretized into $T_0,T_1,\cdots,T_i,\cdots,T_M$. Define a unit step heat flux $q_{step\_t_i}(t)$ at the input position and the temperature history.

$$q_{step\_t_i}(t)=\begin{cases}0 & , \quad t<t_i \\ 1 & , \quad t\geq t_i\end{cases} \tag{1}$$

If $q_{step\_t_0}(t)$ causes $\phi(t)$ increase in $T(t)$, $q_{step\_t_i}(t)$ will causes $\phi(t-(t_i-t_0))$ increase in $T(t)$. So the unit heat flux between t_i and t_{i+1} will cause a temperature increase of $\phi(t-(t_i-t_0))-\phi(t-(t_{i+1}-t_0))$. When $t_{i+1}-t_i$ is small enough, the input heat flux can be assumed constant during it.

$$q(t)=q_{i+1},t\in[t_i,t_{i+1}) \tag{2}$$

And heat flux between t_i and t_{i+1} will cause a temperature increase

$$\Delta T^{i+1}(t)=q_{i+1}\left[\phi(t-(t_i-t_0))-\phi(t-(t_{i+1}-t_0))\right],t\geq t_{i+1} \tag{3}$$

So

$$\Delta T^{i+1}(t_j) = q_{i+1}\left[\phi\left(t_j - (t_i - t_0)\right) - \phi\left(t_j - (t_{i+1} - t_0)\right)\right], j \geq i+1 \tag{4}$$

Using equal time steps Δt

$$\begin{aligned}\Delta T^{i+1}(t_j) &= q_{i+1}\left[\phi\left((j-i)\Delta t\right) - \phi\left((j-i-1)\Delta t\right)\right], j \geq i+1 \\ &= q_{i+1}\left[\phi(t_{j-i}) - \phi(t_{j-i-1})\right]\end{aligned} \tag{5}$$

Define

$$\Delta\phi_i = \phi_{i+1} - \phi_i \tag{6}$$

Then

$$\Delta T^{i+1}(t_j) = q_{i+1}\Delta\phi_{j-i-1} \tag{7}$$

Yields

$$T_j = \Delta T_j + T_0 \tag{8}$$

$$\begin{aligned}\Delta T_j &= \sum_{i=0}^{j-1}\Delta T^{i+1}(t_j) \\ &= \sum_{i=0}^{j-1} q_{i+1}\Delta\phi_{j-i-1}\end{aligned} \tag{9}$$

Define $\Delta \boldsymbol{T} \in \mathbb{R}^{M\times 1}, \boldsymbol{T} \in \mathbb{R}^{M\times 1}, \boldsymbol{q} \in \mathbb{R}^{M\times 1}, \boldsymbol{I} \in \mathbb{R}^{M\times 1}, \boldsymbol{X} \in \mathbb{R}^{M\times M}$

$$\Delta\boldsymbol{T} = \begin{bmatrix}\Delta T_1 & \Delta T_2 & \Delta T_3 & \cdots & \Delta T_M\end{bmatrix}^{\mathrm{T}} \tag{10}$$

$$\boldsymbol{T} = \begin{bmatrix}T_1 & T_2 & T_3 & \cdots & T_M\end{bmatrix}^{\mathrm{T}} \tag{11}$$

$$\boldsymbol{q} = \begin{bmatrix}q_1 & q_2 & q_3 & \cdots & q_M\end{bmatrix}^{\mathrm{T}} \tag{12}$$

$$\boldsymbol{I} = \begin{bmatrix}1 & 1 & 1 & \cdots & 1\end{bmatrix}^{\mathrm{T}} \tag{13}$$

$$\boldsymbol{X} = \begin{bmatrix}\Delta\phi_0 & & & & \\ \Delta\phi_1 & \Delta\phi_0 & & & \\ \Delta\phi_2 & \Delta\phi_1 & \Delta\phi_0 & & \\ \vdots & \vdots & \vdots & \ddots & \\ \Delta\phi_{M-1} & \Delta\phi_{M-2} & \Delta\phi_{M-3} & \cdots & \Delta\phi_0\end{bmatrix} \tag{14}$$

Then

$$\Delta\boldsymbol{T} = \boldsymbol{X}\boldsymbol{q} \tag{15}$$

$$\boldsymbol{T} = \boldsymbol{X}\boldsymbol{q} + T_0\boldsymbol{I} \tag{16}$$

$\boldsymbol{X}$ is the pulse sensitivity coefficient matrix

$$\boldsymbol{X}(r,c) = \frac{\partial T_r}{\partial q_c} = \begin{cases}\Delta\phi_{r-c} & , \quad r \geq c \\ 0 & , \quad r < c\end{cases} \tag{17}$$

$\boldsymbol{X}$ contains M unknown quantities $X_1(t_r) = \dfrac{\partial T_r}{\partial q_1}, r = 1,2\cdots M$.

Considering the heat conduction model in which $q_a(t)$, $q_{loss}(t)$, $T_0(x)$ are irrelevant to q_1 and heat flux is applied at $x=0$.

$$\begin{cases} \rho c\dfrac{\partial T(x,t)}{\partial t}=k\dfrac{\partial^2 T(x,t)}{\partial x^2} & , \quad 0\le x\le L, t>t_0 \\ -k\dfrac{\partial T(x,t)}{\partial x}\Big|_{x=0}=\begin{cases} q_1 & , \quad t_0<t\le t_1 \\ q_a(t) & , \quad t>t_1 \end{cases} \\ -k\dfrac{\partial T(x,t)}{\partial x}\Big|_{x=L}=q_{loss}(t) \\ T(x,t_0)=T_0(x) \end{cases} \tag{18}$$

Eq (18) is differentiated with respect to q_1

$$\begin{cases} \rho c\dfrac{\partial X_1(x,t)}{\partial t}=k\dfrac{\partial^2 X_1(x,t)}{\partial x^2} & , \quad 0\le x\le L, t>t_0 \\ -k\dfrac{\partial X_1(x,t)}{\partial x}\Big|_{x=0}=\begin{cases} 1 & , \quad t_0<t\le t_1 \\ 0 & , \quad t>t_1 \end{cases} \\ -k\dfrac{\partial X_1(x,t)}{\partial x}\Big|_{x=L}=0 \\ X_1(x,t_0)=0 \end{cases} \tag{19}$$

Eq (19) can be solved

$$\begin{aligned} X_1^+(x^+,t^+) &= T^+(x^+,t^+), 0<t^+\le\frac{\alpha}{L^2}\Delta t \\ &= T^+(x^+,t^+)-T^+(x^+,t^+-\Delta t^+), t^+>\frac{\alpha}{L^2}\Delta t \end{aligned} \tag{20}$$

In which

$$T^+(x^+,t^+)=t^++\frac{1}{3}-x^++\frac{1}{2}(x^+)^2-\frac{2}{\pi^2}\sum_{n=1}^{\infty}\frac{1}{n^2}\mathrm{e}^{-n^2\pi^2t^+}\cos(n\pi x^+) \tag{21}$$

$$T^+\equiv\frac{T-T_0}{q_{const}L/k}, t^+\equiv\frac{\alpha t}{L^2}, x^+\equiv\frac{x}{L}, \alpha=\frac{k}{\rho C}, X_1^+(x^+,t^+)=\frac{k}{L}X_1(x,t) \tag{22}$$

Eq (15) is the heat flux identification equation in which $\boldsymbol{X}$ can be solved from Eq (20)~(22) for a specific heat transfer model.

2.2. Regularization methods

Section 2.2.1. analyses the problem of solving this type of equation and the necessity of regularization. In section 2.2.2. Tikhonov regularization method is introduced, and its regularization parameter is determined by the Generalized Cross Validation (GCV) method.

2.2.1. The necessity of regularization. In Eq (15), we can only obtain $\Delta\boldsymbol{T}_{\mathrm{n}}$ due to measurement error in reality

$$\Delta\boldsymbol{T}_{\mathrm{n}}=\Delta\boldsymbol{T}_{\mathrm{exact}}+\boldsymbol{n} \tag{23}$$

In which $\Delta\boldsymbol{T}_{\mathrm{exact}}$ is the exact $\Delta\boldsymbol{T}$, and $\boldsymbol{n}$ denotes the measurement error. Solving $\boldsymbol{q}$ from $\boldsymbol{X}$ and $\Delta\boldsymbol{T}_{\mathrm{n}}$ belongs to the second kind of inverse problem, the direct solution is

$$\boldsymbol{q}_{\mathrm{naive}}=\arg\min_{\boldsymbol{q}}\|\boldsymbol{Xq}-\Delta\boldsymbol{T}_{\mathrm{n}}\|_2 \tag{24}$$

The exact solution satisfies

$$\boldsymbol{Xq}_{\mathrm{exact}}=\Delta\boldsymbol{T}_{\mathrm{exact}} \tag{25}$$

It can be proved [11] that

$$\frac{\left\|\boldsymbol{q}_{\text{exact}}-\boldsymbol{q}_{\text{naive}}\right\|_2}{\left\|\boldsymbol{q}_{\text{exact}}\right\|_2} \le \frac{\sigma_1}{\sigma_n}\frac{\left\|\boldsymbol{n}\right\|_2}{\left\|\Delta\boldsymbol{T}_{\text{exact}}\right\|_2} \tag{26}$$

In which σ_1 and σ_n are the largest and smallest singular value of matrix X respectively. In this problem σ_1 is usually far greater than σ_n, and deviation of the direct solution is intolerable even though there is only tiny measurement noise. That is why we need regularization methods.

2.2.2. Tikhonov regularization method. Tikhonov regularization method works by controlling the magnitude of $\left\|\boldsymbol{q}\right\|_2^2$, and explicitly incorporates the regularity requirement in the formulation of the problem.

$$\boldsymbol{q}_\lambda = \arg\min_{\boldsymbol{q}}\left\{\left\|\boldsymbol{Xq}-\Delta\boldsymbol{T}_{\text{n}}\right\|_2^2+\lambda^2\left\|\boldsymbol{q}\right\|_2^2\right\} = (\boldsymbol{X}^{\text{T}}\boldsymbol{X}+\lambda^2\boldsymbol{I}_M)^{-1}\boldsymbol{X}^{\text{T}}\Delta\boldsymbol{T}_{\text{n}} \tag{27}$$

In which $\boldsymbol{I}_M$ is an $M \times M$ identity matrix and $\lambda(\lambda>0)$ is a positive regularization parameter that controls the weighting between $\left\|\boldsymbol{Xq}-\Delta\boldsymbol{T}_{\text{n}}\right\|_2^2$ and $\left\|\boldsymbol{q}\right\|_2^2$. The first term $\left\|\boldsymbol{Xq}-\Delta\boldsymbol{T}_{\text{n}}\right\|_2^2$ is the residual which measures the goodness of fit, it should be small enough but still larger than $\left\|\boldsymbol{n}\right\|_2$ as we do not want to fit the noise in $\Delta\boldsymbol{T}_{\text{n}}$. The second term $\left\|\boldsymbol{q}\right\|_2^2$ measures the regularity of the solution, we control it to suppress the large amplitude components dominated by high-frequency components in the naive solution.

2.2.3. Generalized cross validation method. In Eq (27), we need to find a suitable λ, such that $\boldsymbol{Xq}_\lambda$ fit $\Delta\boldsymbol{T}_{\text{n}}$ well and $\boldsymbol{q}_\lambda$ is sufficiently regular at the same time.

λ is determined by generalized cross validation (GCV) method. GCV is a statistician's choice, and the goal is to find a λ such that $\boldsymbol{Xq}_\lambda$ predicts $\Delta\boldsymbol{T}_{\text{exact}}$ as well as possible [11].

$$\lambda = \arg\min_\lambda\{G(\lambda)\} = \min_\lambda\left\{\frac{\left\|\boldsymbol{Xq}_\lambda-\Delta\boldsymbol{T}_{\text{n}}\right\|_2^2}{\left(n-\sum_{i=1}^n\varphi_i^\lambda\right)^2}\right\} \tag{28}$$

2.3. Calibration method for three-dimensional structures

The pulse sensitivity coefficient matrix $\boldsymbol{X}$ solved in Eq (20)~(22) is based on one-dimensional hypothesis, and is no longer accurate when the structure has complex three-dimensional features. However, we can still use the approach above by scaling $\boldsymbol{X}$ to $\boldsymbol{X}_{\text{3D}}$

$$\boldsymbol{X}_{\text{3D}} = \gamma\boldsymbol{X} \tag{29}$$

In which $\boldsymbol{X}_{\text{3D}}$ is the estimated $\boldsymbol{X}$ for three-dimensional models and γ is a ratio constant obtained by numerical calibration. The derivation process of real $\boldsymbol{X}_{\text{3D}}$ is much more complicated, and Eq (29) is an approximate formula which turns out to be accurate and efficient enough for engineering estimation.

Given a heat flux of a step signal $\boldsymbol{q}_{\text{ca}}$, input $\boldsymbol{q}_{\text{ca}}$ to the three-dimensional FEA model of the structure and output the temperature rise $\Delta\boldsymbol{T}_{\text{3D}}$,then

$$\gamma \equiv \frac{\text{mean}(\Delta\boldsymbol{T}_{\text{3D}})}{\text{mean}(\boldsymbol{Xq}_{\text{ca}})} \tag{30}$$

In which "mean" is an operation symbol to calculate the average value of the vector components.

3. Numerical studies

In this section two typical structures of high-speed aircrafts are studied, the density ρ, thermal conductivity k and specific heat capacity c of the material are constants and known. (ρ=7850kg/m^3 , $k = 44.5\mathrm{W}/(\mathrm{m}\cdot\mathrm{K})$, $c=475\mathrm{J}/(\mathrm{kg}\cdot\mathrm{K})$).

The basic idea is: given the number of sampling points M , sampling interval Δt, exact surface heat flux $\boldsymbol{q}_{\text{exact}}$, $\boldsymbol{X}$ is solved from Eq (20), then $\Delta \boldsymbol{T}_{\text{exact}}$ is calculated by three-dimensional linear finite element analysis (FEA), and $\Delta \boldsymbol{T}_{\text{n}}$ is obtained by adding noise. The method in section 2. is used to identify $\boldsymbol{q}_{\text{inverse}}$. Compare $\boldsymbol{q}_{\text{inverse}}$ and $\boldsymbol{q}_{\text{exact}}$ by calculating the relative error r .

$$r = \frac{\left\|\boldsymbol{q}_{\text{inverse}} - \boldsymbol{q}_{\text{exact}}\right\|_2}{\left\|\boldsymbol{q}_{\text{exact}}\right\|_2} \times 100\% \tag{31}$$

Substitute $\boldsymbol{q}_{\text{inverse}}$ to the FEA model and obtain $\Delta \boldsymbol{T}_{\text{inverse}}$ inversely, then compare $\Delta \boldsymbol{T}_{\text{exact}}$, $\Delta \boldsymbol{T}_{\text{n}}$ and $\Delta \boldsymbol{T}_{\text{inverse}}$. Mathematical equations for section 3.1. and section 3.2. are not rewritten since the formulas have been presented in section 2.

3.1. Surface heat flux estimation of a thin-wall plate structure

Many typical high-speed aircraft structures like skin, engine wall panel and compression surface can be simplified to thin-wall plate structures. So we study a thin-wall plate structure as the first example: a $0.2\mathrm{m}\times0.1\mathrm{m}\times0.01\mathrm{m}$ steel plate is heated on the $z=0$ surface, the initial temperature is set to be $0\mathrm{K}$ so that $\Delta \boldsymbol{T} = \boldsymbol{T}$ (only the temperature increment is meaningful and the absolute value of temperature is meaningless because the physical parameters of the material do not change with temperature), the measuring point is at $(0,0,0.006)$, set $M=1000$ and $\Delta t=0.1\mathrm{s}$. Temperature rise of the measuring point $\Delta \boldsymbol{T}$ is extracted and a random noise is superimposed. The noise amplitude is a certain percentage of the standard deviation of $\Delta \boldsymbol{T}$.

Simulate a relatively simple case: the heat flux is a sine signal. When the noise amplitude is 2% (which means the amplitude of the random signal is 2% of the standard deviation of $\Delta \boldsymbol{T}$), λ can be obtained by the GCV method as shown in Figure 1, $\boldsymbol{q}_{\text{inverse}}$ and $\boldsymbol{q}_{\text{exact}}$ are shown in Figure 2. It can be seen that the result is very accurate and the relative error $r=3.46\%$. If noise amplitude increases, the relative error will also increase. When noise amplitude reaches 5%, the relative error is 7.27%, the estimation is still quite accurate.

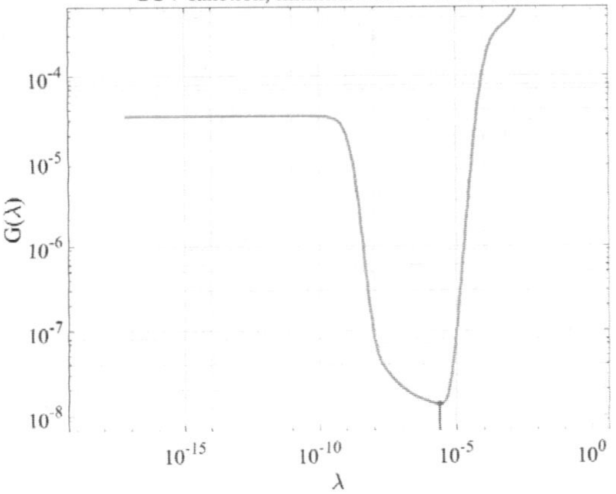

Figure 1. λ based on GCV method (Plate, sine heat flux, 2% noise).

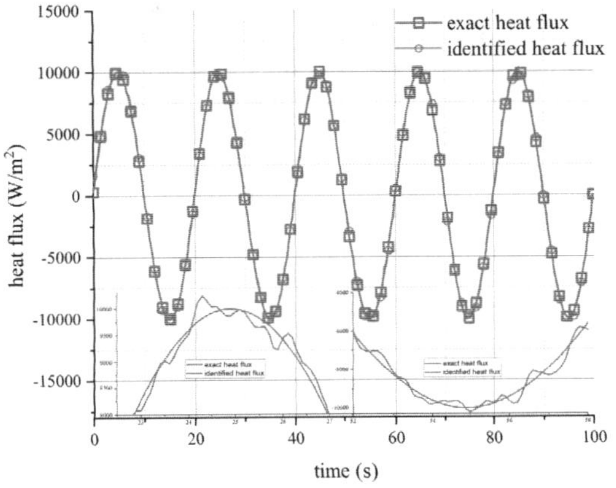

Figure 2. Exact and identified heat flux (Plate, sine heat flux, 2% noise).

Simulate a complex case: the heat flux is a superposition of 3 ramp signals, 1 sine signal with exponential amplitude attenuation, and a random disturbance signal. This is a complex and widely representative heat flux input, the ramp signal represents step change, the sine signal represents regular large fluctuation, and the random signal represents small disturbance noise.

The temperature field result of FEA is shown in Figure 3. The $\boldsymbol{q}_{\text{inverse}}$ and $\boldsymbol{q}_{\text{exact}}$ are shown in Figure 4. When the noise amplitude is 1% and 2%, the relative error is 13.98% and 16.50%. It can be seen that the accuracy decreases when the heat flux history is complex, but we can still identify the general trend of the heat flux, and the accuracy is enough for engineering application. The comparative data for different heat fluxes at different noise levels are shown in Table 1, which shows that the relative error increases with the increase of noise, and λ has to be increased continuously to enhance the degree of regularization in order to obtain a more stable solution. The error increases more quickly for a complex heat flux.

$\Delta\boldsymbol{T}_{\text{exact}}$, $\Delta\boldsymbol{T}_{\text{n}}$ and $\Delta\boldsymbol{T}_{\text{inverse}}$ are plotted in Figure 5, $\Delta\boldsymbol{T}_{\text{inverse}}$ is in good agreement with $\Delta\boldsymbol{T}_{\text{exact}}$, and does not fit the high-frequency components of the noise, which means that the Tikhonov with GCV method have a good regularization effect even though the error is relatively large when the noise is at 2% level.

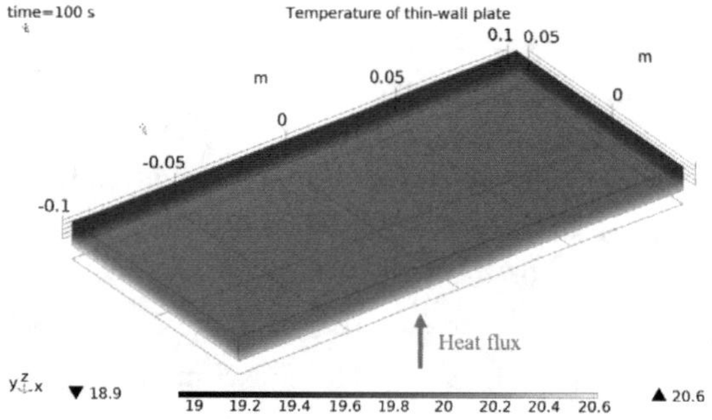

Figure 3. Temperature field result of FEA (Plate, complex heat flux).

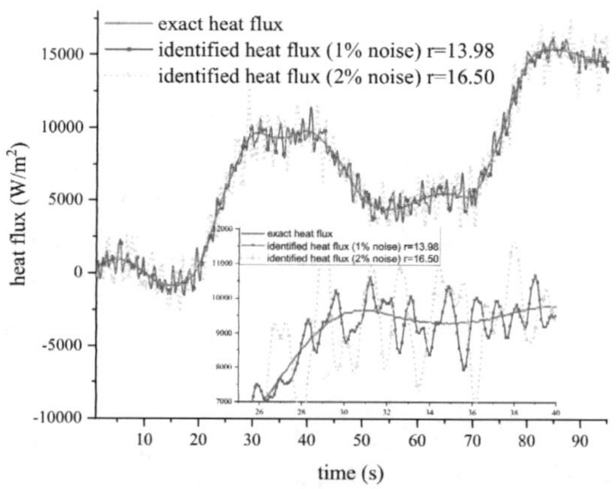

Figure 4. Exact and identified heat flux (Plate, complex heat flux, 1% and 2% noise).

Table 1. Regularization parameter λ and relative error r for different heat fluxes and different noise levels (thin-wall plate).

Noise level		0%	0.1%	0.5%	1%	2%	5%
Sine heat flux	λ	5.20e-8	5.40e-7	1.05e-6	1.47e-6	2.43e-6	3.4e-6
	r (%)	1.58	1.93	3.14	4.27	5.25	6.64
Complex heat flux	λ	8.27e-9	8.92e-7	2.43e-6	4.75e-6	6.63e-6	9.27e-6
	r (%)	6.35	9.73	12.24	13.98	16.50	21.58

Journal of Physics: Conference Series **2313** (2022) 012004 doi:10.1088/1742-6596/2313/1/012004

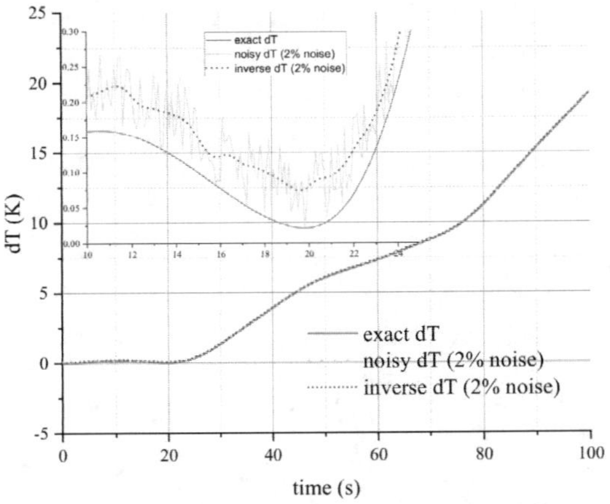

Figure 5. Exact, noisy and inverse dT (Plate, complex heat flux, 2% noise).

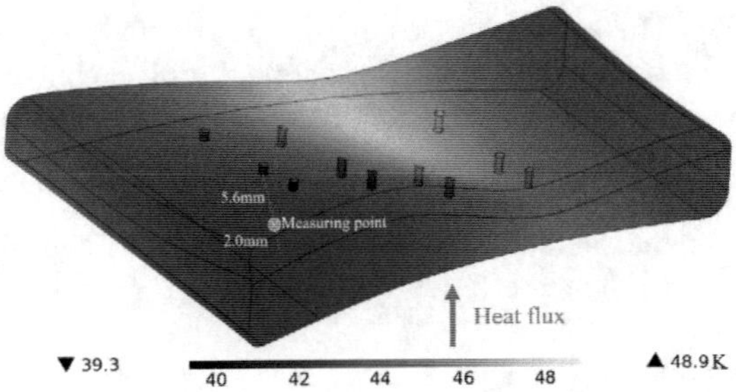

Figure 6. Temperature field result of FEA (3D model, complex heat flux, at 100s).

3.2. *Surface heat flux estimation of a complex 3D structure*

Consider a complex 3D model as shown in Figure 6, the thickness is not uniform and there are small features like holes in the structure, especially there is a hole just above the measuring point. It is generally believed that the error of one-dimensional hypothesis is large in this case, so we need to calibrate the model first. The ratio γ is obtained γ=1.26 by the method in chapter 2.3.

For generality, the complex heat flux in chapter 3.1. is applied. The $\boldsymbol{q}_{\text{inverse}}$ and $\boldsymbol{q}_{\text{exact}}$ are shown in Figure 7, from which we can see the one-dimensional result can only identify the general trend of the heat flux, its average value is larger than the exact heat flux. The accuracy is greatly improved using the method in chapter 2.3. $\Delta\boldsymbol{T}_{\text{exact}}$, $\Delta\boldsymbol{T}_{\text{n}}$ and $\Delta\boldsymbol{T}_{\text{inverse}}$ are plotted in Figure 8, from which we can see the result is a little overfitted, that is why $\boldsymbol{q}_{\text{inverse}}$ has some high-frequency fluctuations.

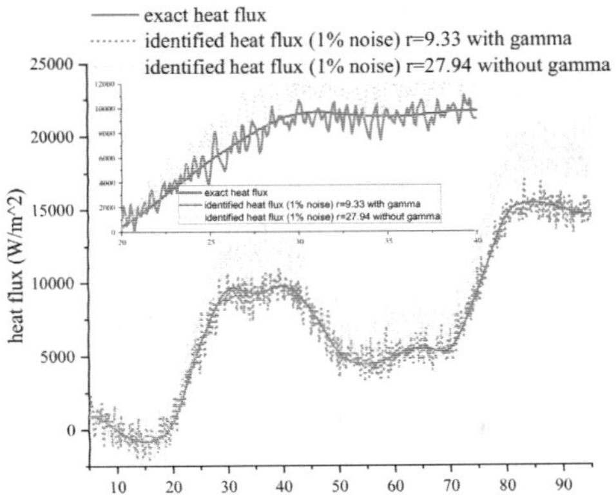

Figure 7. Exact and identified heat flux (3D model, complex heat flux, 1% noise).

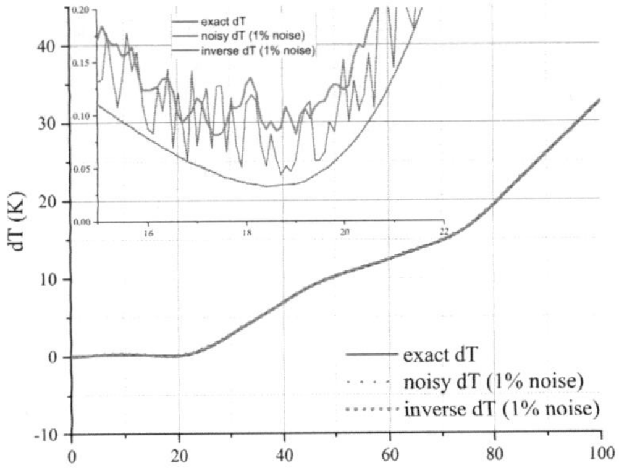

Figure 8. Temperature field result of FEA (3D model, complex heat flux, 1% noise).

4. Conclusion

The following conclusions can be proposed in this article.

1) The heat flux identification equation for typical structures of high-speed aircrafts can be established based on the superposition principle. Tikhonov regularization with parameter determined by GCV can reduce the ill-posedness of the IHCP and solve it stably.

2) The accuracy of estimation is affected by signal feature of the unknown heat flux. For simple sine signal, the error is almost negligible, while precision decreases for complex signals. However, we can still identify the general trend of the heat flux, and the accuracy is enough for engineering application.
3) The pulse sensitivity coefficient matrix in the heat flux identification equation can be corrected by the method in chapter 2.3. , so that complex heat flux on three-dimensional structures can be estimated efficiently.

References

[1] James V. Beck, Ben Blackwell, Charles R. ST. Clair, JR. Inverse heat conduction, ill-posed problems [J]. *Journal of Experiments in Fluid Mechanics*, 2005, 19(1): 1-10.

[2] Stolz, G., Jr. Numerical solutions to an inverse problem of heat conduction for simple shapes[J]. *Heat Transfer*, 1960, 82: 20-26.

[3] Mirsepassi, T. J. Heat-transfer charts for time-variable boundary conditions[J]. *Br. Chem. Eng*, 1959, 4: 130-136.

[4] Mirsepassi, T. J. Graphical Evaluation of a Convolution Integral[J]. *Mathematical tables and other aides to computation*, 1959, 13: 202-212.

[5] James V. Beck. Criteria for comparison of methods of solution of the inverse heat conduction problem[J]. *Nuclear Engineering and Design*. 1979, 53: 11-22.

[6] Sparrow, E. M., Haji-Sheikh, and Lundgren, T. S. The inverse problem in transient heat conduction[J]. *Appl. Mech., Trans. ASME, Series E*, 1964, 86: 369-375.

[7] Grysa, K., Cialkowski, M. J. and Kaminski, H. An inverse temperature field problem of the theory of thermal stresses[J]. *Nuclear Engineering and Design*. 1981, 64: 169-184.

[8] Blackwell, B. F. An efficient Technique for the numerical solution of the one-dimensional inverse problem of heat conduction[J]. *Numerical Heat Transfer*. 1981, 4: 229-239.

[9] Bass, B. R. Applications of the finite element to the inverse heat conduction problem using Beck's second method[J]. *Eng. Ind* 1980, 102: 168-176.

[10] Tikhonov A N, Arsenin V Y . Solutions of ill-posed problems[J]. *Mathematics of Computation*, 1977, 32(144): 491-491.

[11] Hansen P C 2010 Discrete inverse problems: insight and algorithms[M]. Siam Philadelphia.

[12] Ku C Y, Liu C Y, Xiao J E, et al. A collocation method with space-time radial polynomials for inverse heat conduction problems[J]. *Engineering Analysis with Boundary Elements*, 2021, 122: 117-131.

[13] Sun Y, He S. A meshless method based on the method of fundamental solution for three-dimensional inverse heat conduction problems[J]. *International Journal of Heat & Mass Transfer*, 2017, 108: 945-960.

[14] Duda P, Taler J. Numerical method for the solution of non-linear two-dimensional inverse heat conduction problem using unstructured meshes[J]. *International Journal for Numerical Methods in Engineering*, 2015, 48(6): 881-899.

[15] Gu Y, Chen W, Zhang C, et al. A meshless singular boundary method for three-dimensional inverse heat conduction problems in general anisotropic media[J]. *International Journal of Heat & Mass Transfer*, 2015, 84: 91-102.

[16] Wang F, Wen C, Yan G. Boundary element analysis of inverse heat conduction problems in 2D thin-walled structures[J]. *International Journal of Heat and Mass Transfer*, 2015, 91(DEC.): 1001-1009.

[17] Fernandes A P, Santos M, Guimaraes G. An analytical transfer function method to solve inverse heat conduction problems[J]. *Applied Mathematical Modelling*, 2015, 39(22): 6897-6914.

[18] Najafi H, Woodbury K A, Beck J V. Real time solution for inverse heat conduction problems in a two-dimensional plate with multiple heat fluxes at the surface[J]. *International Journal of Heat & Mass Transfer*, 2015, 91(DEC.): 1148-1156.

[19] Najafi H, Woodbury K A, Beck J V. A filter based solution for inverse heat conduction

problems in multi-layer mediums[J]. *International Journal of Heat & Mass Transfer*, 2015, 83: 710-720.

[20] Mohasseb S, Moradi M, Sokhansefat T, et al. A novel approach to solve inverse heat conduction problems: Coupling scaled boundary finite element method to a hybrid optimization algorithm[J]. *Engineering Analysis with Boundary Elements*, 2017, 84: 206–212.

[21] Ran Li, Zhongwei Huang, et al. A modified space marching method using future temperature measurements for transient nonlinear inverse heat conduction problem[J]. *International Journal of Heat & Mass Transfer*, 2017, 106(MAR.): 1157-1163.

Journal of Physics: Conference Series **2313** (2022) 012005 doi:10.1088/1742-6596/2313/1/012005

Numerical visualisation on applying a five-stage multi-scaled square turbulence-generating grid to generating turbulence

Hiroki Suzuki[1, 2]

[1]Graduate School of Natural Science and Technology, Okayama University, 3-1-1 Tsushima-naka, Kita-ku, Okayama-shi, Okayama 700-8530, Japan

[2]Corresponding author's e-mail: h.suzuki@okayama-u.ac.jp

Abstract. This study presents the numerical visualisation of turbulence generated by a five-stage square-type multi-scale turbulence grid. The present study applies a turbulence grid with an increased number of stages to five, in contrast to previous studies using multi-scale turbulence grids with four stages. For the present numerical visualisation, direct numerical analysis is used in this analysis. This analysis uses the high-order central difference schemes and the third-order Runge-Kutta scheme. In addition to the multi-scale turbulence grid with five stages, the present study also covers two turbulence grids with four stages for reference. By using the external force term in the governing equations, the present multi-scale turbulence grids were formed numerically in the computational domain. The spatial homogeneity of the downstream turbulence field is approached in this work. The turbulence generated by the turbulence grids with four stages is characterised by a flow structure with a doughnut-like instantaneous velocity figure in the downstream region. By examining contour diagrams of the streamwise instantaneous velocity, the present study can see that the spatial homogeneity of the turbulence field downstream is improved by increasing the number of stages for the turbulence grid.

1. Introduction

Incompressible turbulent flows are widely found in flows of engineering applications [1]. Because the magnitude of the eddy kinematic viscosity of turbulent flows is generally much larger than that of the molecular kinematic viscosity, heat and mass in turbulent flows can disperse more rapidly than in laminar flows. Therefore, the turbulent flows may contribute significantly to enhancing the mixing of heat and mass in the fluid. By increasing the intensity of turbulence, we expect that the mixing of these species can be further enhanced.

In the wind tunnel experiments, turbulence flows in a test section of a wind tunnel are generated by a turbulence grid installed upstream. The decay characteristics of the turbulence generated by a turbulence grid have been investigated in previous experimental works [2-6] Increasing the intensity of this turbulence has also often been studied in previous research [7-9]. For example, by using an active turbulence grid, previous studies have generated turbulence with high turbulence intensity in a test section of a wind tunnel. On the other hand, previous studies also generate high-intensity turbulence using a static turbulence grid [10-17]. These previous studies have used a multi-scale turbulence grid to generate high-intensity turbulence. The generation of turbulence by means of these multi-scale turbulence grids may also be used to enhance the mixing of heat and mass [18-24]. The turbulence generated by this type of turbulence grid has also been studied numerically. Numerical analysis can provide, for example, instantaneous cross-sectional velocity profiles, difficult to measure experimentally, with sufficient accuracy.

In most previous studies on turbulence generated by a multi-scale turbulence grid, the number of stages of the turbulence grids is limited to four. Turbulence generated by a turbulence grid with five stages has been studied only a little by a previous wind tunnel experiment on this topic [10], however hardly studied by the previous studies. A large number of minimum grid elements in a turbulent grid with five stages can cover space between the grid elements in a turbulence grid with four stages. The role of these many minimal grid elements should be focused on in engineering applications of multi-scale turbulence grids. Previous studies have also pointed out the significant inhomogeneity of turbulence in the downstream from the multi-scale grid [12,13]. The present study expects that a large number of minimal grid elements generated by increasing the number of stages may improve the homogeneity of the generated turbulence downstream.

This study aims to examine turbulence generated by a multi-scale turbulence grid between four and five stages. Specifically, the present study approaches to clarify effects of increasing the number of stages from four to five on turbulence generated by a multi-scale turbulence grid. The present study uses numerical visualisation to approach this issue. By using numerical visualization, I can examine advantages of using a five-stage turbulence grid with a focus on turbulent structures downstream. This work considers that numerical visualization can be used to clearly and concisely focus on the turbulence structure downstream and to examine the advantage of using a five-stage turbulence grid. The numerical visualisation in this analysis is obtained by direct numerical analysis based on previous studies. In this direct numerical analysis, each of the multi-scale turbulence grids used in the present study is numerically simulated in the computational space. The external force term of the governing equations is used to simulate the turbulence grids numerically, based on the previous studies. This direct numerical analysis uses higher-order central difference schemes and a higher-order Runge-Kutta scheme, which are used based on previous studies. Here, a series of spatial discretisation schemes that can conserve the turbulent kinetic energy is used in the state of inviscid limit. The present grid turbulence field is simulated by these analyses, and the present issue is approached using a contour figure of the instantaneous velocity in the streamwise direction.

2. Numerical methods

This study analyzes turbulence generated by a multi-scale turbulence grid. The present multi-scale turbulence grids are installed upstream of the computational domain. The uniform flow inflows normal to the cross-sectional plane, where the multi-scale turbulence grid is installed. The origin of the coordinate system is set at the centre of this turbulence grid plane. In the coordinate system, the streamwise direction is taken as the x-direction, the transverse direction as the y-direction and the spanwise direction as the z-direction. The upstream boundary condition is set at the position $1/4D$ upstream from the streamwise position where the turbulence grid is installed. Here D is the side of the cross-section of the computational domain normal to the streamwise direction. As the inflow boundary condition, a uniform inflow condition is set. As the downstream outflow boundary condition, a convective outflow condition is used. The gradient of the instantaneous pressure for the streamwise direction is set to be zero at the inflow and outflow boundaries. Periodic boundary condition is set for the two directions normal to the streamwise direction. The settings on these boundary conditions are the same as those in the previous studies [12,19].

A multi-scale turbulence grid is installed at the origin of the streamwise direction. In this analysis, two multi-scale turbulence grids with four stages and one turbulence grid with five stages are used. In this work, the square type [10] is used as the present basic grid element for the multi-scale turbulence grids. For the turbulence grids with four stages, the thickness ratio values are set to be 5 and 8.5. For the five-stage turbulence grid, this ratio is also set to 5. For these multi-scale turbulence grids, a value of the fractal dimension is set to two [10]. The geometry of the four stage and five stage square type grids used in this study is set based on the previous studies. Also, the potential advantage of the five stage grids is presented in this study in terms of turbulence structure using numerical visualization. These turbulence grids are numerically reproduced in the computational domain. Here, the direct forcing method using the external force term in the governing equations is used to represent the turbulence grids numerically.

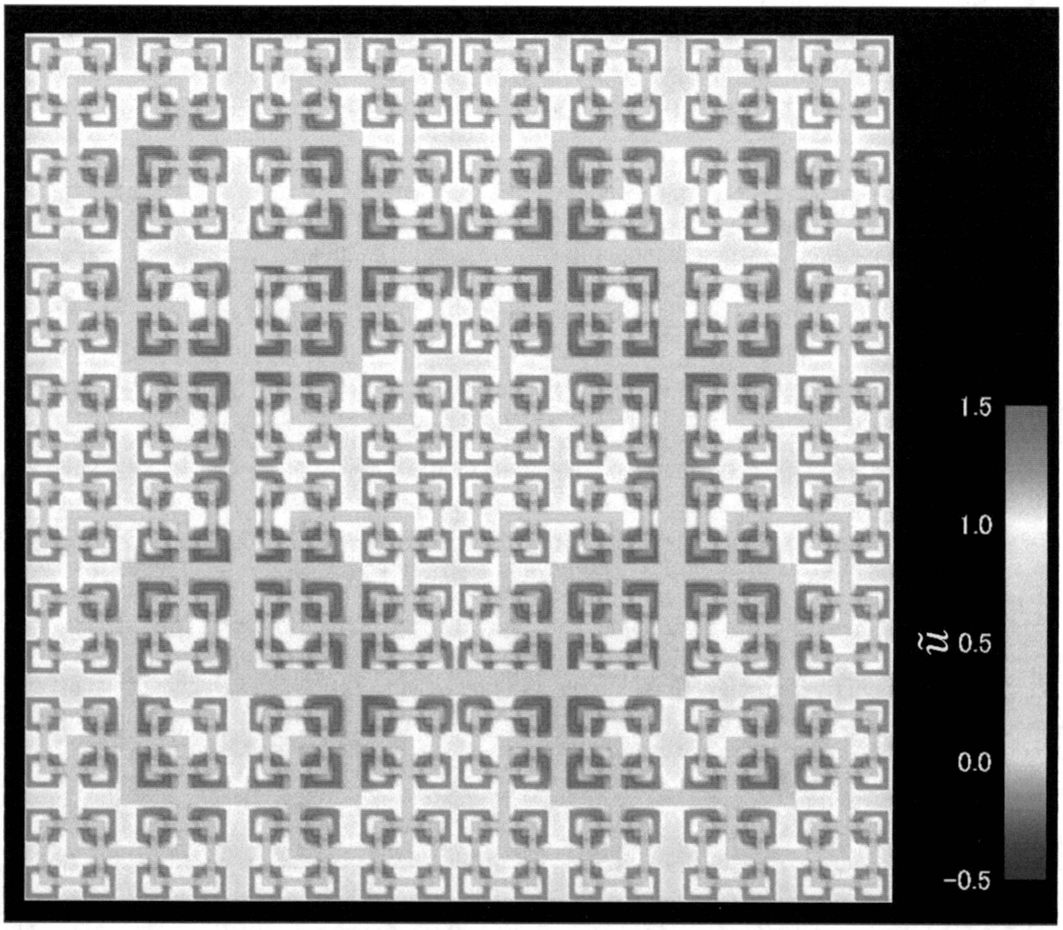

Figure 1. Contour figure of the instantaneous velocity in the mainstream direction in the installed plane of the multi-scale turbulence grid with five stages, where the non-dimensional time is $t = 1$. As shown in the figure, this turbulence grid is reproduced in the computational space.

The computational grid spacing in the streamwise direction is set to be sufficiently small around the turbulence grids. By setting this grid spacing, the flow in the vicinity of each element of the turbulence grids is obtained without the unnatural waves.

The governing equations of the present numerical analysis are the continuity equation and the Navier-Stokes equation for incompressible flows. These equations are non-dimensionalised in the present study. The fractional step method is used to solve the governing equations. The fractional step method in this study is constructed based on the third-order Runge-Kutta method with reduced storage. The Poisson equation given in each fractional step is solved using the direct method. Using this direct method based on the Fourier transform and diagonal matrix algorithm, the error in the continuity equation can be of the order of machine epsilon. The present numerical analysis program developed by our group was used in this study [25].

The staggered grid system is used to discretise the governing equations. The convective terms in the governing equations are discretised using the sixth-order central difference scheme in the streamwise direction and the fourth-order central difference scheme in the two directions normal to the streamwise

Figure 2. A multi-scale turbulence grid with four stages reproduced in the computational domain, where the grid thickness ratio t_r is $t_r = 8.5$. As shown in the figure, this turbulent grid is visualised by an isosurface with zero instantaneous velocity in the streamwise instantaneous velocity.

direction. Here, these higher-order difference schemes are based on the previous studies [12,19]. The viscous terms are analysed using an eighth-order accurate central compact scheme in the streamwise direction, and the Fourier transform in the other two directions. On the basis of a previous study [25], a higher-order compact scheme is used to maintain the accuracy of high-frequency turbulence fluctuations. These discretisation methods are similar to those of the previous study [17].

The side D of the cross-section of the computational domain is taken as the characteristic length in this study. The bulk Reynolds number, defined by the inflow velocity as the characteristic velocity and this characteristic length, is used in this analysis. In the present study, this bulk Reynolds number is set to 40000. For the turbulence grids with four stages, values of the grid thickness ratio tr are set to be $t_r =$ 5 and 8.5. Here, t_r is defined as $t_r = d_0 / d_{N-1}$, where d_0 and d_{N-1} are grid widths of the largest and smallest grid elements, respectively. For the turbulence grids with four stages, grid points in the transverse and spanwise directions are 256 and 320 for $t_r = 5$ and 8.5, respectively. For the turbulence grid with five stages, $t_r = 5$ is set for the grid thickness ratio. The number of grid points in the transverse and spanwise directions is set to 400 for this turbulence grid. For the streamwise flow direction, the number of grid points is set to 768. Temporal and spatial resolution in this study is comparable to that of the previous study [12].

Journal of Physics: Conference Series **2313** (2022) 012005 doi:10.1088/1742-6596/2313/1/012005

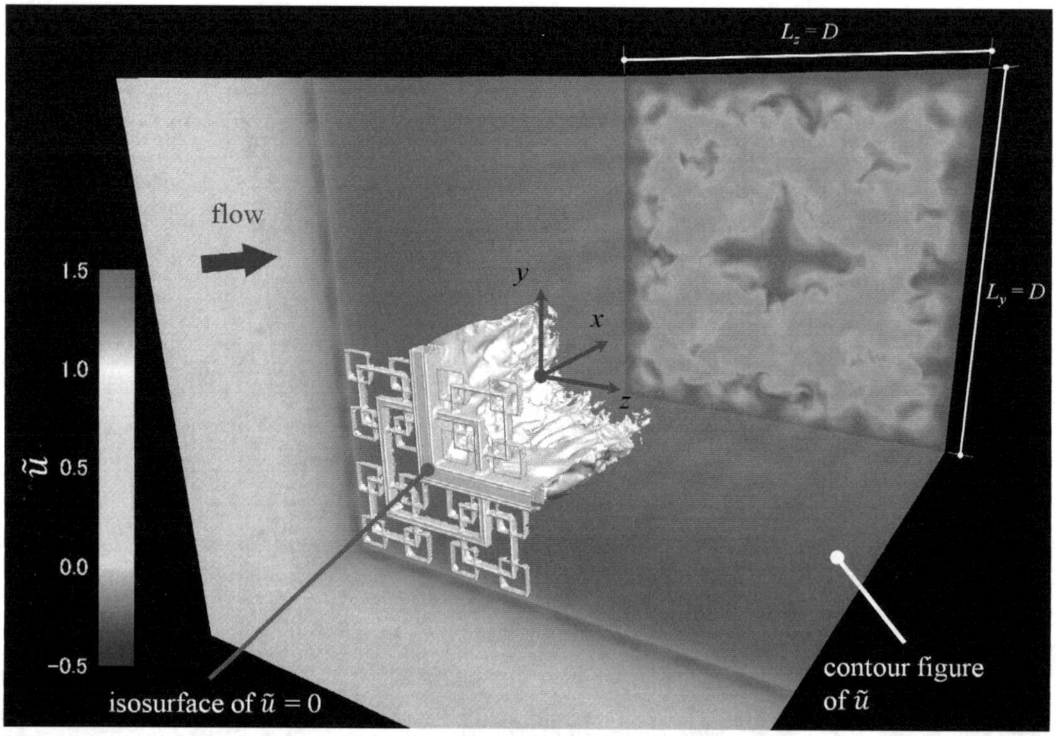

Figure 3. The flow behind the multi-scale turbulence grid with four stages. This turbulence grid is represented by an isosurface with zero velocity in the streamwise direction for a quarter of the total area. The contour figure shows the instantaneous velocity in the streamwise direction. This contour figure of the instantaneous velocity downstream is shown at $x/D = 1.5$. Here the origin is set at the centre of the grid plane.

3. Numerical Results

Figure 1 shows a counter figure of streamwise velocity in the y-z cross-section at $x/D = 0$. The result is shown for the multi-scale turbulence grid with five stages. In the figure, the cross-sectional profile of the streamwise velocity at $t = 1$ is shown. Here, the light blue part represents the value of zero in the streamwise flow velocity. As shown in the figure, we can see that the direct forcing method using the external force term can simulate the shape of the multi-scale turbulence grid with five stages appropriately. The shape of each grid element can be seen from the first grid, where the width of the grid element is the largest, to the fifth grids, where the grid element width is the smallest. As shown in the figure, the fifth grid elements also exist in space between the fourth grid elements correspond to a turbulence grid with four stages. These fifth grid elements may affect the flow in the space between the fourth grid elements. The value of the streamwise flow velocity depends on the position in the turbulence grid. Specifically, regions of a high-velocity value are found near the corners of the first grid. As the time proceeds sufficiently, the velocity increases in the area with large spaces, such as the centre of the first stage grid.

Figure 2 shows the geometry of a multi-scale turbulence grid with four stages. The value of the t_r parameter of this turbulence grid is 8.5. As shown in the figure, we can see that the geometry of each turbulence grid is adequately reproduced over the first to fourth stages. In contrast to the turbulence grid

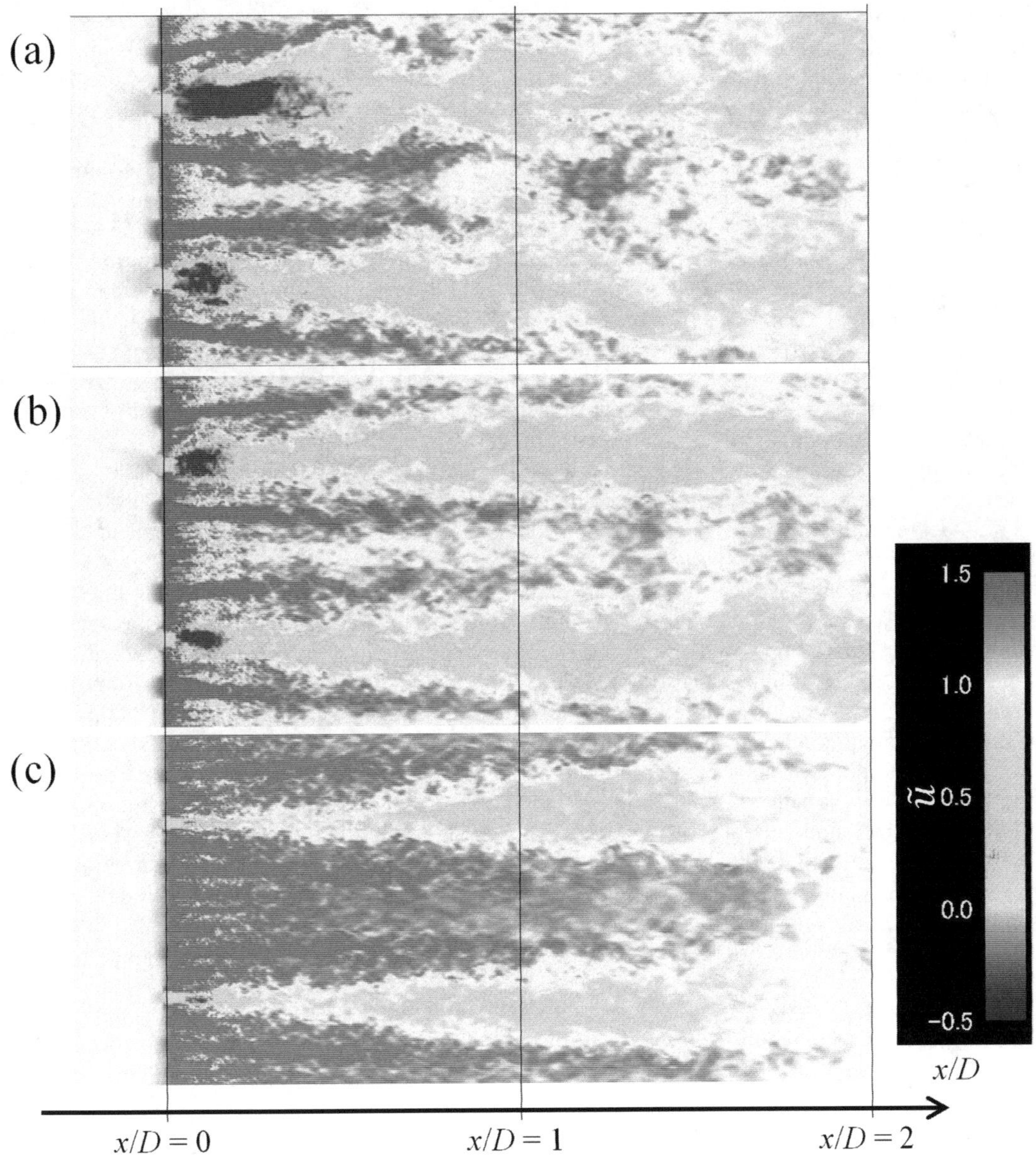

Figure 4. Counter figures of the instantaneous velocity in the streamwise direction for an x-y cross-section, including the central axis. Here, (a): for the multi-scale turbulent grid with four stages under t_r = 8.5. (b): for the turbulence grid with four stages under t_r = 5.0. (c): for the turbulence grid with five stages.

with five stages, the turbulence grids with four stages have a significant space between the grid elements in the fourth stage. We consider that this point is one of the characteristics of the difference in shape between the five-stage turbulence grid and the four-stage grid. Also, in contrast to the turbulence grid with five stages, the width of the first turbulence stage is larger in the turbulence grids with four stages.

This suggests that the effect of turbulence generation by the first-stage turbulence grids is relatively larger in the four-stage turbulence grids than in the five-stage turbulence grid.

Figure 3 visualises a set of contour views of the instantaneous streamwise velocity and the turbulence generated by the turbulence grid with four stages and t_r of 8.5. Here, the generated turbulence due to this turbulence grid is shown by the isosurface of the zero-valued velocity in the streamwise direction. The instantaneous flow field is visualized in this study. The turbulence field varies in time and space. A quarter of the total area of this isosurface is shown to illustrate the results of streamwise velocity in the downstream region. As shown by this zero-valued isosurface of the streamwise velocity, this study observes that turbulence is generated by the present grid with four stages. As shown in the figure, the first stage of the turbulence grid element produces the turbulent flow significantly. The contour of the instantaneous velocity in the streamwise direction is shown at the position $x/D = 1.5$. As shown in the figure, the instantaneous velocity in the streamwise direction is highly inhomogeneous in space. Specifically, there is a region of low velocity in the streamwise direction that forms a doughnut shape. High-velocity regions are observed at the centre and surrounding of the doughnut-like structure. On the other hand, in conventional grid turbulence, which has been widely studied in the previous works, the spatial homogeneity of the downstream streamwise velocity is high. This high spatial inhomogeneity of the instantaneous streamwise velocity is in contrast to the characteristics of conventional grid turbulence. This high degree of inhomogeneity of the main flow is also observed in the results obtained in previous studies (e.g., [13]).

Figure 4 shows contour diagrams of the instantaneous velocity in the streamwise direction obtained on the x-y cross-section at $z = 0$. Here, in this figure, results of the instantaneous velocity in the streamwise direction are compared among the multi-scale grids with four and five stages. For the results obtained by the grid with four stages, the t_r values have been set to 5 and 8.5. As shown in the figure, we can see that turbulence is generated by each multi-scale turbulence grid set at $x/D = 0$. The turbulence generated downstream depends on the shape of the multi-scale grids. The turbulence generated by the turbulence grid with four stages has two large green regions in the downstream. In these green regions, the value of the mainstream velocity is higher than the characteristic velocity. This result indicates that the turbulence obtained by the turbulence grid with four stages has a high spatial inhomogeneity at $x/D = 2$. This non-negligible inhomogeneity is found in the results for both $t_r = 5$ and 8.5. On the other hand, for the turbulence generated by the turbulence grid with five stages, there is higher homogeneity of the instantaneous velocity at the streamwise position of $x/D = 2$. This result implies that the turbulence generated by the turbulence grid with five stages has a higher spatial homogeneity in the downstream region.

4. Discussion and Conclusion

4.1. Discussion

The results of this study are discussed in terms of the spatial homogeneity of generated turbulence. Turbulence generated by the multi-scale grids with four stages may appear to have a relatively insufficient spatial homogeneity, although this homogeneity can be increased in the downstream direction. This insufficient homogeneity is also shown in previous studies (e.g., [10,13]). For example, in the previous studies, the magnitude of this homogeneity is examined using the spatial cross-sectional profiles of turbulence statistics. Also, this observed homogeneity is lower than that of conventional grid turbulence. The present study also shows this lower homogeneity of the turbulence on the multi-scale grid with four stages. As shown in Figures 3 and 4, the instantaneous streamwise velocity in the downstream region is significantly inhomogeneous in space. This homogeneity may be enhanced by increasing the number of stages of a multi-scale grid. This work uses the present multi-scale grid with five stages to generate turbulence. As shown in the figure, the homogeneity of the turbulence generated by the five-stage multi-scale grid is higher than that obtained by the four-stage grid. Therefore, the present study can indicate the possibility of increasing the homogeneity of the turbulence by increasing the number of stages.

The present results are discussed to enhance turbulent mixing for heat and mass. In previous studies [18-21], multi-scale turbulence grids with four stages have been used to enhance turbulent mixing. As seen in the present figure, a turbulence grid with four stages generally has large spaces between the smallest grid elements. Therefore, there are large spaces in the horizontal and vertical planes with the central axis. By reducing these spaces, we expect that the turbulent mixing may be enhanced. In the multi-scale grid with five stages used in this study, these spaces can be reduced. This reduction in the spaces is due to the added stage of the smallest grid elements. As shown in the figure, the addition of the stage of the smallest grid element increases the spatial homogeneity of the downstream velocity in the streamwise direction. Therefore, we can point out that increasing the number of stages of a multi-scale grid from four to five may enhance the turbulent mixing of heat and mass.

4.2. Conclusion

This study aims to clarify effects of increasing the number of stages of a multi-scale turbulence grid on the generated turbulence. In the previous studies, multi-scale grids with four stages are often used. The present study clarifies the effect of increasing the number of stages of a turbulence grid to five on the generated turbulence. In this analysis, direct numerical analysis is used to solve this issue based on the previous studies. This work uses multi-scale turbulence grids with four as well as five stages. Here, for the turbulence grids with four stages, two values of the thickness ratio are used based on the previous studies.

In this analysis, the geometry of the reproduced multi-scale turbulence grid with five stages is firstly seen in the computational space. The spaces in the two-dimensional cross-section with the central axis are reduced in the turbulence grid with five stages than in the grid with four stages. This result can be seen more clearly by comparing the shape of the five-stage turbulence grid with that of the four-stage turbulence grid, which is shown numerically in this study. For the turbulence due to the four-stage turbulence grids, there is a significant spatial inhomogeneity of the instantaneous streamwise velocity downstream. This result for the multi-scale grid with four stages is seen for both conditions of the grid thickness ratio. This inhomogeneity may be reduced by increasing the number of turbulence grid stages from four to five. As shown in the contour figure of the instantaneous velocity in the streamwise direction, the turbulence due to the turbulence grid with five stages is associated with higher homogeneity of the instantaneous velocity in the streamwise direction downstream. Therefore, increasing the number of stages of a turbulence grid from four to five may increase the downstream homogeneity of generated turbulence. This study considers that the advantage of using a five-stage turbulence grid is to enhance the homogeneity of generated turbulence downstream.

As a future study, we consider that turbulence generated by a multi-scale turbulence grid with five stages should first be investigated in more detail. There are fewer previous studies on turbulence generated by turbulence grids with five stages. As shown in the present study, the turbulence generated by the turbulence grids with the number of stages four and five may be significantly different in terms of downstream homogeneity. Applying multi-scale turbulence grids to the enhancement of heat and mass mixing by turbulence has also been investigated in the previous studies. A turbulent grating with five stages has a smaller spatial space between the smallest grid elements than that with four stages. Therefore, the turbulence grids with five stages may enhance turbulent mixing more than those with four stages. We consider that this issue should be approached as future research regarding turbulent mixing enhancement. Specifically, the diffusion of heat and mass in the turbulence produced by a turbulent grid with five stages should be examined.

Acknowledgments

The author is deeply grateful to Emer. Prof. Yasuhiko Sakai (Nagoya University) and Prof. Koji Nagata (Nagoya University) for their support in carrying out this work and their advice. The authors would also like to thank Prof. Toshiyuki Hayase (Tohoku University) for his efforts in the use of the supercomputer. The simulations were carried out using the supercomputer NEC SX-8 at Advanced Fluid Information, Research Center Institute of Fluid Science, Tohoku University. This work was partly supported by the

Japanese Ministry of Education, Culture, Sports, Science and Technology through Grants-in-Aid (Nos. 21K03859 and 22H01684).

References

[1] Pope S B 2000 *Turbulent Flows*. Cambridge University Press
[2] Mohamed M S, LaRue J C 1990 The decay power law in grid-generated turbulence *J. Fluid Mech.* 219 195-214
[3] Kurian T, Fransson J H 2009 Grid-generated turbulence revisited *Fluid Dyn. Res.* 41 021403
[4] Suzuki H, Mochizuki S, Hasegawa Y 2016 Validation scheme for small effect of wind tunnel blockage on decaying grid-generated turbulence *J. Fluid Sci. Tech.* 11 JFST0012
[5] Suzuki H, Mochizuki S, Hasegawa Y 2018 Numerical-based theoretical analysis on effects of weak fluid acceleration of free-stream due to wind-tunnel blockage on grid-generated turbulence *Flow Meas. Instrum.* 62 1-8
[6] Lavoie P, Djenidi L, Antonia R A 2007 Effects of initial conditions in decaying turbulence generated by passive grids *J. Fluid Mech.* 585 395-420
[7] Hideharu M 1991 Realization of a large-scale turbulence field in a small wind tunnel *Fluid Dyn. Res.*, 8 53-64.
[8] Mydlarski L, Warhaft Z 1996 On the onset of high-Reynolds-number grid-generated wind tunnel turbulence *J. Fluid Mech.*, 320, 331-368.
[9] Hearst R J, Lavoie P 2015 The effect of active grid initial conditions on high Reynolds number turbulence *Exp. Fluids* 56 1-20
[10] Hurst D, Vassilicos J C 2007 Scalings and decay of fractal-generated turbulence *Phys. Fluids* 19 035103
[11] Bai K, Meneveau C, Katz J 2012 Near-wake turbulent flow structure and mixing length downstream of a fractal tree *Bound. Layer Meteor.* 143 285-308
[12] Suzuki H, Nagata K, Sakai Y, Hayase T, Hasegawa Y, Ushijima T 2013 Direct numerical simulation of fractal-generated turbulence *Fluid Dyn. Res.* 45 061409
[13] Nagata K, Sakai Y, Inaba T, Suzuki H, Terashima O, Suzuki H 2013 Turbulence structure and turbulence kinetic energy transport in multiscale/fractal-generated turbulence *Phys. Fluids* 25 065102
[14] Weitemeyer S, Reinke N, Peinke J, Hölling M 2013 Multi-scale generation of turbulence with fractal grids and an active grid *Fluid Dyn. Res.* 45 061407
[15] Zhou Y, Nagata K, Sakai Y, Suzuki H, Ito Y, Terashima O, Hayase T 2014 Relevance of turbulence behind the single square grid to turbulence generated by regular-and multiscale-grids *Phys. Fluids* 26 075105
[16] Zhou Y, Nagata K, Sakai Y, Suzuki H, Ito Y, Terashima O, Hayase T 2014 Development of turbulence behind the single square grid *Phys. Fluids* 26 045102
[17] Valente P C, Vassilicos J C 2014 The non-equilibrium region of grid-generated decaying turbulence *J. Fluid Mech.* 744 5-37
[18] Nagata K, Suzuki H, Sakai Y, Hayase T, Kubo T 2008 Direct numerical simulation of turbulent mixing in grid-generated turbulence *Phys. Scr.* T132 014054
[19] Suzuki H, Nagata K, Sakai Y, Hayase T 2010 Direct numerical simulation of turbulent mixing in regular and fractal grid turbulence *Phys. Scr.* T142 014065
[20] Suzuki H, Nagata K, Sakai Y, Ukai R 2010 High-Schmidt-number scalar transfer in regular and fractal grid turbulence *Phys. Scr.* T142 014069
[21] Laizet S, Vassilicos J C 2012 Fractal space-scale unfolding mechanism for energy-efficient turbulent mixing *Phys. Rev. E*, 86, 046302.
[22] Cafiero G, Discetti S, Astarita T 2014 Heat transfer enhancement of impinging jets with fractal-generated turbulence Int. J. Heat Mass Trans., 75 173-183
[23] Gu D, Cheng C, Liu Z, Wang Y. 2019 Numerical simulation of solid-liquid mixing characteristics in a stirred tank with fractal impellers *Adv. Powder Tech.*, 30, 2126-2138.

[24] Teh A L, Siow Y H, Chin W M, Chia C M, Foo J J 2015 Thermal mixing enhancement of a free cooling/heating system with a 2D space-filling plate *App. Therm. Eng.*, 89, 946-957.

[25] Suzuki H, Nagata K, Sakai Y, Hasegawa Y 2013 An attempt to improve accuracy of higher-order statistics and spectra in direct numerical simulation of incompressible wall turbulence by using the compact schemes for viscous terms *Phys. Scr.* 2013 T155 014062

Uncertainty analysis and visualization for a coupled thermohydraulics-neutronics simulation by deterministic sampling

Wen Yang[1, 3], Jun Huang[1], Zhaofei Tian[2] and Guangliang Chen[2]

[1]Institute for Standardization of Nuclear Industry, China Institute of Nuclear Industry Strategy, 43 Fucheng Road, 100048, Beijing, China
[2]College of Nuclear Science and Technology, Harbin Engineering University, 145 Nantong Street, 150001, Harbin, China

[3]Corresponding author's e-mail: yangwen223@outlook.com

Abstract. Uncertainty quantification and sensitivity analysis by deterministic sampling is performed on a thermohydraulics-neutronics coupling simulation of a 5×5 rod bundle in pressurized water reactor. Thermohydraulics calculation is conducted by a CFD simulation and neutronics calculation is conducted by method of characteristic (MOC). The coupling simulation of CFD and MOC, together with large number of calculations needed by random sampling in uncertainty quantification, leads to high computational cost. Therefore, deterministic sampling is introduced in this study and proves its efficiency. Moreover, the calculated results usually have some distribution pattern on the selected section, using mass average value as output is unable to reflect the distribution feature. Therefore, a method realized by Python program is introduced to perform uncertainty quantification and sensitivity analysis of the outputs on the outlet section in fine-mesh level. It is found that the uncertainty of outputs and the inputs' impact on outputs vary in different areas at the outlet section.

1. Introduction

The simulation of pressurized water reactor (PWR) core contains a variety of physics fields, such as neutron physics, thermohydraulics, structural mechanics, and fuel properties. Considering the interactions between different physics fields, a computational fluid dynamics (CFD) simulation is coupled with a three-dimensional neutron physics calculation program based on method of characteristic (MOC). This simulation code [1] enables a more accurate prediction of flow and heat transfer in the PWR core.

In numerical simulations, code computation could contain sources of uncertainty, which could introduce uncertainty in the simulated results. In order to assess the results of the computations, and determine the influence of uncertainty sources on the results, it is necessary to perform uncertain quantification and sensitivity analysis. Common sources of uncertainty [2] are initial and boundary conditions, uncertainties related to the physical models, uncertainties related to the numerical solution, simplification of the geometry, uncertainties due to scaling distortions, uncertainty due to measured previous data, and uncertainty arising from physical instabilities or chaotic behavior.

Uncertainty quantification aims at identifying and quantifying the uncertain inputs related to the simulation purpose and selecting appropriate mathematical methods to quantify the uncertainties presented in the simulation outputs. Uncertainty quantification could be conducted by the uncertainty propagation method. In this method, a set of samples of inputs is selected through sampling techniques, and then the simulation code could generate a set of outputs according to the selected inputs. Afterward, uncertainty information such as mean value and standard deviation could be calculated. Sensitivity analysis aims at determining how the uncertainties of each input affect the overall uncertainty of outputs. Pearson correlation coefficient could provide such information after running a set of simulations mentioned above.

Random sampling is a popular tool to select the samples for inputs according to their probability distribution. At first, Monte Carlo sampling is developed, and it usually needs thousands of samples, which means thousands of times of simulations. In order to reduce the number of calculations, Latin Hypercube Sampling (LHS) is introduced for uncertainty quantification [3]. The number of runs could be reduced to about 100 or even less [4]. However, it is still a problem for uncertainty quantification of CFD/MOC coupling simulations, which would take a large computation cost. Deterministic sampling is developed on the basis of Unscented Kalman Filter [5], it selects samples according to the statistical moments of input variables. The number of runs is significantly reduced when applying deterministic sampling on uncertainty quantification [6].

This paper presents an uncertainty quantification and sensitivity analysis for a PWR 5×5 rod bundle fine-mesh coupled simulation, which couples a CFD simulation for thermohydraulics and a MOC simulation for neutronics. The temperature and velocity of coolant at the outlet section are chosen as outputs considering their importance to the safe circulation of coolant within PWR core. The outputs present different values in different areas of the outlet section, therefore, the distribution pattern is taken into consideration when performing uncertainty quantification and sensitivity analysis. The data are processed by Python after completing the simulations. It is found that the uncertainty of the output varies in different areas. Besides, the inputs' impact on output varies in different areas at the outlet section.

2. PWR 5×5 rod bundle simulation

The coupling model provides a fine mesh level calculation of coupled thermohydraulics and neutronics code. The neutronics calculation is performed by method of characteristics (MOC) and thermohydraulics calculation is a CFD simulation, performed by ANSYS FLUENT© V.16.

2.1. *PWR fuel assembly*

A typical fuel assembly for PWR is composed of fuel rods, guide tubes, an instrument tube, and spacer grids, as shown in Figure 1. The fuel rod is usually around 4 meters long, with 9.5 mm outer diameter and 12.6 mm bundle pitch. Inside the fuel rod, nuclear fuel pallets are made of UO_2, which generates heat. The cladding is made of zirconium alloy. An average fuel rod power density for PWR with 1000MW thermal power is 187.3W/cm. Fuel assemblies are cooled by the coolant, which is water under 15 MPa pressure at the PWR cores.

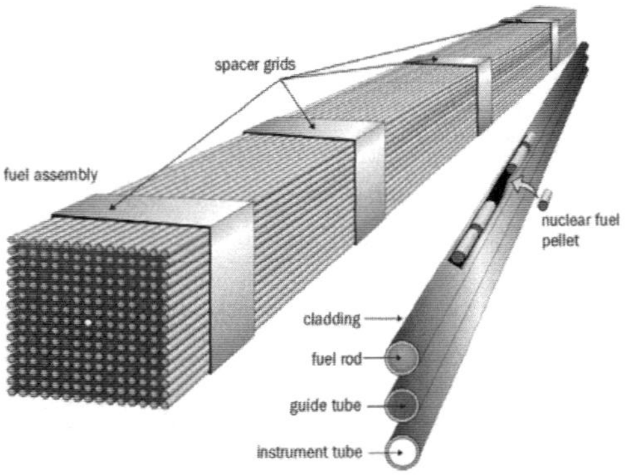

Figure 1. A typical fuel assembly for PWR.

2.2. Simulation model

In this work, a 5×5 rod bundle with clad and coolant channel was modeled. The geometry of the model is shown in Figure 2. The modeled rod bundle is 560 mm long, clad outer diameter is 9.5 mm and bundle pitch is 12.6 mm. The hydraulic diameter for the coolant is therefore 11.78 mm.

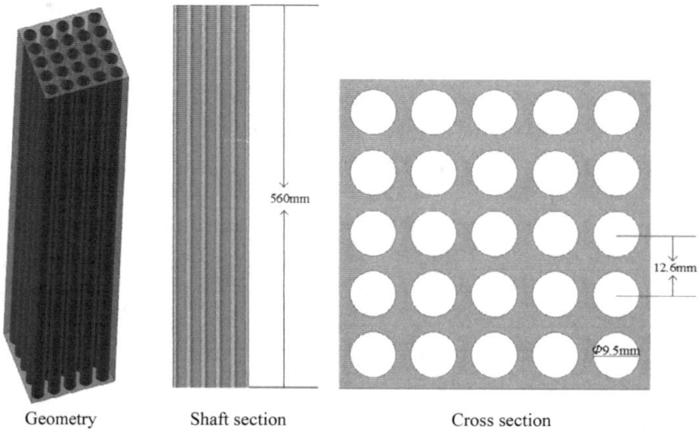

Figure 2. Geometry detail of the simulation model.

2.3. Parameter setting

A previous study [7] has focused on the meshing scheme, numerical calculation method, turbulence model, and boundary condition setting of the coolant area for the 5×5 rod bundle model, main settings are shown in Table 1. For CFD simulation, unstructured mesh was generated, which contains 3172362 cells in total.

Table 1. Parameter setting for CFD simulation.

	Options	Settings
Spatial discretization	Gradient	Least Squares Cell Based
	Pressure	Second Order
	Momentum, turbulent kinetic energy, turbulent dissipation rate, energy	Second Order Upwind
Velocity-pressure coupling algorithm		SIMPLEC
Turbulence model		SKE
Wall function		Standard wall functions
Wall boundary condition		Periodic boundary
Meshing		Unstructured
Inlet boundary condition		Velocity
Property of materials		User-defined, linear Interpolation based on pressure and temperature
Outlet boundary condition		Outflow

Considering that the cross-section is always the same at different axial positions, prism meshing was chosen in this study. Different meshing schemes of coolant subchannel have been examined in previous studies [1], unstructured hexahedral prism is chosen at last, because it can provide the same simulation result with the least number of meshes. Figure 3 shows the meshing scheme examined in previous work and Figure 4 shows the analysis of mesh independency.

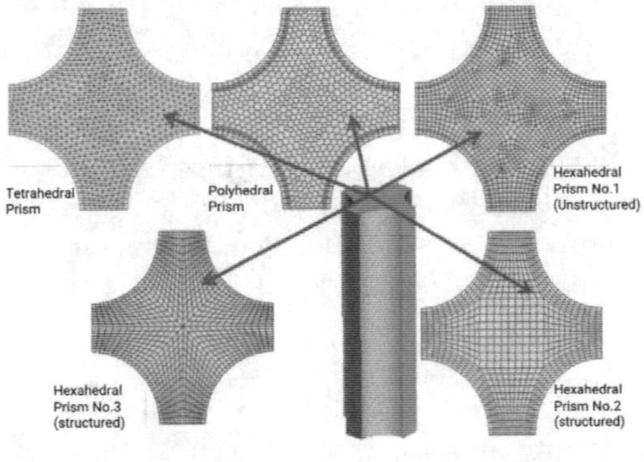

Figure 3. Prim mesh for the simple coolant sub-channel region.

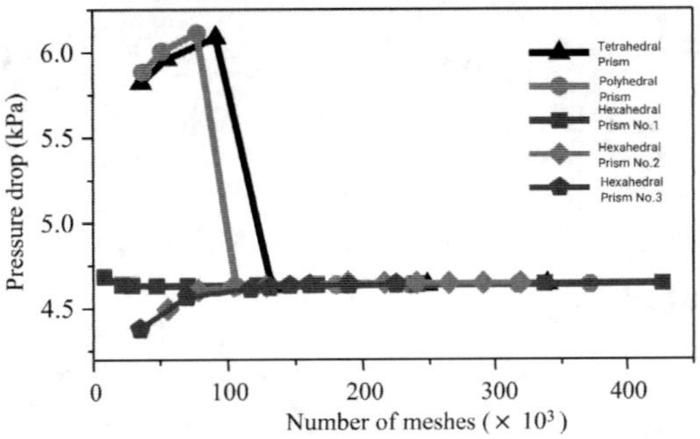

Figure 4. Mesh independency analysis of prim mesh for the simple coolant sub-channel region.

The neutronics calculation uses the three-dimensional neutronics calculation MOC program [8], dividing the 3D transport problem into a coupled 2D-1D problem. The method of characteristics is used for solving the radial 2D equation and Sn diamond difference method is applied in the 1D-axial calculation. Each rod bundle was divided by axial direction (Z1 to Z25) and radial direction (r1 to r3, f1 to f8), which contains 15000 mesh in total. Mesh mapping between CFD and MOC is shown in Figure 5.

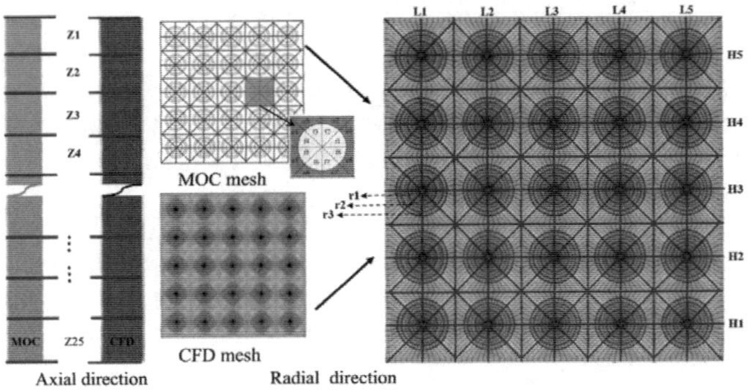

Figure 5. Mesh mapping between CFD and MOC.

2.4. Coupling scheme
During coupled CFD/MOC simulation, on one hand, CFD code transfers fuel temperature, coolant temperature and coolant density to MOC code. On the other hand, MOC code transfers fuel power to CFD code [9]. The transferred data within the mesh mapping scheme was calculated in Eq. 1 and Eq. 2. The symbol $\bar{\rho}$, $\bar{T}$ and V_{tot} are respectively density, temperature, and volume in MOC code. The symbol ρ_i, T_i and V_i are respectively density, temperature, and volume in CFD code. The volume-weighted average method should be used for the density transfer from CFD mesh to MOC cell (Eq. 1), and the mass-weighted average method should be used for temperature transfer from CFD mesh to MOC cell (Eq. 2). The whole process of this coupling simulation is shown in Figure 6.

$$\bar{\rho} = \sum_{i=1}^{n} \frac{\rho_i V_i}{V_{tot}} \tag{1}$$

$$\bar{T} = \sum_{i=1}^{n} \frac{\rho_i V_i T_i}{\bar{\rho} V_{tot}} \tag{2}$$

Monitoring residuals for both CFD and MOC calculation were set at 10^{-6}. The simulation is considered converged when the variation of key parameters, such as maximum fuel temperature, maximum coolant temperature, and maximum fuel power, was less than 10^{-6}.

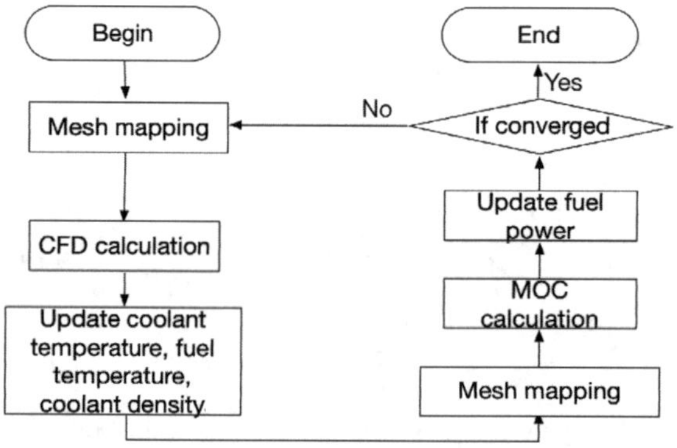

Figure 6. Flowchart for coupled CFD/MOC simulation.

3. Sampling Techniques
The process of uncertainty quantification could be divided into three steps. Firstly, sources of uncertainty should be quantified according to the figure of merit of the simulation. Secondly, sources of uncertainty could be propagated through a set of samples, which are selected by sampling techniques and used as inputs to the simulation. After running simulations for all the samples, a set of outputs could be obtained. Thirdly, post-processing of the obtained outputs is conducted, mean value and standard deviation are usually used to represent uncertainties. In this this study, uncertainty quantification is focused on input uncertainties, such as boundary conditions and heat sources.

3.1. Latin hypercube sampling
Latin hypercube sampling (LHS) is a statistical method to generate random samples. It is among the most popular sampling techniques thanks to its simplicity and projection properties with high-dimensional problems. It divides the cumulated probability function into N different regions, chooses one sample in each region, and finally obtains N samples.

3.2. Taylor expansion for mean value and variance of the output
Uncertainty of the outputs is not related to the number of samples. As is shown in Eq. 3 & 4, a Taylor expansion is performed for the mean value $\langle h(\tilde{q}) \rangle$ and variance $\langle \delta^2 h(\tilde{q}) \rangle$ for an arbitrary function $h(\tilde{q})$, where $\tilde{q}$ is the input parameter, The number of samples is denoted by N, and the ith sample of input is represented by $q^{(i)}$, which could also be represented by its deviation $\delta q^{(i)}$ to the mean value μ. It is

noticed that the uncertainty of outputs is independent to the number of samples. A set of samples that perfectly represents the statistical moments of the inputs shall be enough for uncertainty quantification, regardless of the number of samples. Therefore, the definition of statistical moments and deterministic sampling (DS) are applied to uncertainty quantification.

$$\langle h(\tilde{q})\rangle = \frac{1}{N}\sum_{i=1}^{N} h\left(q^{(i)}\right) = \frac{1}{N}\sum_{i=1}^{N} h\left(\delta q^{(i)} + \mu\right) = \frac{1}{N}\sum_{i=1}^{N}\left[\sum_{j=0}^{\infty}\frac{1}{j!}\delta^{j} q^{(i)}\frac{d^{j}h}{dQ^{j}}(\mu)\right] = \sum_{j=0}^{\infty}\frac{1}{j!}\langle\delta^{j}\tilde{q}\rangle\frac{d^{j}h}{dQ^{j}}(\mu) \quad (3)$$

$$\langle\delta^{2}h(\tilde{q})\rangle = \frac{1}{N}\sum_{i=1}^{N}\left[h\left(q^{(i)}\right) - \langle h(\tilde{q})\rangle\right]^{2} = \sum_{k=0}^{\infty}\sum_{j=0}^{\infty}\frac{1}{j!\,k!}\left[\langle\delta^{j+k}\tilde{q}\rangle - \langle\delta^{j}\tilde{q}\rangle\langle\delta^{k}\tilde{q}\rangle\right]\frac{d^{k}h(\mu)}{dQ^{k}}\frac{d^{j}h(\mu)}{dQ^{j}} \quad (4)$$

3.3. Deterministic sampling

Deterministic sampling (DS) solves the equation of statistical moments to obtain samples. The equation forms as Eq. 5, the samples are denoted by $\{q^{(1)}, q^{(2)}, \cdots q^{(n)}\}$, first and second-order moments of the inputs are respectively denoted by $\langle\delta Q\rangle$ and $\langle\delta^2 Q\rangle$.

$$\begin{cases} \langle\delta Q\rangle = \dfrac{1}{n}\sum_{i=1}^{n}\delta q^{(i)} \\ \langle\delta^{2} Q\rangle = \dfrac{1}{n}\sum_{i=1}^{n}\delta^{2} q^{(i)} \end{cases} \quad (5)$$

This equation is an indefinite equation. Several solutions have been developed using excitation matrix [10], and the calculated samples are in the form of ensembles $\Sigma^{m\times n}$ (shown in Eq. 6), m means the number of inputs and n means the number of samples for each input. $M^{m\times 1}$ is a column vector, it represents the mean value of each input. The symbol $\otimes$ represents Kronecker product. $1^{1\times n}$ is a row vector where all elements are equal at 1. S is an n×n diagonal matrix, it represents the standard deviation for each input. U is a unitary matrix, it represents the correlation between the inputs. And $\hat{V}$ is the excitation matrix. In this study, the DS-Simplex ensemble is chosen as the sampling method because of its ability to reduce the number of samples.

$$\Sigma^{m\times n} = M^{m\times 1} \otimes 1^{1\times n} + U^{T} S U \hat{V} \quad (6)$$

The DS-simplex ensemble uses a unit matrix $I^{m\times m}$ and a column vector $1^{m\times 1}$ in the excitation matrix (shown in Eq. 7). The symbol ⊥ represents Gram-Schmidt orthogonalization. The number of samples is reduced to $n = m + 1$.

$$\hat{V}_{spx} = \sqrt{m+1}\cdot\perp\{I^{m\times m} - 1^{m\times 1}\} \quad (7)$$

3.4. DS compared with LHS

A comparison of DS and LHS method is conducted on the coupled simulation code, peak cladding temperature of the fuel rod is chosen as the figure of merit. Relevant inputs and their uncertainty are recognized, based on a benchmark experiment of OECD/NRC for PWR [11], denoted by Case I in this paper, shown in Table 2.

Table 2. Inputs and their uncertainty for Case I.

Inputs	Mean value	Standard deviation	Probability distribution
Coolant inlet velocity magnitude m/s	4.467	2.234E-02	Normal
System pressure MPa	15.49	5.165E-02	Normal
Total power of fuel rods MW	0.3072	1.024E-03	Normal
Coolant inlet temperature °C	290.0	5.774E-01	Uniform

In Case I, 4 inputs could generate 5 DS-Simplex samples. The number of samples for LHS is determined by Wilks formula, theoretically, 93 samples are needed for 95% quantile and 95% confidence level, therefore, 101 LHS samples were generated in this study. The comparison of results is conducted by calculating the mean value μ and standard deviation V by Eq. 8 & 9, where $h(q^{(i)})$ represents the ith output of the simulation.

$$\mu = \frac{1}{n}\sum_{i=1}^{n} h(q^{(i)}) \tag{8}$$

$$V = \frac{1}{n}\sum_{i=1}^{n}\left(h\left(q^{(i)}\right) - \mu\right)^2 \tag{9}$$

The comparison results are given in Table 3. It is notable that DS provides uncertainty quantification results with the same accuracy as LHS and DS is more efficient than LHS considering the number of simulations. So DS-Simplex method is applied to other simulation cases in this study.

Table 3. Uncertainty quantification for Case I.

Method	Number of samples	Mean value	Standard deviation
DS-Simplex	5	615.41	0.61
LHS	101	615.41	0.57

4. Uncertainty quantification and its visualization

The second case of simulation for PWR steady state operating condition with a lower coolant flow velocity is studied, denoted by Case II in this paper. Two figures of merit are chosen: temperature and velocity of coolant at the outlet section. These parameters are essential to the coolant flow and heat transfer in PWR cores. Four inputs are selected according to those figures of merit, their uncertainties are given in Table 4. The mass average value for the figures of merit on the outlet section is firstly studied, providing an overall description of uncertainty, as is given in Table 5. However, the outputs show a distribution pattern at the outlet section. The areas near fuel rod surfaces tend to have lower velocity and higher temperature.

Python language is used in this study, to calculate the uncertainties of the output parameters grid by grid, and plot them by grid position for all the 14268 meshes at the outlet section. Firstly, simulation results are exported to a TXT file in ASCII format by FLUENT©. Secondly, the data are loaded in the Python program and calculated. Finally, the uncertainty on each grid is represented by a scatter plot according to their coordination. (Shown in Figure 7.)

Table 4. Inputs and their uncertainty for Case II.

Inputs	Mean value	Standard deviation	Probability distribution
Coolant inlet velocity magnitude m/s	1.500	0.7500E-02	Normal
System pressure MPa	15.49	5.165E-02	Normal
Total power of fuel rods MW	0.3072	1.024E-03	Normal
Coolant inlet temperature °C	290.0	5.774E-01	Uniform

Table 5. Uncertainty quantification for Case II.

Figure of merit	Mean value	Standard deviation
Coolant velocity m/s (averaged at outlet section)	1.446	0.007
Coolant temperature °C (averaged at outlet section)	310.991	0.594

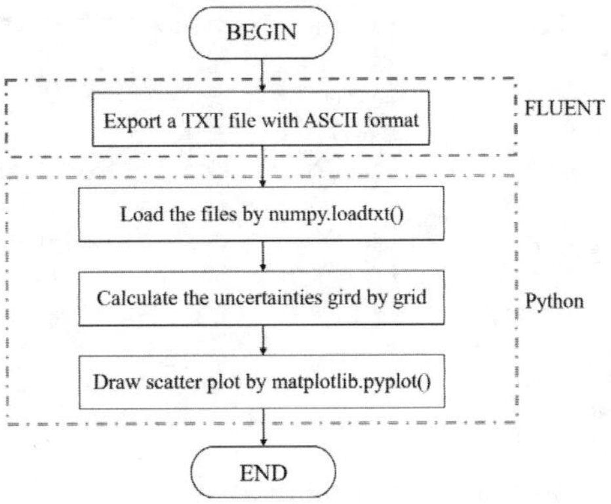

Figure 7. Calculation process for the uncertainty at outlet section.

The mean value of coolant velocity and temperature in each grid is shown in Figure 8. The colors from purple to red indicate the velocity or temperature from small to large, and the blank circular area is the location of fuel rods. Due to the non-slip boundary condition set on the perimeter boundary and at the fuel rod surface, the position close to fuel rod surfaces and coolant perimeter boundaries has a lower velocity (0.66 m/s), while the central area of the coolant channel away from the walls and surfaces has a higher velocity (1.80 m/s). In the area where the velocity is lower and near the fuel rods, the coolant temperature is significantly higher than in other positions, and the maximum value reaches 319.61 °C; while the temperature is lower in the central area of the coolant channel with a large velocity, between 303 °C and 313 °C. Also, at the surrounding boundaries, there isn't any heat source, so the coolant temperature is lower, with a minimum value of 295.97 °C.

As is shown in Figure 9, the distribution of the standard deviation for coolant velocity at the outlet section has the same pattern as its mean value, with a lower standard deviation close to the fuel rod surface and the surrounding walls, a higher standard deviation in the center of the coolant channel. More than half of the grid has a standard deviation greater than the standard deviation of the mass averaged velocity mentioned before. Also, the coolant temperature in the area close to the fuel rods is affected by the fuel power, so the standard deviation in those areas is higher than that of the mass averaged temperature. At the surrounding walls when no heat source is provided, coolant outlet temperature is directly influenced by inlet temperature, so the standard deviation of outlet temperature in those areas is consistent with that of inlet temperature.

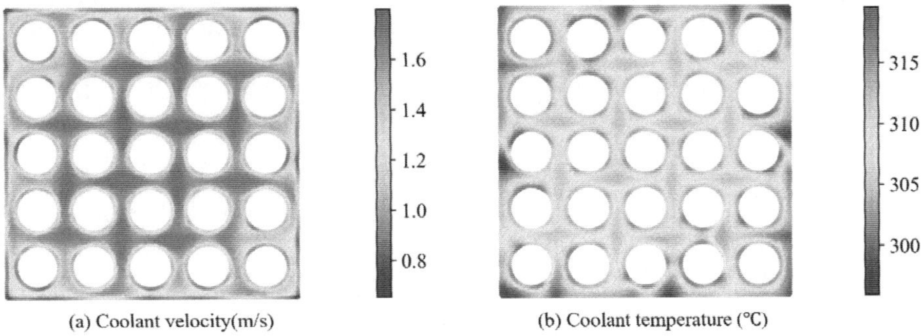

Figure 8. Mean value in each grid at outlet section.

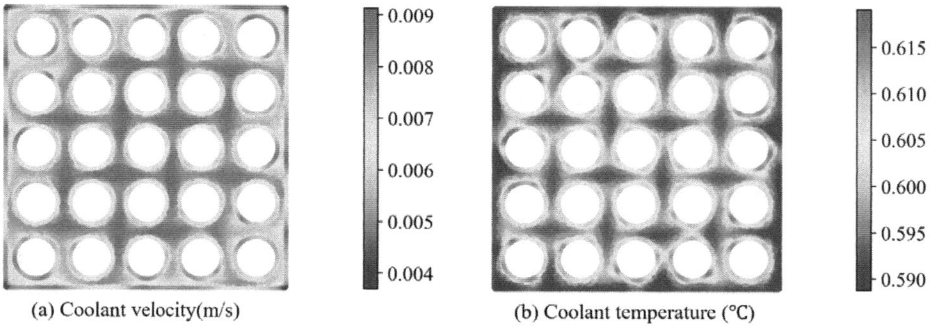

Figure 9. Standard deviation in each grid at outlet section.

In most areas of the outlet section, the uncertainties at each grid are different from uncertainties of mass average values, due to the influence of the coolant flow, inlet temperature, and fuel power in different areas. As a result, the mass average value does not contain all the information for the outputs on the outlet section, and it is necessary to analyze the uncertainties in different areas for fine-mesh level simulations.

5. Sensitivity analysis and its visualization

The sensitivity analysis determines how different inputs affect the output. Pearson correlation coefficient r is usually used to describe the correlation between two parameters, denoted by x and y in Eq. 10. The ith sample and mean value of x and y are respectively represented by x_i, $\bar{x}$ and $y_i, \bar{y}$. The value of r lies between -1 and 1. The negative value shows a negative effect, and the correlation is stronger with a larger absolute value of Pearson correlation coefficient.

$$r = \frac{\sum_{i=1}^{n}(x_i-\bar{x})(y_i-\bar{y})}{\sqrt{\sum_{i=1}^{n}(x_i-\bar{x})^2 \times \sum_{i=1}^{n}(y_i-\bar{y})^2}} \quad (10)$$

The sensitivity analysis of mass average value is given in Figure 10, The blue bar represents the sensitivity analysis for outlet velocity and the red bar with stripe pattern represents the sensitivity analysis for outlet temperature. The horizontal axis presents different inputs, the vertical axis presents the value of person correlation coefficients. It is noticed that outlet coolant velocity is only affected by inlet velocity, and outlet temperature is affected by inlet temperature, fuel power and inlet velocity. System pressure does not affect the figures of merit in this study because the change of pressure is reflected through the change of material properties in this simulation. As a result, four correlations are furtherly studied on the outlet section, including correlations between inlet velocity and outlet velocity, inlet velocity and outlet temperature, inlet temperature and outlet temperature, fuel rod power and outlet temperature. The Pearson correlation coefficients are calculated grid by grid, shown in Figure 8. The correlation is illustrated by grayscale, the darker color represents stronger correlations.

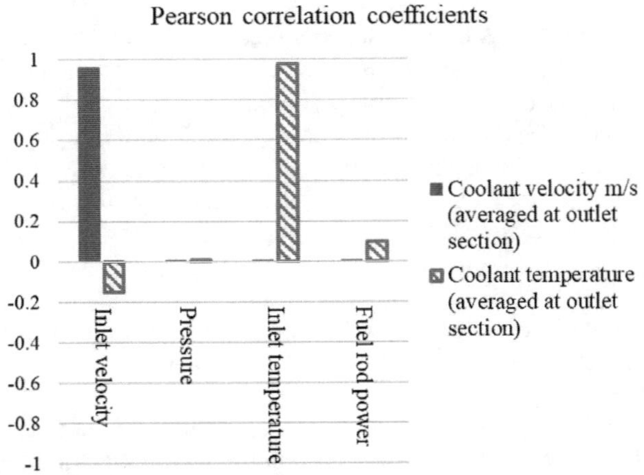

Figure 10. Pearson correlation coefficients between inputs and outputs.

As can be seen from Figure 11, the correlation between inlet velocity and outlet velocity is consistent across the entire outlet section, this is because the velocity in each grid at the outlet section is only affected by the inlet velocity. However, in the areas near the fuel rod surfaces, the outlet temperature is more affected by the inlet velocity and fuel power. In the central area of the coolant channel, the outlet temperature is more affected by inlet temperature, and the influence of inlet velocity and fuel power is reduced. The results of sensitivity analysis grid by grid generally correspond with the results of mass average values. Moreover, it could reveal the difference of correlations in different areas. As a result, performing a sensitivity analysis grid by grid is necessary to study the performance of coolant flow in different areas in PWR cores.

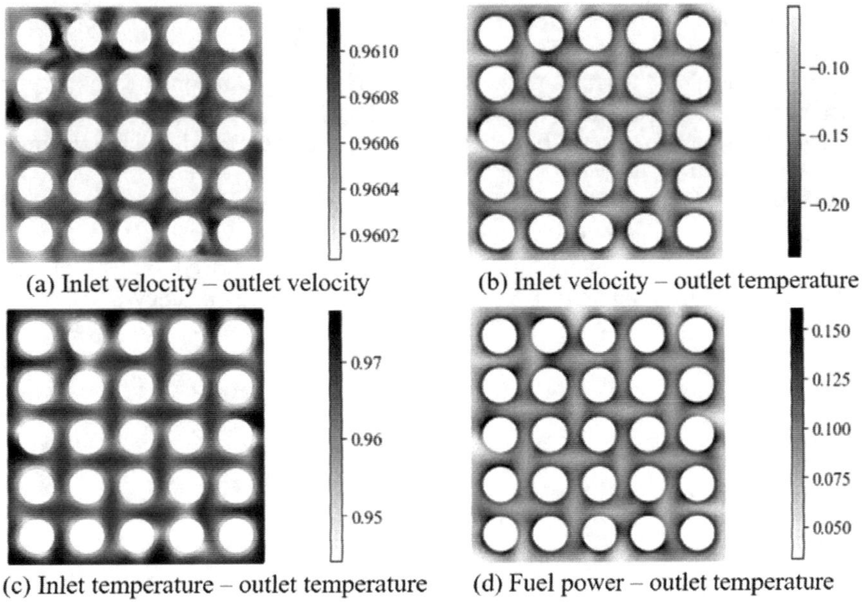

Figure 11. Standard deviation in each grid at outlet section.

6. Conclusions

Uncertainty quantification and sensitivity analysis using deterministic sampling technique for a coupled thermohydraulics/neutronics simulation is studied in this paper. The results prove the accuracy and efficiency of deterministic sampling. A method to analyze and visualize the uncertainties and correlation coefficients for velocity and temperature at the coolant outlet section is proposed in this paper. This new method could better reveal the distribution pattern of those figures of merit, it provides more realistic uncertainty quantification and sensitivity analysis in fine-mesh level than analyzing the mass average values.

References

[1] G. Chen 2017 Research on the Performance of PWR Core via CFD Analysis. Harbin Engineering University.

[2] D. Bestion, A. de Crecy, F. Moretti, R. Camy, A. Barthet, et al. 2016 Review of uncertainty methods for CFD application to nuclear reactor thermalhydraulics. NUTHOS 11-The 11th International Topical Meeting on Nuclear Reactor Thermal Hydraulics, Operation and Safety, Oct 2016, Gyeongju, South Korea.

[3] J.C. Helton, F.J. Davis 2003 Latin Hypercube Sampling and the Propagation of Uncertainty in Analyses of Complex Systems. Reliability Engineering and System Safety, 81(2003): 23–69.

[4] A. Cutrono Rakhimov, D.C. Visser, E.M.J. Komen 2018 Uncertainty Quantification method for CFD applied to the turbulent mixing of two water layers. Nuclear Engineering and Design 333 (2018) 1–15.

[5] S. Julier, J. Uhlmann 2004 Unscented Filtering and Nonlinear Estimation. Proceeding IEEE March, 92 (3) 401-422.

[6] A. Rakhiov, D.C. Visser, K.J Komen 2019 Uncertainty Quantification Method for CFD Applied to the Turbulent Mixing of Two Water Layers–II/Deterministic Sampling for Input Uncertainty. Nuclear Engineering and Design, 348 (2019)146-158.

[7] G. Chen, X. Dong, L. Li, et al. 2018 Research on the Physical Modelling for the Subchannel Analysis of PWR Core. Proceedings of the 2018 26th International Conference on Nuclear Engineering, ICONE26, 1-5.

[8] L. Liang, Z. Liu, Y. Zheng, L. Cao, H. Wu 2017 Leakage reconstruction method for 2D/1D fusion transport calculations. Progress in Nuclear Energy, 2017(97): 60-73.

[9] W. Yang, Z. Tian, G. Chen 2020 Uncertainty Analysis Based on Deterministic Sampling. Nuclear Power Engineering, 2020(S1): 42-45.

[10] J. Hessling 2013 Deterministic Sampling for Propagating Model Covariance. SIAM/ASA Journal on Uncertainty Quantification, 1(1): 297-318.

[11] Nuclear Energy Agency. International Benchmark on Pressurised Water Reactor Sub-channel and Bundle Tests, Volume II. USA: OECD Nuclear Energy Agency, 2016.

Uncertainty analysis of velocity field calibration in wind tunnels based on GUM and MCM

Yufeng Du[1], Neng Xiong[1,2,3] and Jun Lin[1]

[1]China Aerodynamics Research and Development Center, No. 6, South Section, Second Ring Road, Mianyang City, Sichuan Province, P. R. China 621000
[2]College of Aerospace Engineering, Nanjing University of Aeronautics and Astronautics, No. 29 Yudao Street, Nanjing City, Jiangsu Province, P. R. China 210016

[3]Corresponding author's e-mail: xiongneng@cardc.cn

Abstract. The flow quality of wind tunnel is an important index to judge the performance of the wind tunnel, and it directly determines whether the aircraft can obtain high-quality test data, and directly relates to the ability of the wind tunnel to be applied to basic aerodynamic research. Velocity field calibration is necessary in flow field calibration and an important prerequisite for the accurate simulation of Mach number in wind tunnel tests. In this paper, the GUM method and the MCM method are used to compare and analyze the uncertainty of velocity field calibration of the high-speed wind tunnel, and the uncertainty of the Mach number in the range of 0.6 to 4 is solved. The uncertainty evaluation results of the two methods are basically the same, and are equivalent to the wind tunnel Mach number control accuracy value, which verifies the correctness of the evaluation results of the velocity field calibration uncertainty of high-speed wind tunnel, and the MCM method results verify the direct applicability of the GUM method for uncertainty evaluation in similar application scenarios.

1. Introduction

The wind tunnel is a ground test equipment for studying the interaction between aircraft and the atmosphere. The flow quality of wind tunnel directly determines whether the aircraft can obtain high-quality test data, and it is an important indicator for judging the performance of the wind tunnel. Therefore, for the entire life cycle of a wind tunnel from its completion and commissioning to its scrapping and decommissioning, we need to carry out flow field calibration in strict accordance with the requirements of relevant standards, evaluate the flow quality during the entire life of the wind tunnel, and determine whether the wind tunnel can meet the requirements of aircraft development [1-3]. For conventional high-speed wind tunnels, the flow field calibration items mainly include velocity field, direction field, wave-eliminating characteristics of ventilated wall panels, tunnel wall boundary layer, airflow noise, airflow turbulence, airflow instability, etc., and velocity field calibration is a necessary item specified by relevant standards [4]. Through the velocity field calibration, important characteristic parameters such as the root mean square deviation of the Mach number distribution of the high-speed wind tunnel, the Mach number gradient of the flow direction, and the Mach number of the reference point can be evaluated, and the correction relationship between the Mach number of the wind tunnel test and the reference Mach number of the chamber can be established. This is an important prerequisite for the accurate simulation of Mach number in wind tunnel tests. Therefore, accurate calibration of the velocity field is particularly important.

In order to evaluate the accuracy of the velocity field calibration of the high-speed wind tunnel, it is necessary to perform a quantitative analysis on the uncertainty of the Mach number measurement in the calibration process. Uncertainty analysis methods have been widely used in the field of flow field parameter measurement, such as wind tunnel flow field quality measurement uncertainty analysis, rotating machinery unsteady flow parameter test measurement and numerical calculation uncertainty analysis, ship hydrodynamics test parameter uncertainty analysis, etc. [5-15]. However, there are few studies on uncertainty analysis in the field of velocity field calibration which requires extremely high accuracy of test results. Therefore, in order to ensure the accuracy and reliability of the velocity field calibration results of the high-speed wind tunnel, in this paper, we use the Guide to the Expression of Uncertainty in Measurement (GUM) method and Monte Carlo Simulation (MCM) method respectively to evaluate the uncertainty of the velocity field calibration results, and comprehensively analyze the uncertainty results of the GUM method and the MCM method. The results provide a reference for the evaluation of the uncertainty of the velocity field calibration results of the high-speed wind tunnel.

2. Velocity field calibration method and facility

2.1. Velocity field calibration method

For the subsonic and transonic velocity range, the axial detection tube is generally used as the main equipment for velocity field calibration around the world. The axial detection tube is installed on the middle bracket in the test section and coincides with the center line of the test section. The head passes through the test section, the nozzle section to the contraction section, so as to avoid disturbance generated by the head from interfering with the flow field of the test section, thereby ensuring the measurement segment of the axial detection tube to be in almost undisturbed free flow. The core flow Mach number of the wind tunnel test section is calculated by the static pressure measured by the axial detection tube and the total pressure of the settling chamber of the wind tunnel. The specific calculation method is as follows [2]:

$$M = \sqrt{5\left[\left(\frac{p_0}{p}\right)^{2/7} - 1\right]} \tag{1}$$

Where M is the measured Mach number, p_0 is the total pressure in the stable section of the wind tunnel, and p is the static pressure measured by the axial probe tube.

For the supersonic range, the total pressure cross-rake is generally used as the main equipment to carry out velocity field calibration. The total pressure after the shock wave measured by the total pressure cross-rake and the total pressure of the settling chamber of the wind tunnel are used to solve the Mach number of the measuring point by interpolation and iteration according to the following normal shock wave relation [2]:

$$\frac{p_t}{p_0} = \left(\frac{6M^2}{5+M^2}\right)^{3.5}\left(\frac{6}{7M^2-1}\right)^{2.5} \tag{2}$$

Where p_t is the total pressure after shock wave measured by the total pressure cross-rake.

2.2. Facility

The velocity field calibration test is carried out in 1.2m trisonic wind tunnel, which is a cold-flow wind tunnel with a 1.2m×1.2m test section. The Mach number range is from 0.3 to 4.5. The total and static pressure is measured using pressure sensor and scanning valve.

3. Uncertainty evaluation of GUM method

3.1. Introduction to GUM method

The GUM method, as an uncertainty evaluation method based on the transfer law, can directly transfer the uncertainty of the input variable to the output variable through the measurement model and obtain the combined standard uncertainty of the output variable. Because of its simple operation and easy realization, it has been widely used in the field of measurement and effectively solved the uncertainty evaluation problem in various disciplines. It is currently one of the most commonly used measurement uncertainty evaluation methods.

3.2. GUM method evaluation and result analysis

The general expression of combined standard uncertainty is [16]:

$$u_c^2(y)=\sum_{i=1}^{N}\left(\frac{\partial f}{\partial x_i}\right)^2 u^2(x_i) \tag{3}$$

Where $u_c(y)$ is the combined standard uncertainty of the output y, f is the functional relationship between the output y and the input x_i, $u(x_i)$ is the uncertainty of the input x_i, and N is the total number of the input x_i.

According to equations (1) and (3), it can be seen that the combined standard uncertainty of Mach number $u_{c1}(M)$ obtained from the velocity field calibration of the subsonic and transonic range is:

$$u_{c1}^2(M)=\left(\frac{\partial f}{\partial p_0}\right)^2 u^2(p_0)+\left(\frac{\partial f}{\partial p}\right)^2 u^2(p)=\left(\frac{5}{7M}\right)^2\left(\frac{p_0}{p}\right)^{\frac{4}{7}}\left\{\left[\frac{u(p_0)}{p_0}\right]^2+\left[\frac{u(p)}{p}\right]^2\right\} \tag{4}$$

In the same way, according to equations (2) and (3), the combined standard uncertainty of the Mach number $u_{c2}(M)$ obtained from the velocity field calibration of the supersonic range is:

$$u_{c2}^2(M)=\left(\frac{\partial f}{\partial p_t}\right)^2 u^2(p_t)+\left(\frac{\partial f}{\partial p_0}\right)^2 u^2(p_0)=g(M)\cdot\left\{\left[\frac{u(p_t)}{p_t}\right]^2+\left[\frac{u(p_0)}{p_0}\right]^2\right\} \tag{5}$$

Where $g(M)$ is a parameter related only to Mach number, and its expression is as follows:

$$g(M)=\frac{1}{\left\{35\left[\frac{1}{(5+M^2)M}-\frac{M}{7M^2-1}\right]\right\}^2} \tag{6}$$

The $u(p_0)$, $u(p)$ and $u(p_t)$ in equations (4) and (5) can be given according to the accuracy and range of the pressure sensor used in the test, and the combined standard uncertainty of the Mach number can be calculated. The results are shown in Table 1.

Table 1. Results of uncertainty of Mach number.

M	p_0/kPa	$u(p_0)$/Pa	p/kPa	$u(p)$/Pa	M_c	$u_c(M)$
0.6	115	172	91	103	0.588	0.0024
0.8	130	172	86	103	0.792	0.0018
2.0	200	172	143	172	2.013	0.0023
3.0	460	172	150	172	3.008	0.0014
4.0	800	344	111	172	4.000	0.0019

It can be seen from the results in Table 1 that, except for the composite standard uncertainty of Mach number under the condition of M=3 which can be approximated to 0.001, the combined standard uncertainty under the other Mach number conditions can be approximated to 0.002. The magnitude is comparable to the wind tunnel Mach number control accuracy.

Although the GUM method is considered to be very suitable in many cases, in order to ensure the validity of the uncertainty evaluation results, the application of the GUM method is mainly limited by the following premises. Firstly, the measurement model should not contain strong nonlinearity. Secondly, the probability density distribution of the output should be represented by a normal distribution. Thirdly, the magnitude of the output and its uncertainty should not be the same. It can be seen from the above premises that it is not easy to judge whether the conditions for using the GUM method are met.

4. Uncertainty evaluation of MCM method

4.1. Introduction to MCM method

The MCM method provides a general calculation method for measurement uncertainty evaluation, and it shows strong applicability in multiple input quantities and single output variable application scenarios. Compared with the traditional GUM method, the MCM method does not have any restrictions on the measurement model and eliminates the calculation process of the combined standard uncertainty. It is an important alternative to the GUM method when the measurement model is highly nonlinear.

Since the MCM method has a wider range of applications than the GUM method, the MCM method can be considered to investigate the applicability of the GUM method. If the results of the two methods are similar, the GUM method is suitable for the uncertainty evaluation of the velocity field calibration results of the high-speed wind tunnel and similar application scenarios. Otherwise, the MCM method or other more suitable alternative methods need to be considered.

4.2. MCM method evaluation and result analysis

The core idea of the MCM method is to randomly sample within the limited range of the input probability density, and substitute the random samples into the measurement model to obtain the output. Repeat the above process to obtain the output array, and obtain the uncertainty of the output by analyzing the probability density characteristics of the output array. The schematic diagram of the MCM method is shown in Figure 1.

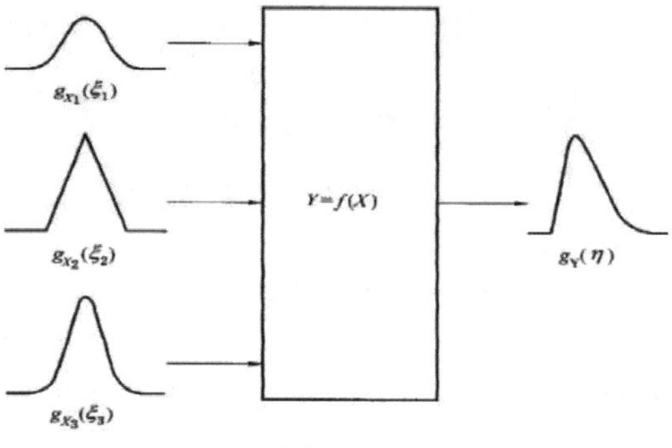

Figure 1. Schematic diagram of MCM.

The specific steps of the MCM method to solve the uncertainty of the velocity field calibration results of the high-speed wind tunnel are as follows.

Firstly, define the input variable, output variable and measurement model. For the measurement model in equation (1), the input is p_0 and p, and the output is M. For the measurement model in equation (2), the input is p_0 and p_t, and the output is M.

Secondly, construct a probability density model of the input. The pressure measurement results can be approximated as a normal distribution with the mean value of the measured value and the standard deviation of 1/3 of the uncertainty, which is conforming to the 3σ principle [17].

Journal of Physics: Conference Series **2313** (2022) 012007 doi:10.1088/1742-6596/2313/1/012007

Thirdly, randomly sample. Use MATLAB to generate a random array that conforms to the above normal distribution, and the size of the array is 10^4. Taking Mach numbers 0.6 and 2 as examples, the distribution histogram of the input quantity is shown in Figure 2 to Figure 5. The vertical axis represents the count of random number which falls in each interval.

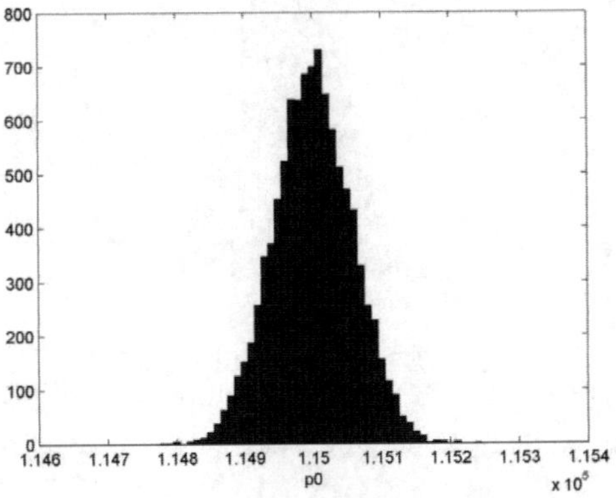

Figure 2. Histogram of p_0 (M=0.6).

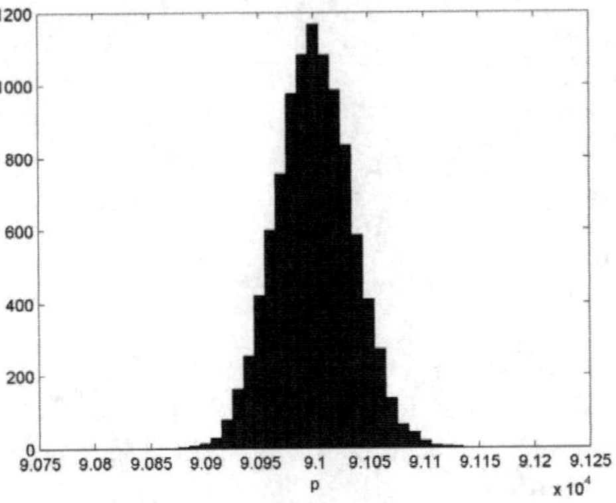

Figure 3. Histogram of p (M=0.6).

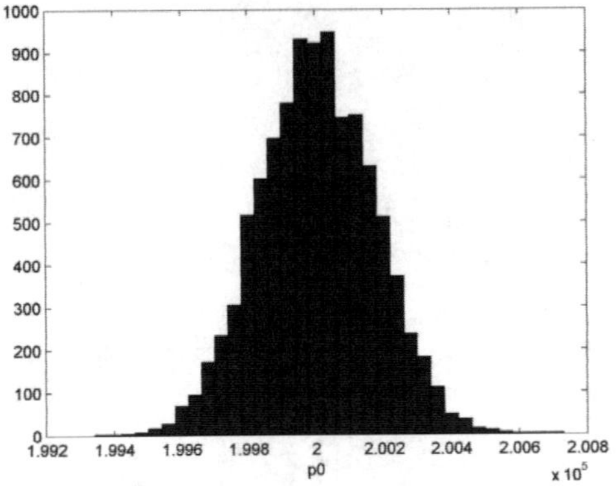

Figure 4. Histogram of p_0 (M=2).

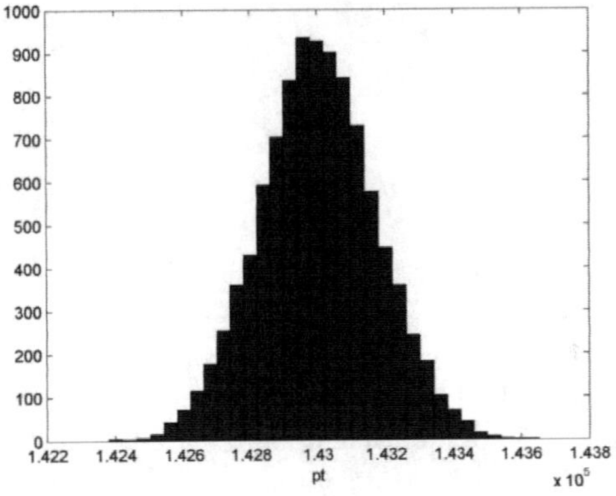

Figure 5. Histogram of p_t (M=2).

Fourthly, calculate the output variable. Taking the generated random array as the input variable and substituting it into the measurement model for calculation, the output variable array is obtained, and the distribution histograms of output variables are shown in Figure 6 and Figure 7. According to the distribution histogram of Mach number, it can be seen that Mach number conforms to the normal distribution.

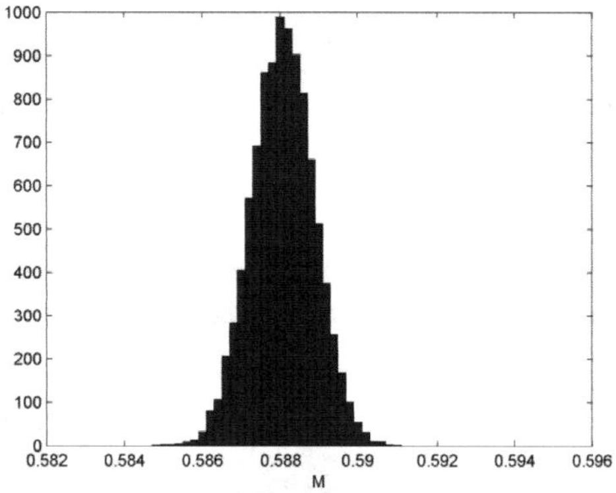

Figure 6. Histogram of M (M=0.6).

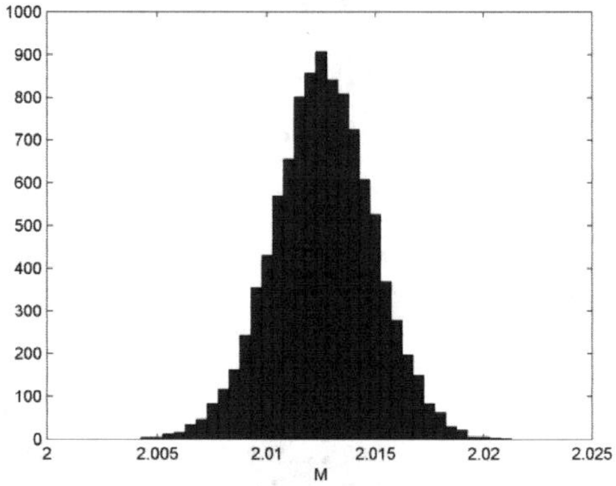

Figure 7. Histogram of M (M=2).

Fifthly, the uncertainty of the output is obtained through the analysis of the statistical characteristics of the output, which is the result of the MCM method evaluating the measurement uncertainty.

The MCM method is used to evaluate the uncertainty of the velocity field calibration of the high-speed wind tunnel, and its comparison with the GUM method is shown in Table 2.

Table 2. Comparison of uncertainty evaluation by MCM and GUM.

M	u_c (GUM)	u_c (MCM)
0.6	0.0024	0.0024
0.8	0.0018	0.0018
2.0	0.0023	0.0022
3.0	0.0014	0.0014
4.0	0.0019	0.0019

From the above comparison results, it can be seen that the uncertainty evaluation results of the MCM method and the GUM method are basically the same, and both are equivalent to the magnitude of the wind tunnel Mach number control accuracy.

5. Conclusion

High-speed wind tunnel velocity field calibration is an important prerequisite for the accurate simulation of Mach number in wind tunnel tests. In this paper, the GUM method and MCM method are used to study the uncertainty of velocity field calibration, and the main conclusions are as follows:

(1) The GUM method and MCM method have basically the same evaluation results of the uncertainty of the velocity field calibration, and both are equivalent to the magnitude of the wind tunnel Mach number control accuracy, which verifies the correctness of the evaluation results of the uncertainty of the velocity field calibration of the high-speed wind tunnel.

(2) The results of the MCM method verify the applicability of directly applying the GUM method for uncertainty evaluation in similar application scenarios.

(3) The applicability of GUM and MCM method in the hypersonic regime will be explored in the future work.

References

[1] Owen F K, Owen A K 2008 Measurement and assessment of wind tunnel flow quality[J]. Progress in Aerospace Sciences, 44: 315-348.

[2] Yun Q L 1991 Experimental aerodynamics[M]. Beijing: National Defense Industry Press, 173-177.

[3] Liu Z C, Liao D X, Dong Y X, et al. 2003 Aerodynamic and structural design of high and low

speed wind tunnels[M]. Beijing: National Defense Industry Press, 282-283.

[4] Chen D H, Wang R B, Liu G Y, et al. 2013 The test and flow field calibration of 2.4m transonic wind tunnel slotted test section[J]. Journal of Experiments in Fluid Mechanics, 27(4): 66-70.

[5] Du Y F, Lin J, Wang X N, et al. 2021 Analysis of modes of disturbances in compressible fluid in subsonic wind tunnel[J]. Acta Aeronautica et Astronautica Sinica, 42(6): 124424.

[6] Du Y F, Lin J, Wang X N, et al. 2019 Measurement technique optimization of turbulence level in compressible fluid by changing overheat ratio of hot wire anemometer[J]. Acta Aeronautica et Astronautica Sinica, 40(12): 123067.

[7] Xu X, Cheng K M, Wang Z J, et al. 2009 Flow field calibration of NUAA φ0.5m hypersonic wind tunnel[J]. Journal of Experiments in Fluid Mechanics, 23(4): 77-81.

[8] Wu Y, Wu D 2020 Calibration and measurement uncertainty evaluation of indication error for wind speed transmitter[J]. Foreign Electronic Measurement Technology, 39(7): 11-16.

[9] Cao J Y 2021 Flow measurement and uncertainty evaluation of pitot static tubes[J]. Instrument Standardization and Metrology, 1: 25-27.

[10] Sun S, Wu R, Liu H 2021 Uncertainty analysis of propeller cavitation fluctuating pressure tests[J]. Journal of Shanghai Ship and Shipping Research Institute, 44(2): 12-17.

[11] Xu Y, Xiong Y, Huang Z 2019Uncertainty analysis of CFD for steady/ transient hydrodynamic performance of propeller[J]. Journal of Experiments in Fluid Mechanics, 31(6): 40-45.

[12] Chang X, Ma K F, Wang C, et al. 2020 Numerical analysis of wake field and tip vortex characteristics of CLT propeller[J]. Journal of Huazhong University of Science and Technology (Natural Science Edition), 48(4): 79-84.

[13] Lei X M, Wu L N, Han Y T, et al. 2020 Uncertainty analysis and evaluation of impeller digital anemometer based on GUM[J]. Foreign Electronic Measurement Technology, 39(3): 30-34.

[14] Wu B S 2007 On geometric parameters in uncertainty analysis of measurement in ship model test[J]. Journal of Ship Mechanics, 11(3): 363-371.

[15] Sun Y, Feng D K, Liu W M, et al. 2020 Uncertainty analysis for sail of submarine based on DDES method[J]. Chinese Journal of hydrodynamics, 35(5): 663-671.

[16] Strutz T 2016 Data fitting and uncertainty: A pratical introduction to weighted least squares and beyond[M]. Springer Fachmedien Wiesbaden, 103-105.

[17] Sheng Z, Xie S Q, Pan C Y 2008 Probability and mathematical statistics[M]. Beijing: Higher Education Press, 42-50.

Journal of Physics: Conference Series **2313** (2022) 012008 doi:10.1088/1742-6596/2313/1/012008

Development of a single-scale initial flow field into steady homogeneous turbulence with validating a constructed Fourier spectral analysis

Hiroki Suzuki[1, 2] and Toshinori Kouchi[1]

[1]Graduate School of Natural Science and Technology, Okayama University, 3-1-1 Tsushima-naka, Kita-ku, Okayama-shi, Okayama 700-8530, Japan

[2]Corresponding author's e-mail: h.suzuki@okayama-u.ac.jp

Abstract. This study presents the development of a single-scale initial velocity field into steady homogeneous isotropic turbulence. Here, we verify a constructed Fourier spectral analysis, which is then used to approach the present issue. The present homogeneous isotropic turbulence is maintained steady by using a linear forcing scheme of Lundgren (2003). Here, two values of the primary coefficient are set here for this scheme. The initial velocity field is also set to maintain the flow globally isotropic. Temporal developments of turbulence statistics, visualization results and structure functions are examined in this study. As shown in a previous study, Rosales and Meneveau (2005), the turbulence length scale at a steady-state could remain independent of a value used for the linear forcing coefficient. The turbulence fields needed 10 to 15 times longer than the turbulence time scale to achieve this steady-state. A larger amount of time was needed for the structure functions to reach their steady-state profiles than for the turbulence statistics.

1. Introduction

Incompressible turbulence is commonly found in industrial equipment, to which the fluids engineering has been relevant [1]. The knowledge of the incompressible turbulent flow has been of fundamental importance in the field of the fluids engineering. Wind tunnel experiments and numerical analyses have been carried out to obtain further the understanding of the incompressible turbulent flows by previous works. These numerical analyses have been characterized as obtaining quantities mainly such as statistics of vorticity and velocity derivatives, which may be rather difficult to obtain by the wind tunnel experiments.

A spectral method [2] is often used to apply numerical analysis to simulating the incompressible turbulent flow. This spectral method can be used in a case where the incompressible turbulent flow is reproduced in a periodic computational domain. In the use of the spectral method, spatial derivatives are simply calculated by multiplying coefficients based on the wavenumber in Fourier space. Therefore, the spectral method can be used to realize high accuracy not only for large-scale flow fields but also for small-scale fields. In contrast, the finite difference methods may underestimate high-frequency fluctuations as shown by a profile of effective wavenumbers obtained for these difference methods. In the spectral method, the pseudo-spectral method has been used to obtain the convective terms. The two-thirds law or the phase shift method is often used to eliminate aliasing errors due to using this method. In the spectral method, the viscous terms can be integrated exactly in time using an integration factor.

Journal of Physics: Conference Series **2313** (2022) 012008 doi:10.1088/1742-6596/2313/1/012008

A multiscale field is used as an initial flow field in the numerical analysis using a three-dimensional periodic domain. Here, the energy density of this multiscale field at each wavenumber is often given by energy spectrum of developed turbulence. For example, this type of initial velocity field has been used in previous studies (e.g., [3]) where steady homogeneous isotropic turbulence was generated using the linear forcing scheme [4-5]. This multiscale initial velocity field seems to have been used recently also in wind tunnel experiments. Here, the initial velocity fields are provided with multiple scales at the same instance using a fractal-shaped turbulence-generating grid [6]. This multiscale turbulence generation using the grids of fractal shape has been studied in succeeding wind tunnel experiments and numerical analysis [7-15].

On the other hand, a single-scale turbulence grid has been widely used in wind tunnel experiments to generate homogeneous turbulence, so-called grid-generated turbulence (e.g., [16-19]). The spatial scale of the shape for this turbulence grid is characterized by the mesh of the grid. We consider the generation of homogeneous turbulence using such the turbulence-generating grids as the use of a single-scale initial field. By taking the grid mesh as the characteristic length, the mesh Reynolds number is defined and used in the previous studies. In the turbulence generation using this kind of turbulence grid, the response to the cross-sectional shape of grid bars has been studied in previous works [20-21]. The single-scale turbulence generation has been maintained over these studies.

In previous works of numerical analysis, a multiscale flow field has often been set as an initial field to generate steady homogeneous turbulence in the analysis using a three-dimensional periodic domain (e.g., [22,23]). In such the cases, a single-scale initial field is not often used to generate this kind of steady turbulence. On the other hand, a single-scale turbulence grid has been widely used to generate homogeneous turbulence in the previous studies of the wind tunnel experiment. Although some previous studies (e.g., [24]) using wind tunnel experiments have compared the nature of turbulence generated between single and multiple scales, few studies appear to have compared the nature of turbulence generation between them.

We considered that the present study should approach this issue based on these perspectives. Therefore, this study aims to investigate the development of steady homogeneous turbulence from an initial flow field with the shape of a single-scale. Specifically, the development of turbulence in a three-dimensional periodic domain from a single-scale field is analysed. Time evolutions of turbulence statistics, visualisation and structure functions in this time domain of developing turbulence are investigated. A numerical code based on the Fourier spectral method [2] is constructed and is used in this study to approach this issue. The Fourier spectral method is expected to be able to analyse not only large-scale turbulence fields but also small-scale fields with sufficient accuracy because operating the spatial derivatives is performed accurately in the Fourier space. The present Fourier spectral code is verified by using the Taylor's analytical solution and the Taylor-Green vortex (TGV) flow [25-26]. The relationship between the generation of turbulence by the multiscale method and the generation by the single scale method has been focused in the previous studies. In addition, previous numerical studies on box turbulence have used multiscale turbulence generation using the spectrum of turbulence. We consider that the results obtained in this study may bridge the gap between these approaches. This study focuses on those two turbulence generation methods. Other turbulence generation methods are presented in previous studies (e.g., [27,28]).

2. Fourier spectral analysis

In this study, incompressible flows are analysed, where, the kinematic viscosity is constant in the flows. The governing equations are the continuity and the Navier-Stokes equations for the incompressible flows [1]. Here, All the physical quantities studied in this work are non-dimensionalised. The governing equations analysed by this study are also non-dimensionalised. The present study aims to analyse the flows in a spatially periodic computational domain. Spatial Fourier transforms can be used in the periodic domain. In the Fourier space, an operation of spatial differentiation for a function can be transformed into that of multiplying the function by coefficients based on the wavenumber. Thus, in the Fourier space, a mathematical form of a differential equation may be simplified. The governing equations for the incompressible flows used in this study can also be transformed to be used in Fourier space. The Fourier-transformed continuity and Navier-Stokes equations [2] are used in this study.

In order to simplify the present governing equations further, these equations are transformed into a pressure-free form using a projection tensor [2]. Here, the governing equations used in this study are the same as those described in previous studies [2-3]. In addition, the governing equations without the viscous terms can be derived by using the integrating factor. By using this integrating factor method, the viscous terms can be analytically integrated in time. The governing equations in Fourier space derived using these two points can then be analysed numerically. For the discretisation of the convection terms, so-called pseudo-spectral method [2-3] is used. This method is considered in this study to be a widely used and a general approach for simulating the homogeneous isotropic turbulence. In this method, the product of the velocity components in the convection terms is obtained in physical space. The values of this product are then also used in the Fourier space by means of the Fourier transform.

The governing equations are developed in time using a highly accurate explicit time integration method in this study. Specifically, in this study, the six-stage fourth-order Runge-Kutta method [29] is used. In this time integration method, the coefficients of this Runge-Kutta method are adjusted so that the numerical dissipation is sufficiently small. Since this pseudo-spectral method [2] is used to discretize the convection terms, aliasing terms are included in the discretised governing equations. These aliasing terms, as error terms, need to be removed and are therefore removed by using the two-thirds rule [2]. Here, this rule is considered in this study to be a widely used method. Therefore, this analysis is free from the aliasing errors, which has been verified in preliminary numerical tests.

3. Verification of the analysis

3.1. Verification using Taylor's analytical solution

This study starts with verifying the present analysis using the analytical solution of the governing equations for the two-dimensional viscous flow. In this study, the widely used Taylor's analytical solution is used. For this analytical solution, time evolutions of the velocity fields are given as follows:

$$u(x,y) = -2\cos(x)\sin(y)\exp(-2t/\mathrm{Re}) \text{ and } v(x,y) = -2\sin(x)\cos(y)\exp(-2t/\mathrm{Re}), \quad (1)$$

where t is the non-dimensional time, u and v are velocity components for x and y directions and Re is the Reynolds number. By using the above forms of the velocity fields, the global kinetic energy K is also given by

$$K = (1/2)\left(\langle u^2\rangle + \langle v^2\rangle\right). \quad (2)$$

Here, $\langle\ \rangle$ denotes the spatial ensemble average. A periodic square computational domain is used in the present simulation, where one of its sides is 2π. The number of grid points for the spatial discretisation N^2 is set to $N^2 = 32^2$. The Reynolds number is set to unity. Based on the time step value of $\Delta t = 0.005$, the governing equations are numerically integrated up to the non-dimensional time of $t = 2$.

Figure 1(a) shows a time evolution of the global kinetic energy in this flow. In this figure, a profile of the kinetic energy obtained in this analysis are compared with that of the analytical solution. The kinetic energy given by the analytical solution can be described by the exponential function. As shown in the figure, this exponential decay of the kinetic energy can also be seen in the present numerical results. Also, the profile obtained in the present analysis are in almost excellent agreement with that of the analytical solution. This result can agree with that the present numerical code has been constructed satisfactorily. Here, we may have to note that a laminar flow is used in this verification. Since the present analysis is intended to be used for analysing turbulent flows, we consider that an additional verification using a turbulent flow field should be performed.

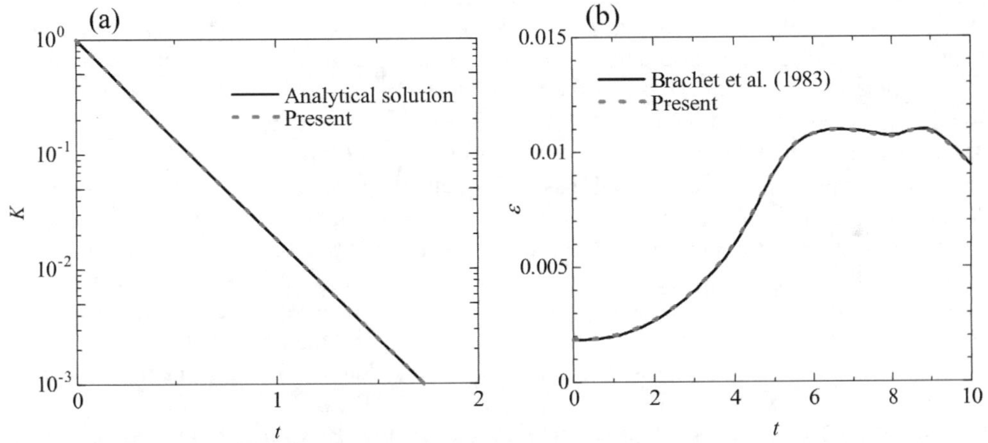

Figure 1. Verification results of the present numerical analysis. (a) Results for Taylor's analytical solution, where K is the global kinetic energy. (b) Viscous dissipation of the turbulent kinetic energy in the developing Taylor-Green vortex (TGV) flow, where $\varepsilon = -dK/dt$.

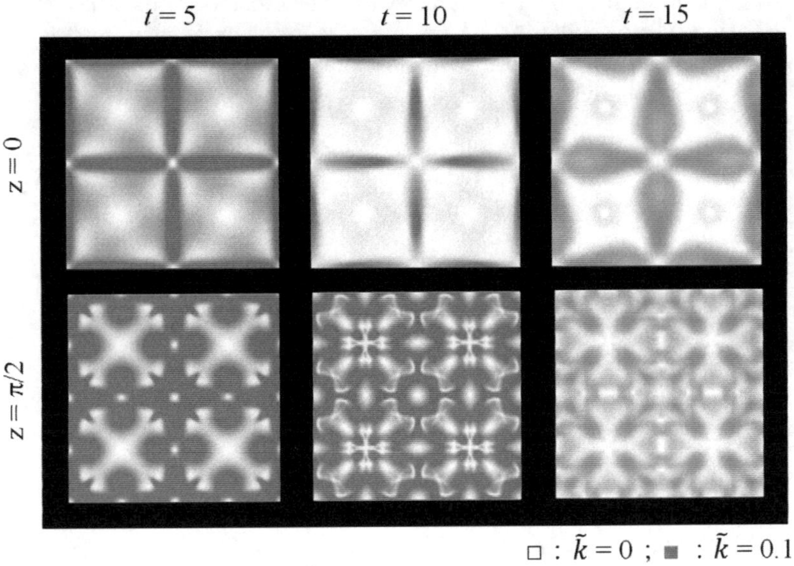

Figure 2. Verification results of this analysis using instantaneous fields of the turbulent kinetic energy in the TGV flow. Here the contour planes are taken at $z = 0$ and $\pi/2$.

3.2. Verification using Taylor-Green vortex flow

The Taylor-Green vortex (TGV) flow [25] is used to verify the present analysis in a turbulence field. This TGV flow has also been widely used in previous studies (e.g., [26]). An initial velocity field of the flow is given as follows.

$$u(x,y,z) = \cos(x)\sin(y)\cos(z),\ v(x,y,z) = -\sin(x)\cos(y)\cos(z) \text{ and } w(x,y,z) = 0. \quad (3)$$

Here, w is a velocity component for z direction. The development of a turbulence field and the decay of the energy from this initial field are analysed. A three-dimensional periodic domain is used as the computational domain, where one side of the domain is 2π. The Reynolds number is set to Re = 400 to compare the present results with those of the previous study. The number of grid points is set to $N^3 = 128^3$ to analyse this flow based on this previous study. For a value of the time increment Δt, $\Delta t = 0.005$ [25], the flow is integrated temporally up to the non-dimensional time $t = 10$ [25].

Figure 1(b) shows a time evolution of dissipation rate of the turbulent kinetic energy, ε. Here, this dissipation rate is given as follows: $\varepsilon = -dK/dt$, where K is the turbulent kinetic energy averaged over the entire computational domain at each instant. The results obtained in this study are verified by comparing with those of this previous study. As shown in the figure of the previous study, a value of the dissipation rate increases with time up to about $t = 5$. Then, the dissipation rate is not likely to change significantly from the time $t = 5$ to $t = 10$. As shown in the figure, the time evolution of this dissipation rate in the present study agrees well with that in the previous study. This agreement is maintained over the time range from the initial time to $t = 10$. The agreement of the present data with that of the previous study can be considered to support that the present analysis is able to analyze the turbulence field with sufficient accuracy.

Figure 2 shows time evolutions for the spatial contour of instantaneous turbulent kinetic energy. Here, an x-y plane of the spatial contour is used. In the previous figure, the spatial mean of the turbulent kinetic energy is used, whereas, in this figure, the instantaneous spatial contour is shown. This study focuses on the instantaneous distribution to verify the symmetry of instantaneous fields. The instantaneous contour figures are shown at $z = 0$ and $z = \pi/2$. As shown in the figure, there is a sufficiently high symmetry in the instantaneous turbulence contour figures. Fine instantaneous structures are found with a sufficiently high symmetry at $z = \pi/2$. These results also support, in the present opinion, the appropriate construction of the present analysis.

4. Numerical results

4.1. Numerical conditions

The present numerical analysis is used to analyze a steady homogeneous isotropic turbulent flow field. The continuous supply of energy using an external force term is needed to maintain a turbulence field steady. Here, velocity components u, v and w are taken for the directions, x, y and z, respectively, and in this study, the external force term is provided by using a linear forcing scheme [3-5]. Specifically, the components of the external force term, F_x, F_y and F_z for x, y and z directions, in the governing equation are set as follows [3].

$$(F_x, F_y, F_z) = (Au, Av, Aw). \tag{4}$$

Here A is a coefficient of the scheme. For comparing the present results with those of the previous study [3], $A = 0.1333$ and $A = 0.0667$ [3] are set. This steady turbulence is reproduced in the three-dimensional periodic domain. One side of this periodic domain is 2π. Physical quantities are discretised by 128^3 grid points in this domain. The Reynolds number is set to Re = 222.67 based on the previous work. This study validates the spatial resolution of the analysis using the Kolmogorov scale. For the two analyses of the present study, the minimum grid width normalized by the Kolmogorov scale is larger than 1.4 [1]. This result indicates that the present analyses are carried out with sufficient spatial resolution.

This study uses a single-scale field as the initial velocity field. This single-scale field is given as follows:

$$\begin{aligned} u(x,y,z) &= (2/3^{1/2})\,(-\cos(x)\sin(y) + \cos(x)\sin(z)),\text{and} \\ v(x,y,z) &= (2/3^{1/2})\,(-\cos(y)\sin(z) + \cos(y)\sin(x)) \text{ and} \\ w(x,y,z) &= (2/3^{1/2})\,(-\cos(z)\sin(x) + \cos(z)\sin(y)).\text{ nd} \end{aligned} \tag{5}$$

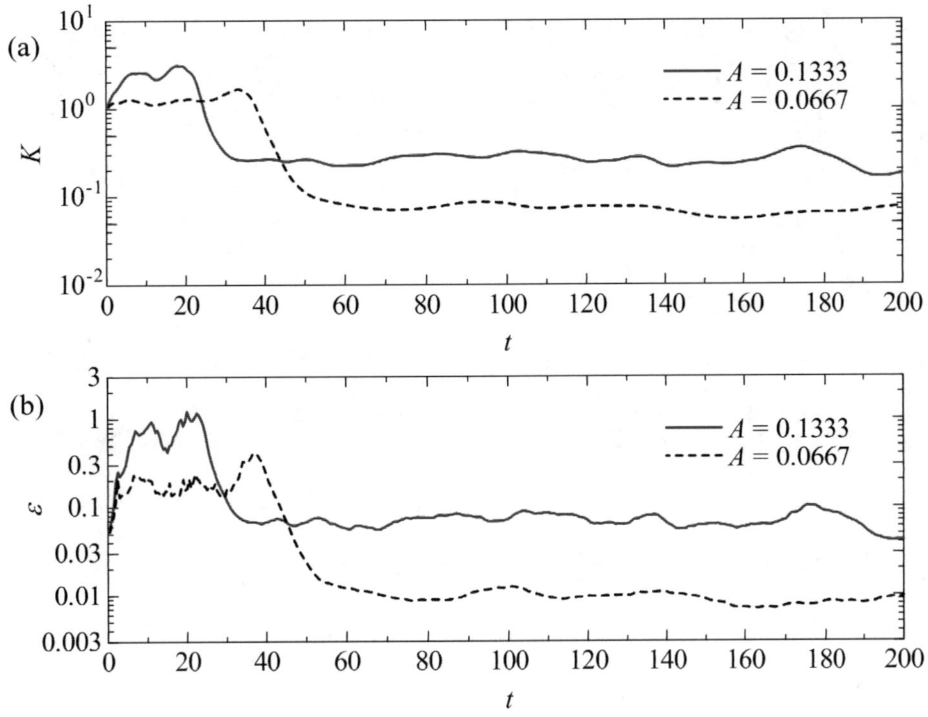

Figure 3. Temporal evolutions of the turbulent kinetic energy and its isotropic dissipation, (a) and (b), respectively. Here these quantities are spatial mean values at each instant.

This field is globally isotropic, and the divergence of this field is analytically zero. In addition, the global kinetic energy at the initial time is set to be unity. In the previous work, the initial velocity field is given by using multiple scales with energy spectra. On the other hand, the present study focuses on developing a single-scale initial velocity field into steady homogeneous turbulence. From this initial stage, the present study develops the flow field in time up to the non-dimensional time $t = 200$.

4.2. Results on fundamental turbulence statistics
Figure 3(a) shows time evolutions of the turbulent kinetic energy. Here, this turbulent kinetic energy is a spatially averaged value at each instant of the time. As shown in the figure, the turbulent kinetic energy is large in the region where the non-dimensional time is small. Specifically, in the initial developing region, the turbulent kinetic energy may be larger than the initial value. This larger value of the turbulent kinetic energy tends to reduce as the time progresses. For a sufficiently large time, the turbulent kinetic energy is found to vary under a constant mean value. As can be seen in the figure, for a small value of the coefficient, the steady-state seems to start at a larger time. The time required to achieve the steady-state may depend on the magnitude of the coefficient. As this coefficient is increased, the turbulent kinetic energy produced per unit time by the forcing terms is increased. The magnitude of the turbulent kinetic energy in this steady-state is larger in conditions where the coefficient is large.

Figure 3(b) shows temporal evolutions of viscous dissipation rate of the turbulent kinetic energy, ε. Here, this viscous dissipation rate is calculated using the intensity of the longitudinal velocity derivatives for the velocity fluctuations by assuming the isotropic behaviour of the turbulence. This value is also spatially averaged at each time. As shown in the figure, a value of the dissipation rate is large in the

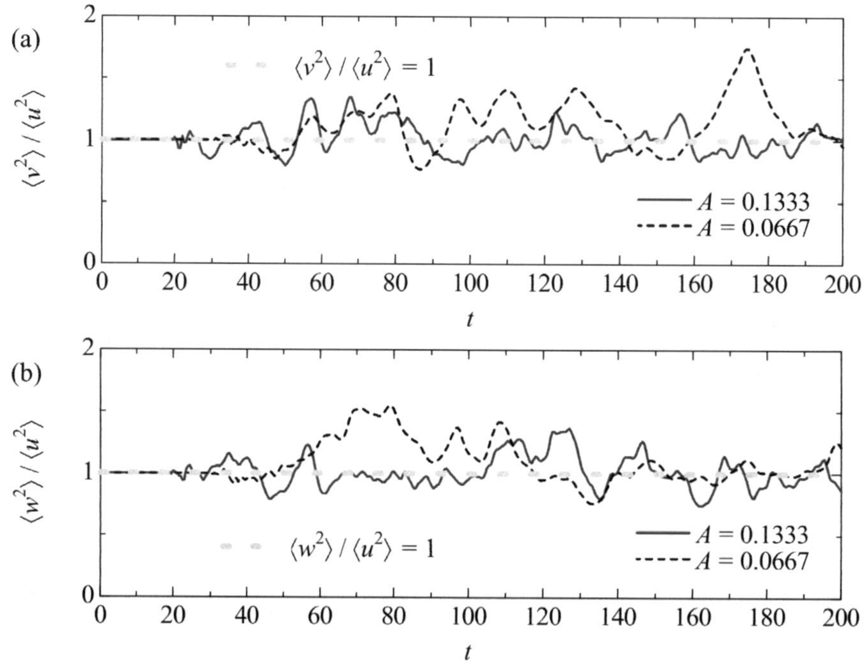

Figure 4. Temporal evolutions of the large-scale anisotropy values defined as $\langle v^2\rangle/\langle u^2\rangle$ and $\langle w^2\rangle/\langle u^2\rangle$, (a) and (b). Here these values are obtained as spatial mean values at the time.

region of the initial time. However, a value of the dissipation rate decreases as the time increases. The temporal profiles of the dissipation rate in the initial time region are similar to those of the turbulent kinetic energy. At a sufficiently large time, a value of the dissipation rate appears to be statistically steady. The time at which the statistically steady dissipation rate starts is comparable to that of the steady turbulent kinetic energy. The value of the steady-state dissipation rate appears to depend on the coefficient. For a smaller value of the coefficient, a value of the dissipation rate decreases. This decrease in the dissipation rate appears to be larger than that of the turbulent kinetic energy. This result may imply that the dissipation rate is more sensitive to the coefficient than the turbulent kinetic energy.

Figure 4 shows temporal profiles of the large-scale anisotropy values [16-19]. Here, the present values of the large-scale anisotropy are given as follows: $\langle v^2\rangle/\langle u^2\rangle$ and $\langle w^2\rangle/\langle u^2\rangle$, where, when a turbulence field is isotropic, a value of the large-scale anisotropy factors is exactly unity. As shown in the figure, obtained values of the large-scale anisotropy are seen to be unity in the region of the initial time domain. This result implies that the initial isotropic velocity field has not developed into turbulence in this time domain. At a non-dimensional time larger than $t = 50$, values of the large-scale anisotropy factors are varied with respect to time. The amplitude of these variations is comparable between both of the values for the coefficient in the linear forcing. On the other hand, when a value of the coefficient is small, the variations of the anisotropy with time appear to be slower. This difference may point to the possibility that a time scale value of the turbulence is different due to the different values of the coefficient.

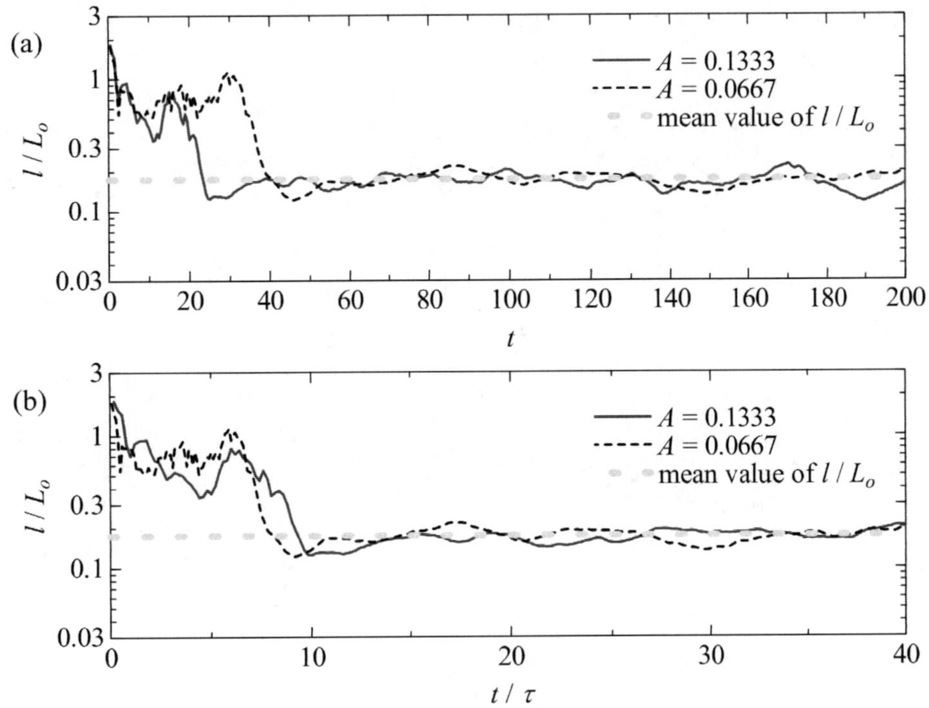

Figure 5. Temporal evolutions of the turbulent length scale normalized by the side of the computational domain, $L_o = 2\pi$. Here quantities for the horizontal axis are nondimensional time and the time normalized by the turbulent time scale. Blue-dashed line denotes the mean value of 0.175 calculated from the previous study [3].

4.3. Results on turbulent length scale

Figure 5(a) shows variations for length scale of turbulence with time. Here, this length scale l is given as follows [3]: $l = q^3 / \varepsilon$, where $q^2 = 2K/3$, and $L_o = 2\pi$. As shown in the figure, the length scale has a large value at the initial time period. In the time region for the sufficiently developed turbulence, a value of this length scale can be regarded as statistically steady. This value of the length scale is approximately in agreement between the two values of the coefficient. The previous study [3] has pointed out that the length scale of homogeneous turbulence obtained by the linear focusing is independent of the coefficient. Also, a value of the length scale normalised by L_o is obtained to be about 0.175. Values obtained in the present study for the length scale are of similar magnitude to that of the previous study. This agreement suggests that the steady-state turbulent flows simulated in the present study have identical characteristics with that of the previous study. Therefore, the present study considers that this agreement can validate the present turbulence analysis. In this study we have focused on time series of turbulence statistics. Although the time series data are indeed varied temporally, we consider that the points obtained in this study are significant.

As shown in Figure 5(b), a time at which the steady statistical value of the length scale is reached is earlier when the coefficient A is larger. This difference in the coefficient may affect a value of the time scale of turbulence. We focus on the time scale of turbulence to clarify this difference. Here, this turbulence time scale τ is obtained as follows [3]: $\tau = q^2/\varepsilon$. The present values of the turbulence time scale are $\tau = 2.5$ and $\tau = 5$ for the values of the coefficients, $A = 0.1333$ and $A = 0.0667$, respectively. In

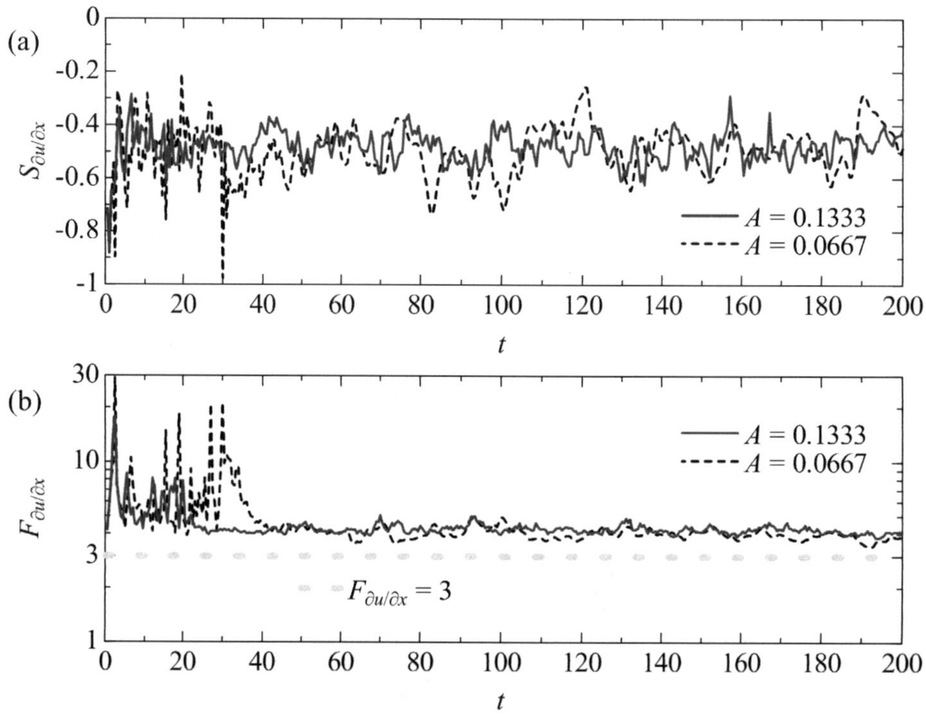

Figure 6. Temporal evolutions of skewness and flatness factors, (a) and (b), of the longitudinal velocity derivative.

Figure 5(b), the horizontal axis is taken as the non-dimensional time normalized by this turbulence time scale. Using the time normalized by the time scale, as shown in this figure, the present study can find that, between the two values of the coefficient, values of the time at which the steady value of the length scale starts is almost in agreement. This result implies that the time from the initial time to the time when the turbulence field becomes statistically steady is about ten times larger than the turbulence time scale.

4.4. Results on spatial derivatives of the flow field

The present study then focuses on turbulence statistics related to spatial derivatives of velocity fluctuations. In particular, the skewness and flatness factors are calculated for the longitudinal velocity derivatives of the velocity fluctuations [30,31] in this study. Previous studies (e.g., [30,31]) have pointed out that these quantities deviate from values for the Gaussian behavior of a fluctuation. This deviation can be considered to correspond to the intermittency of small-scale turbulence fluctuations. Figure 6(a) shows time profiles of the skewness and flatness factors of the longitudinal velocity derivative, $S_{\partial u/\partial x}$ and $F_{\partial u/\partial x}$. Here, these quantities are also spatially averaged at each time instant. As shown in the figure, obtained values of the skewness factor is smaller than zero, which is considered as the value for a Gaussian fluctuation. Similarly, obtained values of the flatness factor are also larger than three, which is also the value for the Gaussian. These deviations are also quantitatively comparable with those of the previous studies (e.g., [30,31]). This agreement of these values agrees with that characteristics of the small-scale turbulence field obtained in this study are similar to those shown by the previous studies.

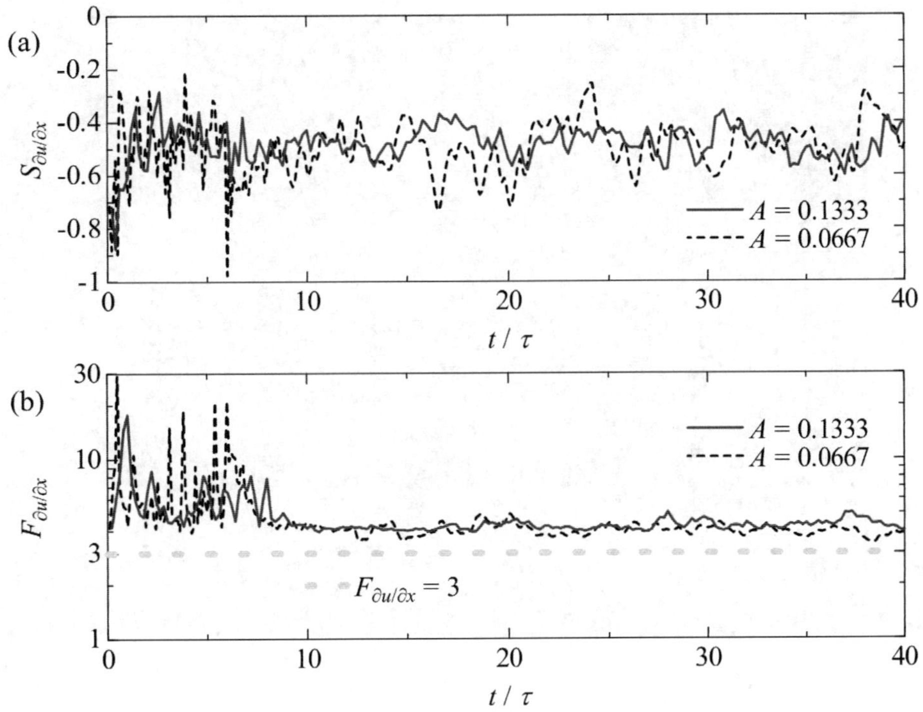

Figure 7. Temporal evolutions of skewness and flatness factors, (a) and (b), of the longitudinal velocity derivative, where the horizontal axis is taken to be the non-dimensional time normalized by the turbulent time scale.

The present values of the skewness and flatness factors can also be considered to be statistically steady in regions where the non-dimensional time is large. When the non-dimensional time is not large, these values may be different from those in the steady-state. For the flatness factor of the velocity derivative, we can clearly see that values in the initial time region are different from those in the steady-state. The non-dimensional time at which values of the flatness factor for the steady-state begin seems to depend on the coefficient of the external force term. Based on this point and the results for the length scale shown in the previous figure, the non-dimensional time giving the profiles of the skewness and flatness factors is also normalised using the time scale. Figure 6(b) shows time profiles of the skewness and flatness factors normalised by the time scale, taking the non-dimensional time as the horizontal axis. As shown in the figure, values of the times at which the flatness factor values for the steady-state start are roughly coincident between the two values of the coefficient.

4.5. Results on visualization and structure functions

Figure 7 visualises flow fields in the initial developing temporal domain. Here, results for the instantaneous turbulent kinetic energy in the *x*-*y* plane at $z = 0$ are shown. Based on the results obtained previously in this study, the results are shown for non-dimensional times $t/\tau = 4$, 8 and 12, normalized by the turbulence time scale. As shown in the results for $t/\tau = 4$, the flow fields are less homogeneous spatially during the developing time domain, with visible remnants of the initial flow field. At $t/\tau = 12$, at the end of the developing time domain, the statistically spatial homogeneity of the flow field appears

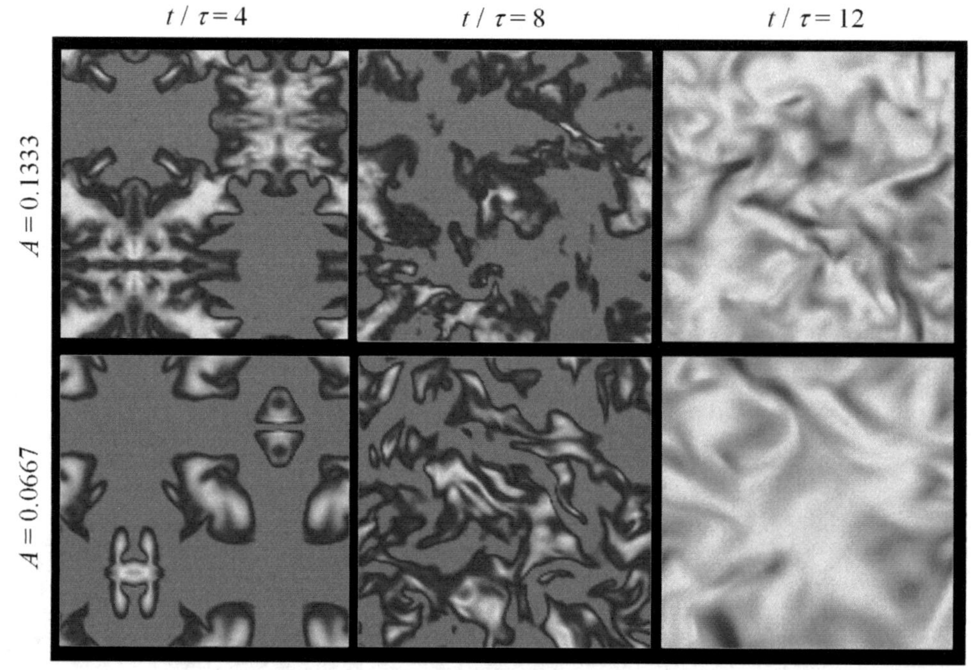

Figure 8. Evolution of the instantaneous turbulent kinetic energy field. Here the values of the time are taken by using the turbulent time scale.

to be enhanced. At this time, the different values of the coefficients can provide different flow fields. We consider that this difference is due to that in the turbulent Reynolds number caused by the difference in values for the coefficient. The turbulent Reynolds number value of the turbulence field due to the coefficient $A = 0.1333$ is higher than that for $A = 0.0667$. At $t/\tau = 8$, which is considered in the developing time domain, the spatial homogeneity of the turbulence field is lower than that at $t/\tau = 12$.

At this time instant, the appearance of the turbulence field can be seen to be different depending on the value of the coefficient. For the large value of the coefficient, the remnants of the initial flow field are still visible, whereas the remnants cannot be clearly found for small values of the coefficient. As shown in the Figure 8, for $t/\tau = 8$, there is little qualitative difference between the values of the coefficient in terms of the length scale of the turbulence field and the skewness and flatness factors of the velocity derivative obtained in this study. This study considers that turbulence structures at each time should be approached in terms of the magnitude of a spatial scale to investigate the difference in the visualization results at $t/\tau = 8$.

Structure functions are often used to examine turbulence structures characterised by using velocity components, where the longitudinal second and third-order structure functions for the distance between two spatial points r [1], $D_{11}(r)$ and $D_{111}(r)$, used in this study are expressed as follows, respectively.

$$D_{11}(r) = \langle (u(x+r) - u(x))^2 \rangle \text{ and } D_{111}(r) = \langle (u(x+r) - u(x))^3 \rangle. \quad (6)$$

Additionally, to avoid a difference caused by the magnitude difference of the second and third-order intensities for the velocity fluctuation to affect results obtained in this study, the structure functions have been normalised using these quantities. Figure 9 shows time evolutions of the second-order structure-function in the developing time range. Here, the results are shown at times t/τ, $t/\tau = 4$, 8 and 12, which

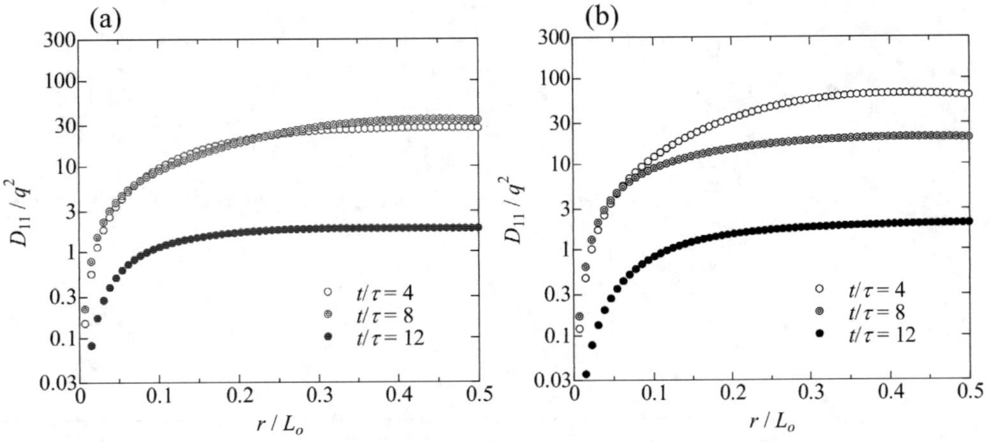

Figure 9. Evolution of the second-order structure function normalised by the intensity q^2 for the values of the coefficient $A = 0.1333$ (a) and $A = 0.0667$ (b).

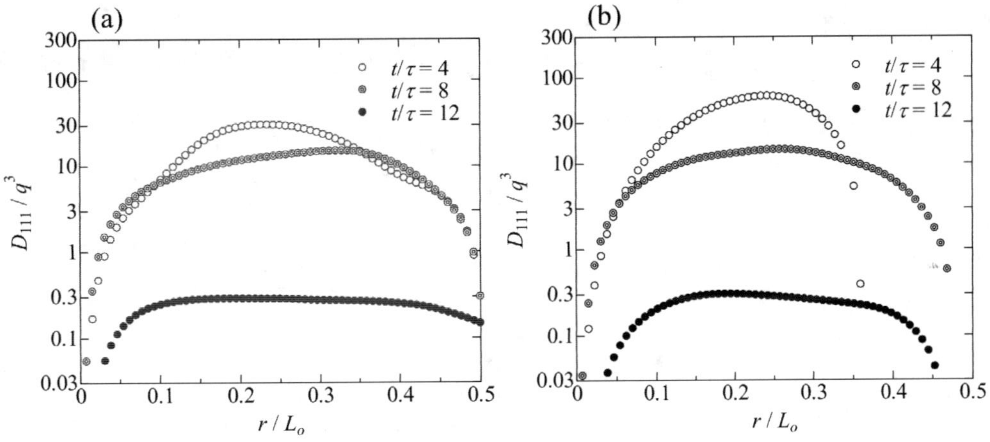

Figure 10. Evolution of the third-order structure function normalised by the cubic of q for the values of the coefficient $A = 0.1333$ and $A = 0.0667$, (a) and (b).

are normalised on the time scale, where $q^2 = 2/3\ K$. For smaller values of this normalised time, values at larger distance of the second-order structure function are larger. In the case for the coefficient with $A = 0.1333$, the shape of the profiles for the structure-functions are similar between $t/\tau = 4$ and 8, whereas, for the coefficient $A = 0.0667$, the shape is different between these two time instants. This difference may reflect the difference in the visualisation results shown in the previous figure. As a value of the normalised time increases, a value of the structure function may decrease. At $t/\tau = 12$, the profiles of the structure functions are similar between the two values of the coefficients, except for the result for small values of the distance.

Figure 10 shows profiles of the third-order structure function normalized by the cube of q in the developing time domain. Here, these profiles are shown at $t/\tau = 4$, 8 and 12. As shown in the figure, a value of the profiles for the third-order structure function decreases as t/τ increases. At $t/\tau = 12$, the

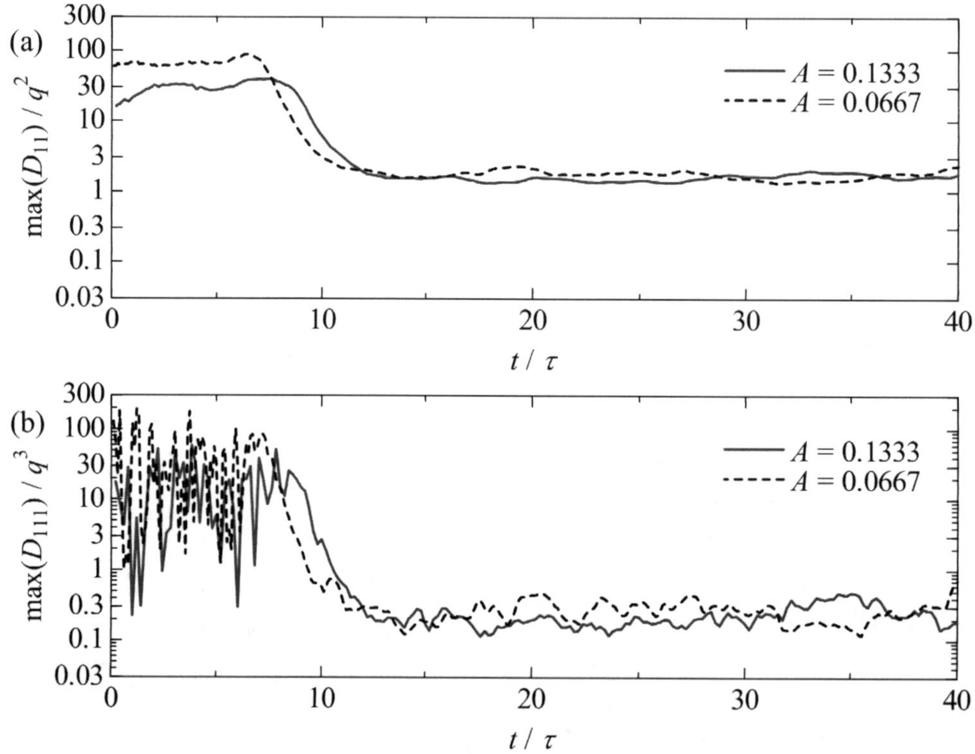

Figure 11. Temporal evolution of the maximum values of the normalized second and third-order structure functions, (a) and (b).

profiles of the third-order structure function are similar between the two values of the coefficient, except for those at a large distance value of r. As shown in the results for $t/\tau = 4$ and 8, the difference in the profile between the two values of the coefficient is larger for the value at a large distance r. This difference may be caused by the difference in energy supply per unit time due to the different magnitude of the external force term.

This study examines time evolutions of these structure functions by showing those of the maximum values of the normalised structure functions. The time evolutions of the maximum values are shown in Figure 11. As shown in the figure, the maximum values for the second-order structure function are large in the developing time domain and decrease to steady-state values at a sufficiently large time region. A value of the normalised time at which the maximum becomes steady is about $t/\tau = 15$. The time evolutions of the maximum value for the normalised third-order structure function are also shown in the figure. The time at which the normalised third-order structure function becomes steady is also approximately $t/\tau = 15$ as similar with those for the second-order structure function. On the other hand, the normalised time at which the turbulence statistics become steady is about $t/\tau = 10$, as shown in the previous figures. This difference implies that a time required for profiles of the second and third-order structure functions to become steady may be larger than that needed for the statistics to become steady.

5. Discussion and Conclusion

5.1. Discussion

This study discusses the present results in terms of the dependence of an initial flow field. In this study, a kind of the single-scale initial velocity field has been used, whereas, in the previous study [3], a multi-scale velocity field was used with an energy spectrum of developed turbulence. We consider that the initial velocity field in the present study is less close to the developed steady turbulence field than that in the previous study. On the other hand, the time at which the turbulence field becomes steady in the present study is comparable to that in the previous study. This result suggests that, when the initial velocity field is provided in a multi-scale way, a part of the scale may be able to develop steady turbulence as in the previous study. In the present study, the wavenumber of the single-scale velocity field is taken as that corresponding to the size of the computational domain. A further study should approach examining whether increasing the wavenumber of a single-scale velocity field still results in the development of steady turbulence.

The set magnitude of the turbulent Reynolds number, while set in the light of the previous study [3], may not be set sufficiently high to involve the inertial subrange [1] in the present study. The dependence of the present obtained results on this Reynolds number may be raised to be an issue which should be investigated in the future. As shown in the present study, the length of the developing time domain is about $t/\tau = 10$ - 15. A further issue that needs to be examined may also be whether this region may be affected by this Reynolds number. One of the most important key words assumed in this study is the isotropy of turbulence. The turbulence field and the initial field analyzed in this study are globally isotropic. Although the turbulence field found in actual engineering devices may often be anisotropic, the present study analyzes the isotropic field based on previous studies.

5.2. Conclusion

The purpose of this study is to examine temporal evolutions of turbulence fields in the time domain, where a single-scale initial velocity field develops into steady homogeneous isotropic turbulence. In contrast to the previous study, a single-scale field is set as the initial velocity field. Here, this initial velocity field is globally isotropic. In the developing time region, turbulence statistics and structure functions are investigated here. The present turbulence fields are reproduced in the three-dimensional periodic domain, where the turbulent flows are maintained steady by the linear focusing approach based on the corresponding previous study. Two values of the coefficient for the external force term in the governing equations set in the linear focusing method are used. A constructed Fourier spectral method based on the fourth-order Runge-Kutta method is used to analyze the present turbulence fields. This study uses an analytical solution and the Taylor-Green vortex (TGV) flow to verify this constructed spectral method. In the verification using the TGV flow, the spatial symmetry of the instantaneous field is also examined using the instantaneous fields.

The turbulent kinetic energy, its viscous dissipation rate and the large-scale anisotropy are firstly used to indicate that the flow fields in the present study are sufficiently developed to be each turbulent state. The steady-state turbulent flow fields in the present study are validated by comparing values of the skewness and flatness factors of a longitudinal derivative of a velocity fluctuation. Also, the turbulent length scale of the present turbulence fields is also compared with those of the previous study. Then, results of the visualisation for the turbulence field in the developing time domain and the results of the second and third-order structure functions are presented. From these results, the present study has seen that the initial time domain in which the turbulence field reaches the steady-state is 10 to 15 times as large as the turbulence time scale. Also, values of the turbulence statistics converged to the steady-state values in a shorter time than the second and third-order structure functions. Therefore, these results imply that turbulence statistics converge more rapidly than second- and third-order structure functions in the present study. Moreover, the initial velocity field can converge smoothly to the developed turbulence if the initial velocity field is on a single scale.

As a future study, the wavenumber dependence of a single-scale field set as an initial velocity field may need to be investigated. For example, whether a turbulence field still reaches a steady state when the wavenumber value is changed should be examined. Also, the dependence of the turbulence field on the turbulent Reynolds number in the developing region needs to be examined. In this study, this Reynolds number is set based on the previous studies to compare the obtained results with those of these studies. There may be the need for further analysis of temporal developments of turbulence statistics in the future outlook. Further examination of the results obtained in this study in terms of spectra may also be one of the future studies. The present study is interested in examining the impact of increasing this Reynolds number on the turbulence field in the developing region.

Acknowledgments
This work was partly supported by the Japanese Ministry of Education, Culture, Sports, Science and Technology through Grants-in-Aid (Nos. 20H02069, 21K03859 and 22H01684).

References
[1] Pope S B 2000 *Turbulent Flows*. Cambridge University Press
[2] Canuto C, Hussaini M Y, Quarteroni A, Zang T A 1988 *Spectral Methods in Fluid Dynamics*. Springer Berlin Heidelberg
[3] Rosales C, Meneveau C 2005 Linear forcing in numerical simulations of isotropic turbulence: Physical space implementations and convergence properties *Phys. Fluids* 17 095106
[4] Carroll P L, Blanquart G 2013 A proposed modification to Lundgren's physical space velocity forcing method for isotropic turbulence *Phys. Fluids* 25 105114
[5] De Laage de Meux B, Audebert B, Manceau R, Perrin R 2015 Anisotropic linear forcing for synthetic turbulence generation in large eddy simulation and hybrid RANS/LES modeling *Phys. Fluids* 27 035115
[6] Seoud R E, Vassilicos J C 2007 Dissipation and decay of fractal-generated turbulence *Phys. Fluids* 19 105108
[7] Nagata K, Suzuki H, Sakai Y, Hayase T, Kubo T 2008 Direct numerical simulation of turbulent mixing in grid-generated turbulence *Phys. Scr.* T132 014054
[8] Suzuki H, Nagata K, Sakai Y, Ukai R 2010 High-Schmidt-number scalar transfer in regular and fractal grid turbulence *Phys. Scr.* T142 014069
[9] Suzuki H, Nagata K, Sakai Y, Hasegawa Y 2013 An attempt to improve accuracy of higher-order statistics and spectra in direct numerical simulation of incompressible wall turbulence by using the compact schemes for viscous terms *Phys. Scr.* 2013 T155 014062
[10] Suzuki H, Nagata K, Sakai Y, Hayase T, Hasegawa Y, Ushijima T 2013 Direct numerical simulation of fractal-generated turbulence *Fluid Dyn. Res.* 45 061409
[11] Nagata K, Sakai Y, Inaba T, Suzuki H, Terashima O, Suzuki H 2013 Turbulence structure and turbulence kinetic energy transport in multiscale/fractal-generated turbulence *Phys. Fluids* 25 065102
[12] Zhou Y, Nagata K, Sakai Y, Suzuki H, Ito Y, Terashima O, Hayase T 2014 Relevance of turbulence behind the single square grid to turbulence generated by regular-and multiscale-grids *Phys. Fluids* 26 075105
[13] Zhou Y, Nagata K, Sakai Y, Suzuki H, Ito Y, Terashima O, Hayase T 2014 Development of turbulence behind the single square grid *Phys. Fluids* 26 045102
[14] Valente P C, Vassilicos J C 2014 The non-equilibrium region of grid-generated decaying turbulence *J. Fluid Mech.* 744 5-37
[15] Zheng S, Bruce P J K, Cuvier C, Foucaut J M, Graham J M R, Vassilicos J C 2021 Nonequilibrium dissipation scaling in large Reynolds number turbulence generated by rectangular fractal grids. *Phys. Rev. Fluids* 6 054613
[16] Mohamed M S, LaRue J C 1990 The decay power law in grid-generated turbulence *J. Fluid Mech.* 219 195-214

[17] Kurian T, Fransson J H 2009 Grid-generated turbulence revisited *Fluid Dyn. Res.* 41 021403
[18] Suzuki H, Mochizuki S, Hasegawa Y 2016 Validation scheme for small effect of wind tunnel blockage on decaying grid-generated turbulence *J. Fluid Sci. Tech.* 11 JFST0012
[19] Suzuki H, Mochizuki S, Hasegawa Y 2018 Numerical-based theoretical analysis on effects of weak fluid acceleration of free-stream due to wind-tunnel blockage on grid-generated turbulence *Flow Meas. Instrum.* 62 1-8
[20] Lavoie P, Djenidi L, Antonia R A 2007 Effects of initial conditions in decaying turbulence generated by passive grids *J. Fluid Mech.* 585 395-420
[21] Hearst R J, Lavoie P 2015 The effect of active grid initial conditions on high Reynolds number turbulence *Exp. Fluids* 56 1-20
[22] Orszag, S A, Patterson, G S 1972 Numerical simulation of three-dimensional homogeneous isotropic turbulence *Phys. Rev. Lett.* 28 76
[23] Rogallo R S 1981 Nasa technical memorandum *NASA Ames Res. Center* 81315
[24] Nagata K, Saiki T, Sakai Y, Ito Y, Iwano K 2017 Effects of grid geometry on non-equilibrium dissipation in grid turbulence *Phys. Fluids* 29 015102
[25] Brachet M E, Meiron D I, Orszag S A, Nickel B G, Morf R H, Frisch U 1983 Small-scale structure of the Taylor–Green vortex *J. Fluid Mech.* 130 411-452
[26] Brachet M 1991 Direct simulation of three-dimensional turbulence in the Taylor–Green vortex *Fluid Dyn. Res.* 8 1-8
[27] Hideharu M 1991 Realization of a large-scale turbulence field in a small wind tunnel. *Fluid Dyn. Res.* 8 53-64
[28] Mydlarski L, Warhaft Z 1996 On the onset of high-Reynolds-number grid-generated wind tunnel turbulence *J. Fluid Mech.* 320 331-368
[29] Berland J, Bogey C, Bailly C 2006 Low-dissipation and low-dispersion fourth-order Runge–Kutta algorithm *Comput. Fluids* 35 1459-1463
[30] Sreenivasan K R, Antonia R A 1997 The phenomenology of small-scale turbulence. *Ann. Rev. Fluid Mech.* 29 435-472
[31] Zhou T, Antonia R A 2000 Reynolds number dependence of the small-scale structure of grid turbulence *J. Fluid Mech.* 406 81-107

Flow structure of pitched and skewed microjet vortex generator array

Xingya Da[1, 3], Yuan Yi[1], Yue Zhang[2], Longkai Guo[1] and Yang Tao[1]

[1]China aerodynamics research and development center, Mianyang, China;
[2]Beihang University, Beijing, China

[3]Corresponding author's e-mail: Daxingya@cardc.cn

Abstract. To understand the micro-vortex structure of Microjet Vortex Generators, a plate with pitched and skewed jet holes was established. Simulations using SST-DDES method based on WENO scheme were conducted to analyse the effect of jet pressure, hole interval and incoming boundary layer thickness on the flow structure. The pitch angle and skew angle are 55 degrees and 45 degrees respectively. The total pressure of microjets ranges from 2.5 to 3.5 times the mainstream pressure. The interval of microjets ranges from 6.75 to 13.5 diameters of the hole, and the thickness of incoming boundary layer ranges from 2 to 4 diameters of the hole. The flow field is dominated by three vortices, including a main vortex and two secondary vortices which are significantly influenced by jet pressure and interval. The interaction between secondary vortices and main vortex plays an important role in the process of boundary layer mixing.

1. Introduction

Vortex generator jet(VGJ) is a kind of flow control method of using pitched and/or skewed jet to enhance the mixing of the boundary layer and the mainstream [1]. Compared with solid-state vortex generators, this method does not bring additional resistance and structural fracture risk, can be adjusted according to the flow field, and has a wide application range.

Since it was proposed, the principle of microjet vortex generator has been widely known. It enhances the mixing of the boundary layer and the mainstream by generating streamwise vortices, and provides the ability of the boundary layer to resist the adverse pressure gradient. Johnston et al. [2] realized through low-speed wind tunnel tests that the use of jet array can produce pairs of counter-rotating vortex, which can significantly affect the cross flow. This understanding has played an important role in promoting the development of inlet flow control. For a large S-bend inlet, because of the curvature of the profile, secondary flow (cross flow perpendicular to the flow direction) naturally occurs in the inlet. Secondary flow leads to the accumulation of the boundary layer to the low-pressure side, and further development leads to flow separation. Therefore, the use of micro jet array to generate a vortex field can strengthen the mixing of the boundary layer and the mainstream, and offset the secondary flow of the mainstream to achieve the effect of 'four or two pulling a thousand catties'. The research of Rabe, Da Xingya et al. [3-7] shows that even if there is a high-intensity separation, the flow separation can be restrained by the microjet array, and the jet flow is only about 0.6% of the inlet flow.

Compton et al. [8] measured the flow field structure of a single-hole jet with 45-deg pitch angle and the same skew angle, in which the velocity ratio of jet to incoming flow is about 1.3. In this study, the diameter of the jet is 6 mm, which belongs to the flow category with large scale and high Re number. The flow field structure is measured by a 3.2mm five-hole probe, which can identify the large vortex structure and cannot capture small-scale vortex structures. The study reveals that the optimum jet skew angle is between 45 ° and 90 °. Compton pointed out that the physical mechanism of jet flow vortex is not clear, which is speculated to be related to flow vortex.

Zhang et al. [9-11] used laser Doppler to measure the flow field structure of a single jet with pitch and skew angles in low-speed turbulent flow field, and also only captured large-scale vortex structure. Coussement et al. [12] used large eddy simulation method to study the flow field structure of pulsed jet. The research object is a low-speed incompressible jet with vertical blowing, and the incoming velocity is 1.2m/s. The revealed flow field characteristics are quite different from those of microjet vortex generator. Fernandez et al. [13] measured the flow field structure of the vertical microjet array using PIV. The incoming velocity was 1.47m/s, and the maximum jet velocity was only 14.7m/s. The study revealed the variation that the smaller the interval, the stronger the turbulence pulsation, and found two transmission mechanisms of large-scale contour-rotating vortices and turbulent transport.

For inlet flow control cases, the jet velocity reaches the speed of sound and even higher. The flow structure for such sonic jets has not been well studied. In order to further understand the mechanism of microjet generators, this paper investigates the flow structure and key parameters through CFD computations. The simulation conditions is extracted from an existing inlet model for which the effectiveness of the microjet flow control has been proved. Through unsteady CFD simulations, a structure containing three-vortices is identified which shows differences from existing studies.

2. Simulation method

2.1. Governing equations

In this work, an in-house code, named MAP and developed by Ning of Beihang University is used for all simulations [14]. A delayed detached eddy simulation method based on SST turbulence model is used. Reynolds averaged NS equation based on SST Turbulence model is

$$\frac{\partial}{\partial t}\int_{\Omega}\boldsymbol{Q}d\Omega+\oint_{\partial\Omega}\left(\boldsymbol{F}_c-\frac{1}{Re_\infty}\boldsymbol{F}_v\right)dS=\int_{\Omega}\boldsymbol{G}d\Omega \tag{1}$$

where

$$\boldsymbol{Q}=\begin{bmatrix}\rho\\ \rho u\\ \rho v\\ \rho w\\ \rho E\\ \rho k\\ \rho\omega\end{bmatrix}\quad \boldsymbol{F}_c=\begin{bmatrix}\rho V\\ \rho uV+n_x p_t\\ \rho vV+n_y p_t\\ \rho wV+n_z p_t\\ \rho HV\\ \rho kV\\ \rho\omega V\end{bmatrix}\quad \boldsymbol{F}_v=\begin{bmatrix}0\\ n_x\tau_{xx}+n_y\tau_{yx}+n_z\tau_{zx}\\ n_x\tau_{xy}+n_y\tau_{yy}+n_z\tau_{zy}\\ n_x\tau_{xz}+n_y\tau_{yz}+n_z\tau_{zz}\\ n_x\Theta_x+n_y\Theta_y+n_z\Theta_z\\ (\mu+\sigma_k\mu_T)\boldsymbol{n}\cdot\boldsymbol{\nabla}k\\ (\mu+\sigma_\omega\mu_T)\boldsymbol{n}\cdot\boldsymbol{\nabla}\omega\end{bmatrix}\quad \boldsymbol{G}=\begin{bmatrix}0\\0\\0\\0\\0\\ \tilde{P}_k-D_k\\ P_\omega-D_\omega+P_{cross}\end{bmatrix}$$

The high-order scheme construction method based on WENO reconstruction is used for the convection term, and the preprocessed LDFSS (low diffusion flux splitting schemes) scheme is used for the flux calculation; the central difference is used for the viscous term; for the time accurate unsteady simulation, the virtual time stepping method with high computational efficiency is used, and the numerical format is the three-point Euler backward difference method. Details of the governing equation can be found in Ref. 14 as well.

2.2. Simulation model

The CFD simulation domain contains three microjets. Figure 1 shows only one of them, which means that the total simulation domain contains three domains in Figure 1. The mainstream flow direction is along the X-axis direction, the diameter of the jet hole is D which is 2 mm. The spacing of holes is L.

The pitch angle with the XOZ plane is 55 degrees and the skew angle with the XOY plane is 45 degrees. The height of the domain is 12.5 D. The distance between the upstream boundary and the hole center is 10 D, and the distance between the downstream boundary and the hole center is 40 D. Structural grid which is more accurate for shear layer simulation is used, and the amount of grid point is about 15 million for a jet domain. In order to reduce the amount of unsteady simulation, periodic symmetric boundary conditions are used for the left and right sections. The height of the first layer to the wall is 5×10^{-6} which makes the y+ is about 2. The grid numbers in the X, Y and Z directions are 359, 198 and 53 respectively.

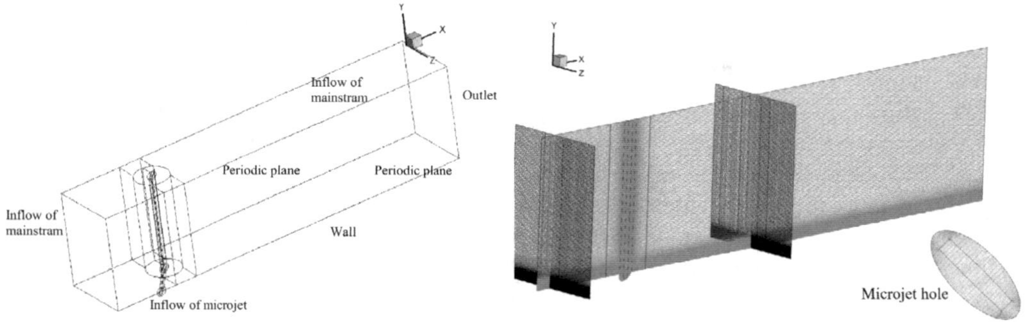

Figure 1. Simulation model and mesh.

2.3. Boundary conditions and simulation cases

The subsonic inflow boundary condition is given at the mainstream inlet for which the total temperature T0 is 288.2 K, the total pressure P0 is 104107 Pa, and the thickness of the boundary layer is D. The back pressure condition is given at the mainstream outlet which ensure that the mainstream Mach number is 0.5 with the velocity about 165m/s. The time step is taken as 1e-6s which ensures that about 12 timesteps are required for the mainstream passing through the jet hole. As shown in Table 1, different jet inlet total pressure, jet hole spacing and inlet boundary layer thickness are given for the analysis. The jet total pressure ratio from case 1 to case 3 is reduced from 3.5 to 2.5, the hole spacing is adjusted in case 4 and case 5, and the inlet boundary layer thickness is adjusted in case 6 on the basis of case 2. For all jet total pressure conditions, the jet velocity is guaranteed to be sonic speed of about 310m/s. The inflow boundary layer of the microjets, as well as the mainstream, is set to be turbulent. The Re number for the jet ranges from 77500 to 108500 based on the jet pressure, velocity and diameter, while the Re number of the mainstream keeps at 1×10^6 based on the inflow pressure, velocity and the length of the simulation tunnel.

Table 1. Simulation cases for the plate microjet.

case	Total pressure ratio P_j/P_0	Spacing	Boundary layer thickness	Re_D
1	3.5 P_0	9D	2D	108500
2	3.0 P_0	9D	2D	93000
3	2.5 P_0	9D	2D	77500
4	3.0 P_0	13.5D	2D	93000
5	3.0 P_0	6.75D	2D	93000
6	3.0 P_0	9D	4D	93000

3. The main vortex structure analysis

As shown in Figure 2, it is isosurface map with $Q = 1 \times 10^9$ for the simulation case 1. From the map, it can be seen that the flow produces three main vortices. As shown in Figure 3 are the vorticity contours at different sections. Figure 3(a) is the section at the center of the jet outlet at an angle of 45 degrees with the z-axis. The mainstream is coming from the right side and passing through the jet flow. A significant vortex is generated behind the jet. As compared with Figure 2, it can be identified that this vortex is just the vortex 1 in Figure 3.

Figures 3(b) to 3(e) shows the vorticity contours at section X = 0.002 to X = 0.04. As shown in Figure 3(b), a pair of reverse vortex pairs can be seen. The left vortex in Figure 3(b) is just the vortex 1 in Figure 2. Close to the wall in the right side, a vortex is developing from the near wall. This vortex will evolve into the vortex 2 in Figure 2.

As shown in Figure 3(c), the vortex 2 is gradually lifted off and the vortex 3 is developing near the wall. As shown in Figure 3(d), the vortices 1 and 2 two are blending into one vortex, while the vortex on the right is further lifted, but its intensity is weakened. The vortex near the wall is further lifted off to form the vortex 3 in Figure 2. As shown in Figure 3(e), the vortices 1 and 2 formed previously evolve by rotating in a counter-clockwise direction, while the vortex 2 and vortex 3 near the wall mentioned above are further strengthened.

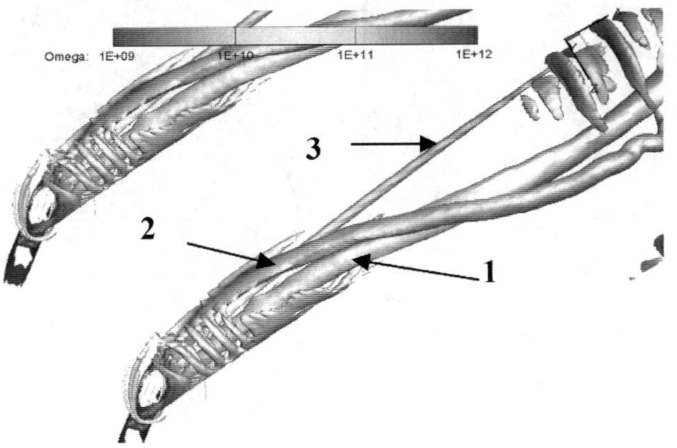

Figure 2. Vortex structures.

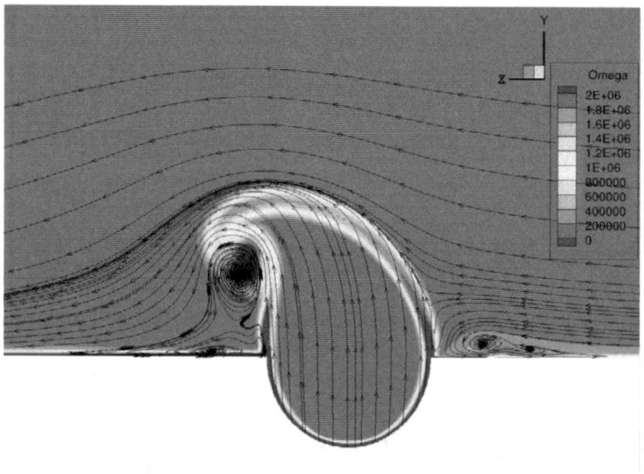

(a). Vorticity contours vertical to the jet hole

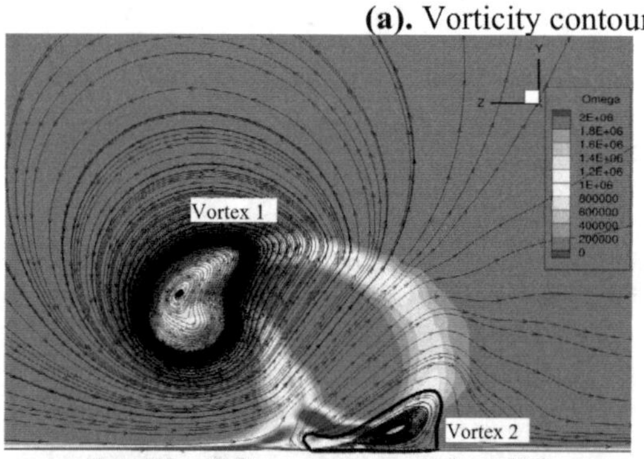

(b). Vorticity contours at X=0.002

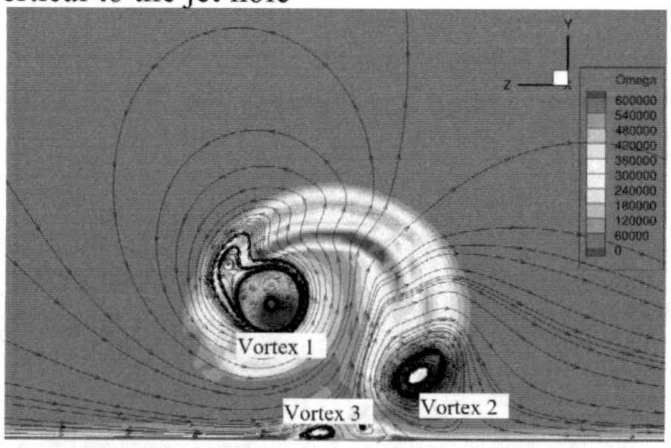

(c). Vorticity contours at X=0.008

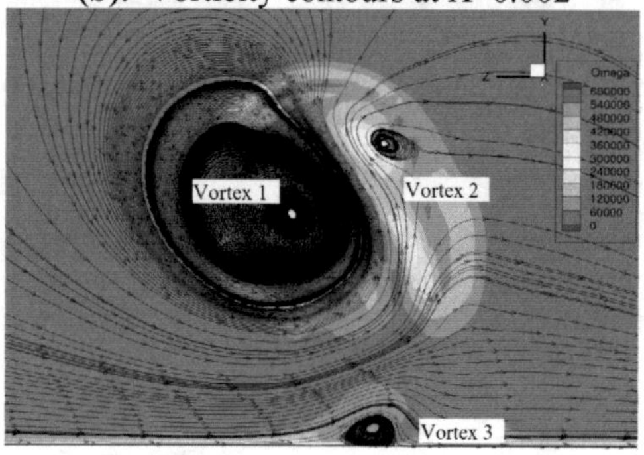

(d). Vorticity contours at X=0.02

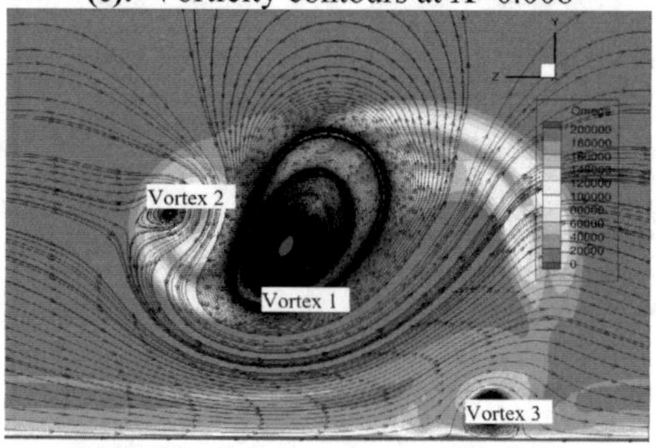

(e). Vorticity contours at X=0.04

Figure 3. Vorticity contours.

4. Effect of jet pressure on the vortex structure

4.1. Vortex structure analysis

Figure 4 shows the upper and lower views of the vortex structure. In this figure, in order to show more details, Q is set as 10^8. For vortex 1, when the total pressure is higher, the strength of the jet is larger, which makes the vortex core rise quickly. However, as the jet moves downstream, the vortex core is further closer to the wall. For vortex 2, while the total pressure is smaller, its strength is also smaller, and thus, it breaks down earlier. For vortex 3, when the total pressure is larger, the larger velocity difference between the jet and the mainstream makes it break down very early.

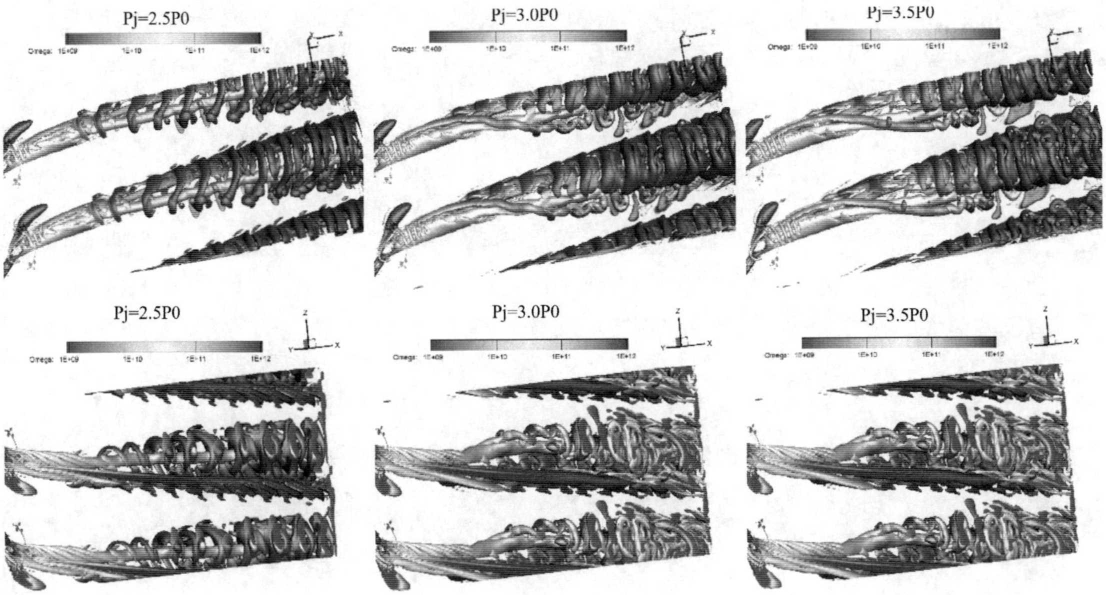

Figure 4. Vortex structures at different jet pressure ratio of 2.5D, 3.0 and 3.5 (upper: top view, lower: bottom view).

4.2. Downstream flow structure

Figure 5(a) shows the vorticity contours at x = 0.008. The vorticity structure at different total pressures at this section is similar, but with the increase of jet total pressure, the intensity of vorticity increases, and the vortices tends to lift off. Figure 5(b) shows the vorticity contours at x = 0.02. Here, when the total jet pressure ratio is 2.5, the second vorticity peak at the upper right corner disappears. Figure 5(c) shows the vorticity contours at x = 0.04. There is a large difference in the flow structure between flows with different total jet pressure. The most obvious part is that when the total pressure ratio is 2.5, there is only one concentration point of the vorticity, so there is only one vortex structure. When the total jet pressure is larger, a relatively abundant vortex structure can be formed and a weak vortex can be found near the concentration point.

Figure 5(d) shows the vorticity contours at x = 0.06. Here, with the increase of the total pressure, the position of the maximum vorticity shows a trend of gradually approaching the wall. This is because when the total jet pressure is larger, the two vortices interact with each other, so that the vortex core of vortex 1 is suppressed to the wall. At the same time, the maximum vorticity first increases and then decreases. This is because when the total pressure ratio is 2.5, the intensity of vortex 2 is too small, resulting in the dissipation in advance, and there is no process of two vortices interacting. Therefore, the maximum vorticity is larger at this time. When the total pressure is 3.0 and 3.5, there is a process of two vortices interacting. When the total pressure of the jet is larger, it can be clearly seen that vortex 3 near the wall is rolled up and interacts with vortex 1.

In Figures 5(e) and 5(f), the vorticity and velocity contours at X = 0.08 are shown. With the increase of the total pressure ratio, the vortex structure formed by the interaction between the jet and the mainstream is more complex, and the mixing between the jet and the main stream is more uniform. As shown in the figure, when the total pressure ratio is 2.5, a relatively clear vortex 3 can be seen near the wall. When the total pressure is larger, vortex 3 has already broken down and mixed with vortex 1.

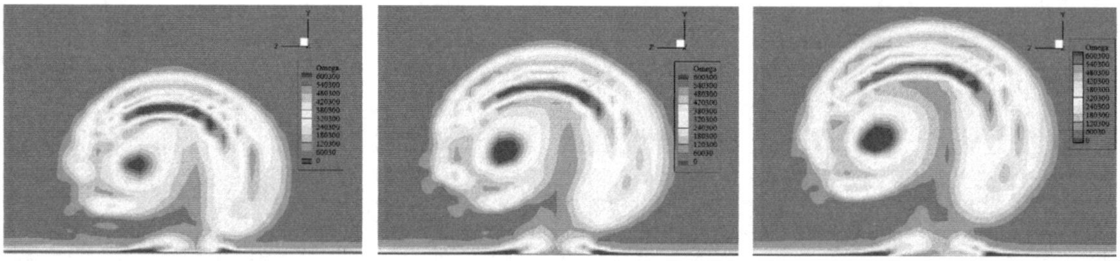

(a). Vorticity contours at X=0.008 with pressure ratio of 2.5, 3.0 and 3.5 from left to right

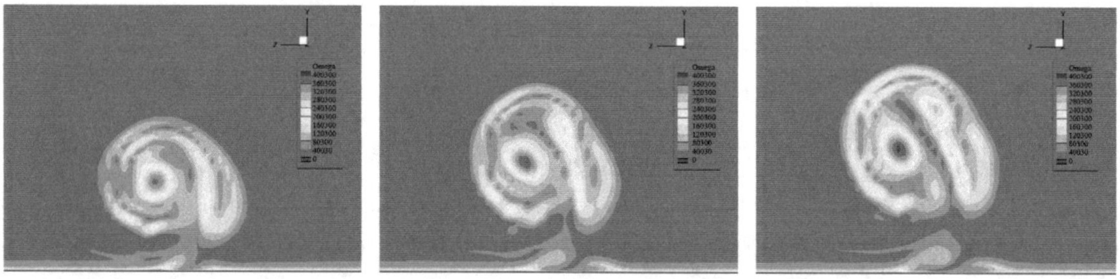

(b). Vorticity contours at X=0.02 with pressure ratio of 2.5, 3.0 and 3.5 from left to right

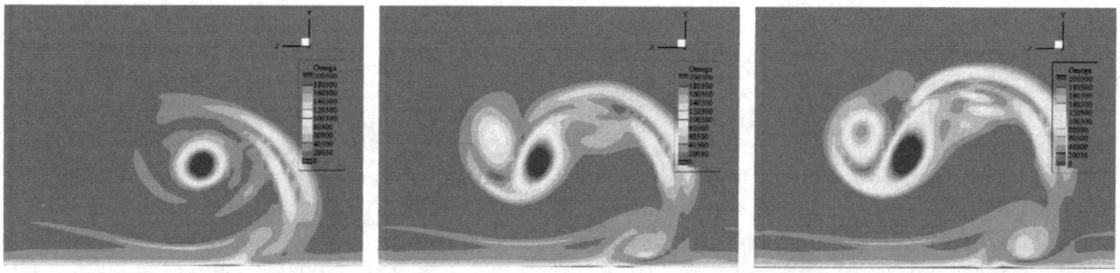

(c). Vorticity contours at X=0.04 with pressure ratio of 2.5, 3.0 and 3.5 from left to right

(d). Vorticity contours at X=0.06 with pressure ratio of 2.5, 3.0 and 3.5 from left to right

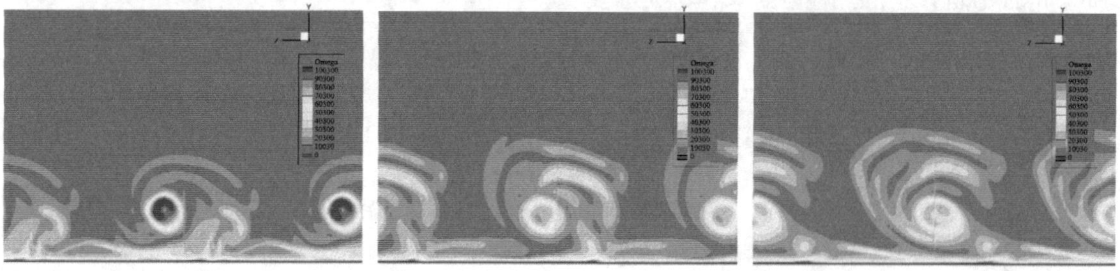

(e). Vorticity contours at X=0.08 with pressure ratio of 2.5, 3.0 and 3.5 from left to right

Journal of Physics: Conference Series **2313** (2022) 012009 doi:10.1088/1742-6596/2313/1/012009

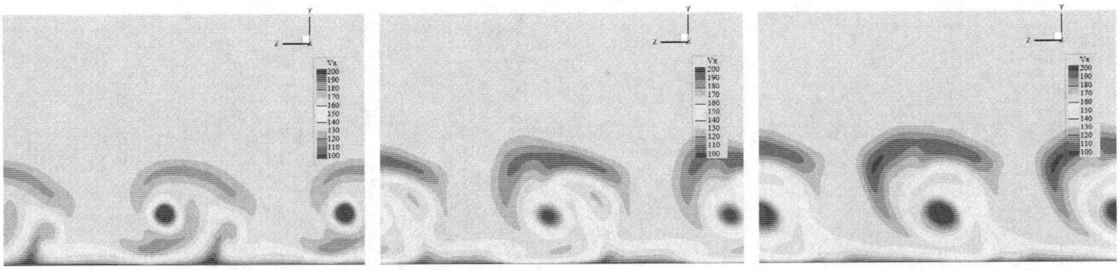

(f). Velocity contours at X=0.08 with pressure ratio of 2.5, 3.0 and 3.5 from left to right

Figure 5. Vorticity contours with different pressure ratios.

5. Effect of jet spacing on the vortex structure

Comparation between cases 2, 4 and 5 shows the effect of jet intervals. It can be seen in Figure 6 that structures in three cases are basically the same. However, in the downstream, when the spacing of the jet holes decreases, the influence of the adjacent jet becomes greater, which gradually enhances the interaction between neighboring vortices. Therefore, the breaking down of the large vortex is advanced. In this figure, Q is selected as 10^8.

Figure 7(a) is the vorticity contours and streamlines at x = 0.06. When the hole spacing is smaller, the vortices formed by adjacent jet holes interact with each other. When the jet hole spacing is 6.75, the fluid between adjacent holes is also rolled up, making the mixing be more uniform. When the distance is enlarged, the vortex interaction generated by adjacent holes decreases.

Figures 7(b) and 7(c) shows the vorticity and velocity contours at x = 0.08. It can be seen that, when the hole spacing is smaller, the fluid in the boundary layer is more involved in the vortex. This makes the jet be more evenly mixed with the mainstream and the boundary layer contain more high-speed fluid, resulting in a fuller velocity distribution.

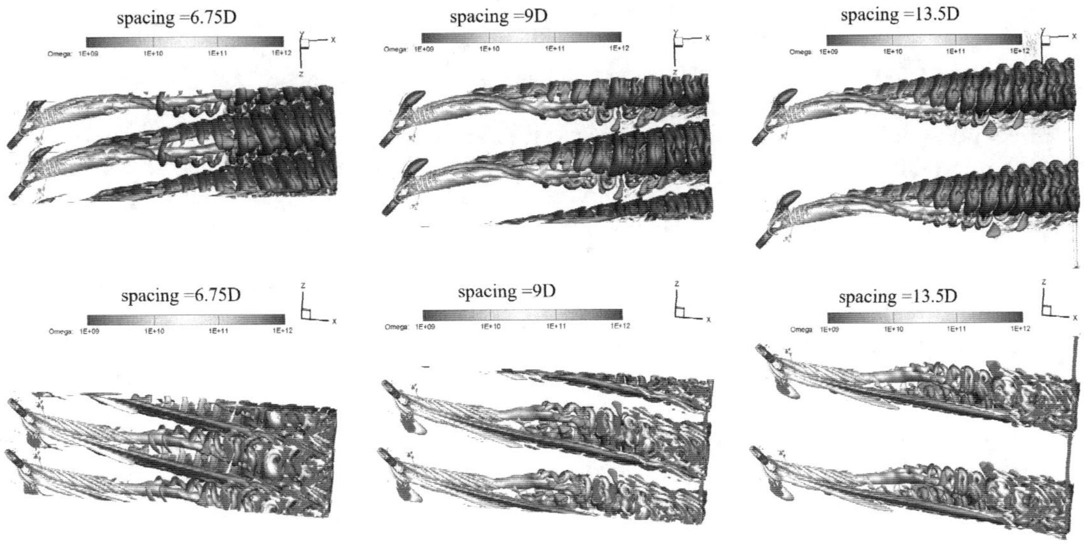

Figure 6. Vortex structures at different spacing of 6.75D, 9D and 13.5D from left to right.

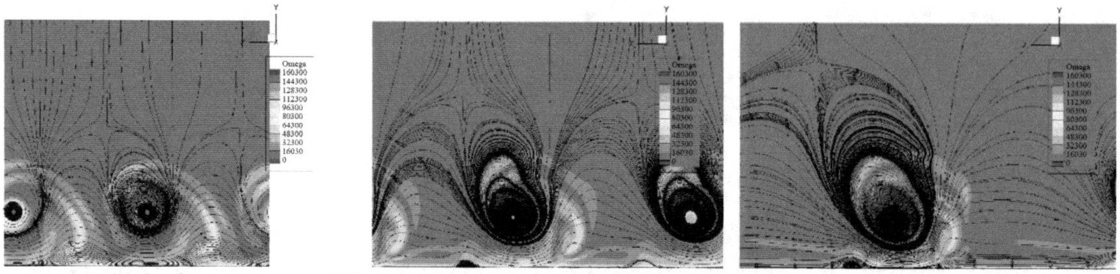

(a). Velocity contours at X=0.06 with spacing of 6.75D, 9D and 13.5D from left to right

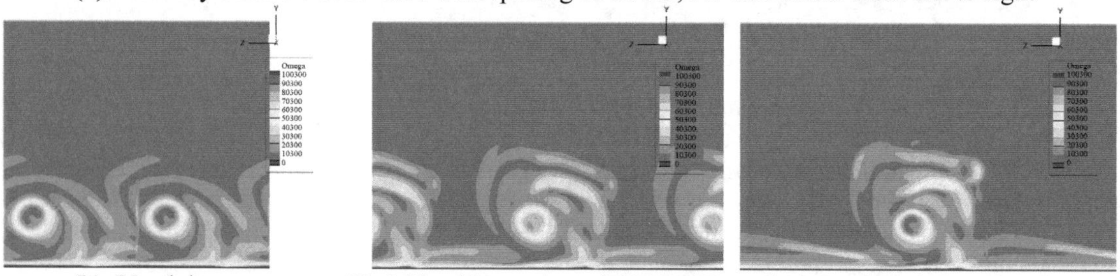

(b). Vorticity contours at X=0.08 with spacing of 6.75D, 9D and 13.5D from left to right

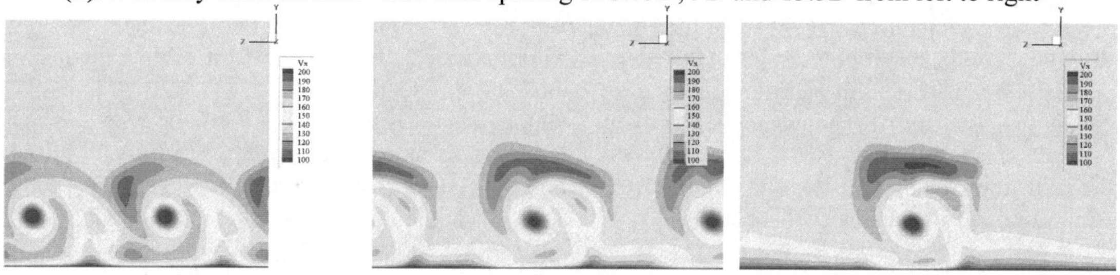

(c). Velocity contours at X=0.08 with spacing of 6.75D, 9D and 13.5D from left to right

Figure 7. Vorticity contours at different spacings.

6. Effect of boundary layer thickness on the vortex structure

Figure 8 shows the comparison of vortex structures with different boundary layer thicknesses. In this figure, Q is set as 10^8. From the perspective of vortex structure, the difference is neglective. However, when we see from the bottom side, we can find a relatively significant difference for the vortex-ring structure which is formed by the vortex 3. Figure 9 shows the outlet velocity profile compared with inlet velocity profile. Obviously, the two cases have fuller velocity profile than that of no microjet blowing. The case for the 4D case has a similar velocity profile with that of the 2D case. However, the difference for the 4D case is more significant, indicating that the improvement is promising.

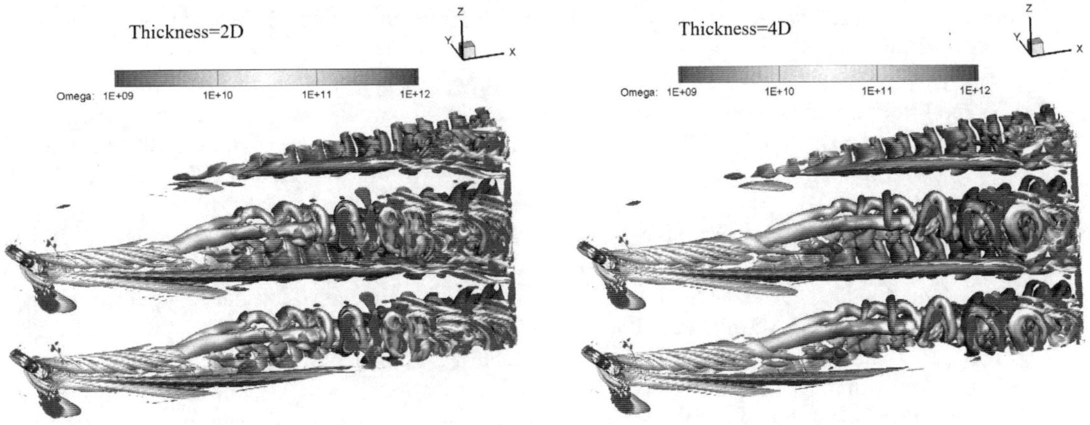

Figure 8. Vortex structures at different inlet boundary layer thickness.

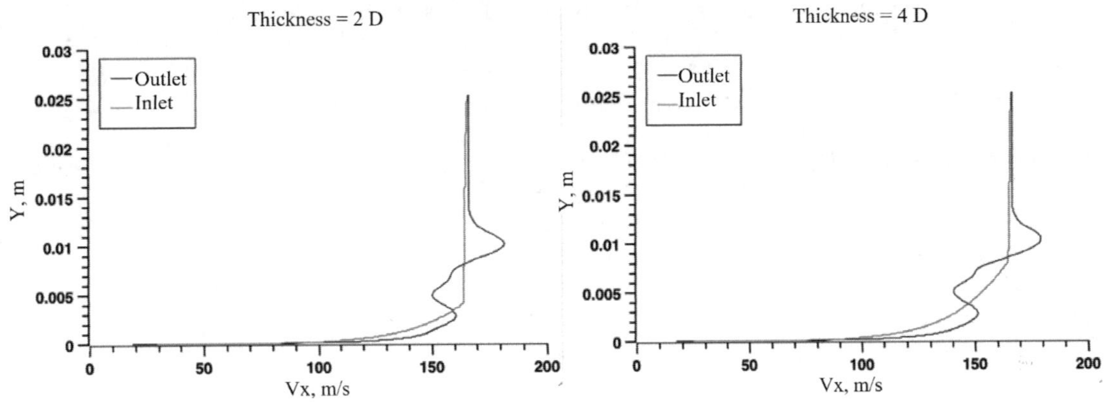

Figure 9. Velocity profile with different inlet boundary layer thickness (left: 2D, right: 4D).

7. Conclusions

A flat plate model with microjet array is established to understand the flow structure of vortices. By setting different working conditions, the effects of jet total pressure, incoming boundary layer thickness and jet hole spacing on jet action are compared and analyzed.

The flow field generated by pitched and skewed microjet array mainly includes one main vortex, two secondary vortices and some vortex rings. The main vortex is generated by the strong shear layer between jet and mainstream, and the other two vortices are the secondarily induced by the interaction between the main vortex and the mainstream. The development of vortices is highly related to the jet pressure. With the increase of jet pressure, the secondary vortices interact with the main vortices, resulting in more dispersed vorticity of the main vortices far behind, and better mixing effect of the flow field.

In the range of studied spacing, with the decrease of spacing, the stronger the vortex structure interaction is, and the more uniform of the mixing is. The thicker the boundary layer is, the more fluid the jet rolls up and the larger the secondary vortex structure is.

References

[1] Ball W H 1984 Tests of Wall Blowing Concepts for Diffuser Boundary Layer Control[R]. AIAA-84-1276.

[2] Johnston J P and Nishi M 1990 Vortex Generator Jets—Means for flow separation Control[J]. AIAA Journal, 28(6): 989-994
[3] Hamstra JW, Miller DN, Truax PP 2000 Active inlet flow control technology demonstration[R]. 22nd Internal Council of Aeronautical Sciences; 2000 Aug 27-Sep 1; Harrogate, UK.
[4] Rabe AC 2003 Effectiveness of a serpentine inlet duct flow control scheme at design and off-design simulated flight conditions[D]. Virginia: Virginia Polytechnic Institute and State University.
[5] Anderson JM 2003 Non-intrusive sensing and feedback control of serpentine inlet flow distortion[D]. Virginia: Virginia Polytechnic Institute and State University.
[6] Da X, Fan Z, Fan J, et al. 2015 Microjet flow control in an ultra-compact serpentine inlet[J]. Chinese Journal of Aeronautics; 28(5): 1381-90.
[7] Da X, Fan Z, Fan J 2020 Closed-loop flow control of an ultra-compact serpentine inlet based on nondimensional model[J]. Chinese Journal of Aeronautics; 33(10): 2555-2562.
[8] Compton D A and Johnston J P 1992 Streamwise vortex production by pitched and skewed jets in a turbulent boundary layer[J]. AIAA Journal, 30(3): 640-647
[9] Zhang X, Collins M W 1997 Measurements of a longitudinal vortex generated by a rectangular jet in a turbulent boundary layer [J]. Physics of Fluids, 9(6): 1665-1673.
[10] Zhang X, Collins M W 1997 Nearfield evolution of a longitudinal vortex generated by an inclined jet in a turbulent boundary layer [J]. Journal of Fluids Engineering, 119(4): 934-939.
[11] Zhang X 1998 Turbulence Measurements of a Longitudinal Vortex Generated by an Inclined Jet in a Turbulent Boundary Layer[J]. Journal of Fluids Engineering. 120(4):765-771
[12] Coussement Axel, Gicquel O. and Degrez G 2012 Large Eddy Simulation of a pulsed jet in cross-flow. Journal of Fluid Mechanics, 695:1-34
[13] Fernandez E. J., and Alvi F. S. 2014 Vorticity and Turbulence Properties of Microjet Arrays for Active Flow Control[R]. AIAA-2014-0769.
[14] Wu J and Ning F 2013 Development of hybrid RANS/LES Code and Its Validation[J]. Transactions of Nanjing University of Aeronautics and Astronautics, 30(1): 24-32.

Journal of Physics: Conference Series **2313** (2022) 012010 doi:10.1088/1742-6596/2313/1/012010

Improved theory of the elastic-plastic response of spherical containment vessels subjected to internal blast loading and its application

Zeng He[1], Zhihuan Wang[1, 2], Baihua Zhang[1], Leiming Jia[1] and Zhou Tian[1, 3]

[1]Northwest Institute of Nuclear Technology, Xi'an 710024, China
[2]Department of Engineering Physics, Tsinghua University, Beijing 100084, China

[3]Corresponding author's e-mail: tianzhou@nint.ac.cn

Abstract. Baker built the displacement equations of motion for a thin spherical shell to spherically symmetric internal blast loading, and further provided the analytical solution of the elastic-plastic response under a triangular decaying pressure, neglecting shell thinning and radius variation [1], which was widely used in the fields of analysis and design of explosion containment vessels. However, it appears that some improvements and corrections are needed when we revisit this classic problem. In this paper, the elastic-plastic response of thin spherical shells is investigated mathematically. The complete equations of motion for three models are developed. Then the main ideas of the analytical and numerical methods are presented, as well as some important procedures and results by using Laplace transform techniques, which are convenient to use and generalize. Three typical examples are solved to exhibit differences between our improved theory and Baker's [1]. Furthermore, our models are employed to gain some insights into response regularities, including the effect of various degrees of strain-hardening on maximum displacements under impulsive loading, and the effect of loading duration on the shell displacement for a perfectly plastic material.

1. Introduction

Explosion containment vessels have been widely used in the fields of manufacturing industry, public security, and national defense [2-4]. In order to assess the capability of containment shells to withstand the effects of internal blast, it is necessary to study its dynamic behaviour in both theoretical and practical senses. According to the vessel shapes, they can be classified into spherical shells, cylindrical shells, and cylinders with hemispherical or ellipsoidal end caps. According to the application scenarios, they can be classified into single-use and frequent-use.

Baker is the first to discuss the elastic [5-7] and elastic-plastic responses [1,8] of thin spherical shells, and observes rich phenomena in experiment and achieves excellent knowledge in theory. Because the spherical vessel's motion under spherical symmetric loading is of single degree-of-freedom (SDOF), Baker has obtained analytical solutions of elastic and elastic-plastic responses of spherical shells undergoing a triangular decaying pressure, and compared the theory with experiment [1,5,8]. Besides, Baker also considers the response of a spherical shell to blast loading from an eccentric source [6,7]. He has determined the vibration modes of a thin spherical shell and obtained infinite-series solution of elastic response, both of which are verified by comparison with experimental data.

Journal of Physics: Conference Series **2313** (2022) 012010 doi:10.1088/1742-6596/2313/1/012010

Based on the work of Baker, extensive researches in the aspects of theory, experiment and numerical simulation have been carried out for the response of spherical explosion vessels [2-4, 9-25]. For example, Liu has monitored the pressure and temperature at the internal surface of a vessel as well as the strain of its external surface, which provides experimental data for understanding the whole process of flow field evolution and vessel response [22]. Hu has calculated the normally reflected pressure acting at the inner wall of the vessel by using the DYTRAN program, and then analyzed the vessel's dynamic response by an analytical method and the LS-DYNA software, which has practical significance in engineering design and safe operation [11-14]. In addition, the strain growth phenomenon of explosion vessels has attracted much attention since it is introduced by Buzukov [9]. For example, Dong has studied the mechanisms of strain growth by using the LS-DYNA software, and put forward corresponding protection suggestions [17,18].

Most of the above studies are limited within the range of elastic deformation. As pointed out by White [2], due to more explosive energy can be absorbed by plastic strain, the shell thickness of the vessel that allows certain plastic deformation is much smaller than that of the elastic design criteria, thereby substantially reducing difficulties in manufacture, transport, storage and disposal [4]. In addition, we discover that some improvements or corrections are needed for the Baker's theory [1]: (1) Analytical solutions in the plastic response phase (including perfect plasticity and linear plasticity) are incorrect after the transient pressure ends and before the elastic motion resumes, neglecting radius variation and shell thinning; (2) No equations of motion and the corresponding solutions are given after the elastic motion resumes, neglecting radius variation and shell thinning; (3) Only equation of motion is given in the plastic deformation phase, considering radius variation and shell thinning. However, in so far as the author has known, Baker's solutions are still widely accepted, such as they are almost completely quoted in the monograph of Yang, where only ε is altered to 2ε for the stress-strain relation in the plastic regime [26]. Therefore, it is necessary to reinvestigate the elastic-plastic response of a thin spherical shell subjected to internal blast loading.

Our improvements mainly include: (1) Correct analytical solutions are derived in the plastic response phase after the transient pressure ends and before the elastic motion resumes, neglecting radius variation and shell thinning, and the possible reason of Baker's mistake is given; (2) Equations of motion and the corresponding solutions are provided after the elastic motion resumes, neglecting radius variation and shell thinning; (3) Equations of motion in the elastic loading phase and in the elastic resuming phase are added, considering radius variation and shell thinning, and our definition of strain may be more suitable for large deformation. Besides, some general analytical solutions are also given, which can be easily employed to other pressure waveforms.

2. Equations of motion

The equation of motion in the radial direction of a thin spherical shell to the transient pressure $p(t)$ is

$$h\left(\sigma_{\theta\theta}+\sigma_{\phi\phi}\right)+\rho hR\frac{\mathrm{d}^2u}{\mathrm{d}t^2}=Rp(t) \tag{1}$$

where h and R are shell thickness and radius, respectively, $\sigma_{\theta\theta}$ and $\sigma_{\phi\phi}$ are the membrane stresses of shell element, ρ is density of the shell material, u is the radial displacement. The two membrane stresses $\sigma_{\theta\theta}$ and $\sigma_{\phi\phi}$ should be equal due to spherical symmetry and can be designated by $\sigma=\sigma_{\theta\theta}=\sigma_{\phi\phi}$. Then the above equation may be simplified as

$$\frac{\mathrm{d}^2u}{\mathrm{d}t^2}+\frac{2\sigma}{\rho R}=\frac{p(t)}{\rho h} \tag{2}$$

This is the basic equation of motion for thin spherical shells subjected to symmetric pressure. The complete specific equations of motion are presented below for three cases, i.e., elastic response, elastic-plastic response neglecting radius variation and shell thinning, and elastic-plastic response considering these effects, while Baker only gave partial specific equations of motion [1].

2.1. *Model 1: elastic response*
Let us start from a mild loading. The spherical shell always vibrates within the elastic range and obeys Hooke's law, as portion *OA* displayed in Figure 1,

$$\sigma = \frac{E}{1-\nu}\epsilon = \frac{E}{1-\nu}\frac{u}{R_0}, \qquad \sigma \le Y \tag{3}$$

where *E*, ν and *Y* represent Young's modulus, Poisson's ratio and yield strength of the shell material, $\varepsilon = \varepsilon_{\theta\theta} = \varepsilon_{\phi\phi} = u/R_0$ is spherical membrane strain with R_0 the initial value of the shell radius, due to the fact that the radius variation can be neglected in most cases.

Substituting equation (3) into equation (2), we obtain the equation of motion in the elastic phase,

$$\frac{\mathrm{d}^2u}{\mathrm{d}t^2} + \omega_e^2 u = \frac{p(t)}{\rho_0 h_0} \tag{4}$$

where ρ_0 and h_0 denotes the initial values of shell material and shell thickness respectively, and ω_e is the circular frequency of the elastic response defined as

$$\omega_e = \frac{1}{R_0}\sqrt{\frac{2E}{\rho_0(1-\nu)}} \tag{5}$$

Equation (4) has the same form as the forced vibration of a SDOF system without damping [3], thus it is stated that shell is at the breathing mode [17].

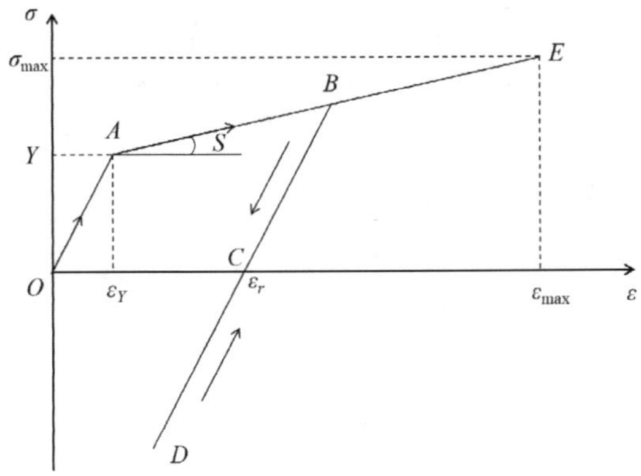

Figure 1. Bilinear elastic-plastic stress-strain relations for the spherical shell material.

2.2. *Model 2: elastic-plastic response, neglecting radius variation and shell thinning*
As the loading becomes more intense, equation (4) holds until the stress reaches yield. The shell material then enters the plastic flow regime, and it is then assumed to obey the linear strain-hardening relation shown in Figure 1. The stress beyond yield can be written as

$$\sigma = Y + S(\epsilon - \epsilon_Y) \tag{6}$$

where ε_Y is the yield strain of the shell material (*P*oint A), and the slope *S* of the plastic portion is called the strain-hardening coefficient, while perfect plasticity is regarded as a special case $S = 0$. Denote the time of yielding as t_A, and the time of elastic motion resuming as t_B, we get the equation of motion in the plastic phase from equations (2) and (6)

$$\frac{\mathrm{d}^2u}{\mathrm{d}t^2} + \omega_p^2 u = \frac{p(t)}{\rho_0 h_0} - \frac{2(Y - S\epsilon_Y)}{\rho_0 R_0}, \qquad t_A < t \le t_B \tag{7}$$

where ω_p is the circular frequency of the plastic response defined as

$$\omega_p = \frac{1}{R_0}\sqrt{\frac{2S}{\rho_0}} \tag{8}$$

The end of plastic phase, *P*oint B in Figure 1, is characterized by maximum radial displacement, i.e., the velocity $\dot{u} = 0$. The corresponding stress and strain of *P*oint B are denoted as σe_B and ε_B, respectively. Then the material experiences elastic motion along path *BCD* parallel to path *OA*, where the stress is given by

$$\sigma = \sigma_B + \frac{E}{1-\nu}(\epsilon - \epsilon_B) \tag{9}$$

and the equation of motion in the elastic unloading phase is

$$\frac{\mathrm{d}^2 u}{\mathrm{d}t^2} + \omega_e^2 u = \frac{p(t)}{\rho_0 h_0} - \frac{2}{\rho_0 R_0}\left(\sigma_B - \frac{E}{1-\nu}\epsilon_B\right), \qquad t > t_B \tag{10}$$

The second term (including negative sign) on the right hand of equation (10) stands for an equivalent loading originating from plastic deformation, which is always positive and constant. Therefore, it produces a permanent strain in the shell material, as *P*oint C shown in Figure 1. Setting $\sigma = 0$ in equation (9), the permanent strain ε_r is given by

$$\epsilon_r = \epsilon_B - \frac{1-\nu}{E}\sigma_B \tag{11}$$

Baker didn't give the equation after elastic motion resumes [1]. We can see that when is loading is intense enough, the shell material goes through elastic loading-plastic loading-elastic unloading successively, following the stress-strain path $O \rightarrow A \rightarrow B \leftrightarrows C \leftrightarrows D$ in the sequence of events occurring in Figure 1.

2.3. *Model 3: elastic-plastic response, considering radius variation and shell thinning*

For typical brittle materials ($\epsilon \ll 1$), the above models are sufficient to predict the dynamic response. If the shell material is ductile, maximum allowable strain can be around 0.1, and the shell thickness and radius may vary noticeably. These effects will now be accounted for.

The basic equation of motion still applies, whereas shell thickness *h* and radius *R* can change over time in equation (2). As a spherical shell undergoes arbitrary deformation, the relations between displacement, radius and strain are as follows [27],

$$u = R - R_0, \qquad \epsilon = \frac{u}{R} + \frac{1}{2}\left(\frac{u}{R}\right)^2 \approx \frac{u}{R} = 1 - \frac{R_0}{R} \tag{12}$$

The second term in the definition of strain is usually at least one order of magnitude smaller than the first term, and thus can be omitted. The specific equations of motion are established next in the same process as Model 2. For super-ductile materials allowing much larger strain, the full definition of strain should be employed.

In the elastic loading phase, the shell material is compressible but shell thickness is almost constant [1,8], yielding

$$h = h_0, \qquad \rho R^2 = \rho_0 R_0^2 \tag{13}$$

Combined with equations (2) and (3), we can obtain the evolution equation of shell radius,

$$\frac{\mathrm{d}^2R}{\mathrm{d}t^2}+\frac{2E}{\rho_0R_0^2(1-\nu)}R-\frac{2E}{\rho_0R_0(1-\nu)}=\frac{p(t)}{\rho_0h_0R_0^2}R^2,\qquad t\le t_A \tag{14}$$

During plastic deformation, the shell material remains essentially incompressible so that the volume of the thin spherical shell is the same at any time [1,8]. Similarly, the evolution equation of shell radius becomes:

$$\frac{\mathrm{d}^2R}{\mathrm{d}t^2}+\frac{2\left[Y+S\left(1-\epsilon_Y\right)\right]}{\rho_A}\frac{1}{R}-\frac{2SR_0}{\rho_A}\frac{1}{R^2}=\frac{p(t)}{\rho_Ah_AR_A^2}R^2=\frac{p(t)}{\rho_0h_0R_0^2}R^2,\quad t_A<t\le t_B \tag{15}$$

In the elastic motion resuming phase, one can still assumes that the shell material is compressible with shell thickness being constant. Hence the evolution equation of shell radius is

$$\frac{\mathrm{d}^2R}{\mathrm{d}t^2}+\frac{2}{\rho_BR_B^2}\left[\sigma_B+\frac{E}{1-\nu}\left(1-\epsilon_B\right)\right]R-\frac{2ER_0}{\rho_BR_B^2(1-\nu)}=\frac{p(t)}{\rho_0h_0R_0^2}R^2,\quad t>t_B \tag{16}$$

Baker only gave a differential equation of spherical shell radius in the plastic phase under the assumption $\epsilon=u/R_0$ [1], which is different from equation (14).

3. Solutions of the equation of motion

In this section both the analytical and numerical methods to solve the preceding equations of motion are presented briefly. In mathematics, this work is transferred to handle a series of second-order nonhomogeneous ordinary differential equations with initial values. To be more specific, the equations for Model 1 and Model 2 are linear, while that for Model 3 is nonlinear. As a consequence, Model 1 and Model 2 can be dealt with in both analytical and numerical approaches. Solutions of Model 3, however, must be accomplished by numerical computation.

The main ideas of the analytical and numerical methods are given below, as well as some key processes and results by using Laplace transform techniques for the analytical method. Detailed derivation of the complete solutions of Model 1 and Model 2 by method of characteristic root is somewhat tedious, and will be published in another paper.

Before going on, let us assume an appropriate form of the transient pressure $p(t)$. In the preliminary analysis of containment vessels, blast loading is normally simplified as a triangular pulse [1,8,18], i.e.,

$$p(t)=\begin{cases}P\left(1-\dfrac{t}{T}\right), & 0\le t\le T\\ 0, & t>T\end{cases} \tag{17}$$

where P is the peak overpressure, T is the equivalent duration of the transient loading which is determined by impulse I as $T = 2I/P$.

3.1. *Analytical method*

Laplace transform is a common method to solve the initial value problem of ordinary differential equation, with the advantage of simple conduct and convenient generalization [28].

3.1.1. Model 1. Consider the general form of the transient pressure p(t) for the moment. Let us implement Laplace transform on equation (4), insert the initial conditions, and then use the convolution theorem [28]. Eventually, we can obtain the solution of equation (4) as below,

$$u(t)=\frac{1}{\rho_0h_0\omega_e}\int_0^t p(\tau)\sin\omega_e\left(t-\tau\right)\mathrm{d}\tau \tag{18}$$

Baker didn't this general solution of elastic response in Ref. [1]. The specific solution for the triangular decaying pressure of equation (17) is

$$u(t)=\begin{cases}\dfrac{P}{\rho_0 h_0\omega_e^2}\left(1-\dfrac{t}{T}-\cos\omega_e t+\dfrac{\sin\omega_e t}{\omega_e T}\right), & t\le T\\ \dfrac{P}{\rho_0 h_0\omega_e^2}\left[-\cos\omega_e t+\dfrac{\sin\omega_e t-\sin\omega_e(t-T)}{\omega_e T}\right], & t>T\end{cases} \tag{19}$$

which agrees with Baker's solution [1].

3.1.2. Model 2. Here we mainly focus on the situation where the loading continues after the stress reaches yield but ends before the elastic motion resumes, i.e., $t_A < T \le t_B$. The first formula of equation (19) still holds for $t < t_A$. With regard to response after t_A, the shell material for perfect plasticity and linear plasticity are discussed below.

We begin with a perfectly plastic material. When $t_A < t \le t_B$, the shell material is in plastic phase, and the equation of motion, equation (7), reduces to

$$\frac{d^2u}{dt^2}=\frac{p(t)}{\rho_0 h_0}-\frac{2Y}{\rho_0 R_0}, \qquad t_A<t\le t_B \tag{20}$$

Perform the standard procedures of Laplace transform by invoking the continuity conditions that radial displacement and velocity are continuous at $t = t_A$. The resulting solution is,

$$u(t)=\frac{1}{\rho_0 h_0}\int_{t_A}^{t} p(\tau)(t-\tau)\,d\tau-\frac{Y}{\rho_0 R_0}(t-t_A)^2+\dot{u}_A(t-t_A)+u_A \tag{21}$$

which is missing in Ref. [1]. Substituting the transient pressure, equation (17), into above integration, we can obtain the specific solution,

$$u(t)=\begin{cases}\dfrac{P}{\rho_0 h_0}\left[-\dfrac{t^3}{6T}+\dfrac{t^2}{2}+\left(\dfrac{t_A^2}{2T}-t_A\right)t+\dfrac{t_A^2}{2}-\dfrac{t_A^3}{3T}\right]-\dfrac{Y}{\rho_0 R_0}(t-t_A)^2+\dot{u}_A(t-t_A)+u_A, & t_A<t\le T\\ \dfrac{P}{\rho_0 h_0}\left[\dfrac{(T-t_A)^2}{2T}t-\dfrac{T^2}{6}+\dfrac{t_A^2}{2}-\dfrac{t_A^3}{3T}\right]-\dfrac{Y}{\rho_0 R_0}(t-t_A)^2+\dot{u}_A(t-t_A)+u_A, & T<t\le t_B\end{cases} \tag{22}$$

When $t > t_B$, the shell material is in the elasticity resuming phase, and the equation of motion, equation (10), may be written as

$$\frac{d^2u}{dt^2}+\omega_e^2 u=-\frac{2}{\rho_0 R_0}\left(Y-\frac{E}{1-\nu}\epsilon_B\right), \qquad t>t_B \tag{23}$$

The solution is obvious by invoking the continuity conditions at $t = t_B$,

$$u(t)=u_B\cos\omega_e(t-t_B)-\frac{2}{\rho_0 R_0\omega_e^2}\left(Y-\frac{E}{1-\nu}\epsilon_B\right)\left[1-\cos\omega_e(t-t_B)\right], \qquad t>t_B \tag{24}$$

In contrast, the first formula of equation (22) is the same as in Ref. [1], whereas the second formula is wrong and solution after the elastic motion resumes, equation (24), is missing therein.

The same procedure is also applicable to a linearly plastic material. Therefore, below we directly list the solution at $t_A < t \le t_B$

$$u(t)=\frac{1}{\omega_p}\int_{t_A}^{t}\left[\frac{p(\tau)}{\rho_0 h_0}-\frac{2(Y-S\epsilon_Y)}{\rho_0 R_0}\right]\sin\omega_p(t-\tau)\mathrm{d}\tau+u_A\cos\omega_p(t-t_A)+\frac{\dot{u}_A}{\omega_p}\cos\omega_p(t-t_A) \quad (25)$$

which is not presented in Baker's theory [1]. And the specific solution for the triangular decaying loading is

$$u(t)=\begin{cases} u_A\cos\omega_p(t-t_A)+\frac{\dot{u}_A}{\omega_p}\sin\omega_p(t-t_A)-\frac{2(Y-S\epsilon_Y)}{\rho_0 R_0\omega_p^2}\left[1-\cos\omega_p(t-t_A)\right] \\ +\frac{P}{\rho_0 h_0\omega_p^2}\left[1-\frac{t}{T}-\left(1-\frac{t_A}{T}\right)\cos\omega_p(t-t_A)+\frac{\sin\omega_p(t-t_A)}{\omega_p T}\right], \quad t_A<t\le T \\ u_A\cos\omega_p(t-t_A)+\frac{\dot{u}_A}{\omega_p}\sin\omega_p(t-t_A)-\frac{2(Y-S\epsilon_Y)}{\rho_0 R_0\omega_p^2}\left[1-\cos\omega_p(t-t_A)\right] \\ +\frac{P}{\rho_0 h_0\omega_p^2}\frac{1}{\omega_p T}\left[\sin\omega_p(t-t_A)-\sin\omega_p(t-T)-\omega_p(T-t_A)\cos\omega_p(t-t_A)\right], T<t\le t_B \end{cases} \quad (26)$$

The equation of motion in the elasticity resuming phase has the same expression as equation (23), except for that Y should be replaced with σ_B. Thus the corresponding solution can be gained readily. Similarly, the first formula of equation (26) is the same as in Ref. [1], whereas the second formula is wrong and solution after the elastic motion resumes is not provided by Baker.

Hence, we can see that Baker didn't give the solutions the elasticity resuming phase, for either a perfectly plastic material or a linearly plastic material [1,8]. Besides, his analytical solutions in the plastic phase after the pressure duration are incorrect. Taking equation (26) as an example. Baker only obtained the solution at $t_A < t \le T$, which is identical to the first formula therein, and he further stated that it "can be modified to give displacement after the pulse by substituting $t = T$ in the last term" [1]. We notice, however, this substitution couldn't lead to the second formula of equation (26). Possible reasons responsible for this mistake can be inferred from equation (25). When calculating the integration of the first term on the right side by using the triangular pressure, we only need to replace the upper integral limit t with T, i.e.,

$$\int_{t_A}^{t} p(\tau)\sin\omega_p(t-\tau)\mathrm{d}\tau=\int_{t_A}^{T} P\left(1-\frac{\tau}{T}\right)\sin\omega_p(t-\tau)\mathrm{d}\tau \quad (27)$$

However, Baker may also replace t in the integrand $\sin\omega_p(t-\tau)$ with T [1,8], thereby making the wrong statement.

3.2. Numerical method

From above we have found that the analytical solutions are rather complicated, even though the transient pressure takes a simple form. We deduce that generally it is difficult in acquiring explicit analytical solutions for complex or irregular loadings. Moreover, there exists no analytical solutions for Model 3. These disadvantages prompt us to solve the equation of motion in a numerical manner.

Recall different models in Section 2, we find that all the equations of motion belong to second-order differential equations with respect to time, with the dependent variable being displacement u for Model 1 and Model 2, and radius R for Model 3. Write the equation of motion in a more general form,

$$\frac{\mathrm{d}^2 y}{\mathrm{d}t^2}=f(t,y) \quad (28)$$

where $y = u$ corresponds to Model 1 and Model 2, while $y = R$ to Model 3. In order to exploit the common numerical methods for first-order ordinary differential equation, one is required to rewrite the above second-order differential equation as first-order differential equations. Let us define $Y_1 = y$, $Y_2 = \dot{y}$ (which is equal to $\dot{u}$ for Model 1 and Model 2, and $\dot{R}$ for Model 3), so equation (28) can be modified as [29]

$$\begin{cases} \dfrac{\mathrm{d}Y_1}{\mathrm{d}t} = Y_2 \equiv F_1(t, Y_1, Y_2) \\ \dfrac{\mathrm{d}Y_2}{\mathrm{d}t} = f(t, Y_1) \equiv F_2(t, Y_1, Y_2) \end{cases} \tag{29}$$

It is convenient to express equation (29) and the initial conditions in a vector form,

$$\begin{cases} \dfrac{\mathrm{d}\boldsymbol{Y}}{\mathrm{d}t} = \boldsymbol{F}(t,\ \boldsymbol{Y}) \\ \boldsymbol{Y}(t = 0) = \boldsymbol{Y}_0 \end{cases} \tag{30}$$

where $\boldsymbol{Y} = (Y_1, Y_2)^{\mathrm{T}}$, $\boldsymbol{F} = (F_1, F_2)^{\mathrm{T}}$, $\boldsymbol{Y}_0 = (y_0, \dot{y}_0)^{\mathrm{T}}$. Equation (30) can then be solved by using various famous numerical algorithms, such as trapezoid method, Runge-Kutta method and so on [29].

4. Examples and discussions

Baker has computed the responses of spherical shells subjected to different blast loadings, and showed the results for elastic deformation, elastic-plastic deformation neglecting radius variation and shell thinning, as well as elastic-plastic deformation considering these effects [1,8]. These typical examples are recalculated by analytical solutions and numerical method developed in this paper, where the fourth-order classical Runge-Kutta method is employed to give numerical solutions [29], and then differences are recognized between our results and those from Ref. [1].

The initial shell dimensions are $R_0 = 15$ in, $h_0 = 1/16$ in. The properties of the shell material is as follows: density $\rho_0 = 0.283$ lb/in^3, Young's modulus $E = 3\times10^7$ psi, Poisson's ratio $\nu = 1/3$, yield stress $Y = 4\times10^4$ psi, strain-hardening coefficient $S = E/(1-\nu)/2$. Radial displacement at yielding, u_A, can be computed to be 13.33×10^{-3} in.

4.1. Results of elastic response

Blast Loading 1 has peak overpressure $P = 120$ psi and loading duration $T = 13.8\times10^{-5}$ s. The total time is 45×10^{-5} s, with time step $\Delta t = 4.5\times10^{-7}$ s. Calculated results are shown in Figure 2, which illustrate that the analytical and numerical solutions match each other quite well. Maximum displacement $u_{\max} = 5.82\times10^{-5}$ in, indicating that Loading 1 is too weak to cause plastic flow. Result of shell displacement extracted from Ref. [1] is also plotted for comparison. It can be observed that displacement curves differ slightly during $(20\text{~}25)\times10^{-5}$ s. Maximum displacements is 5.85×10^{-3} in from Ref. [1], with relative difference around 0.5% away from our result. Because the analytical solutions of the elastic response are the same, this small discrepancy may arise from computational precision or round-off error.

4.2. Results of elastic-plastic response, neglecting radius variation and shell thinning

Blast Loading 2 has peak overpressure $P = 1000$ psi and loading duration $T = 6.65\times10^{-5}$ s, which may give rise to plasticity (we will come back to this problem later). Calculated results of Model 2 are shown in Figure 3, where the analytical and numerical curves still overlap completely, with the same time step $\Delta t = 4.5\times10^{-7}$ s as in Figure 2. We discover that the deformation already exceeds the elastic range due to $u_{\max} > 30\times10^{-5}$ in. Baker's displacement results are illustrated again for comparison [1]:

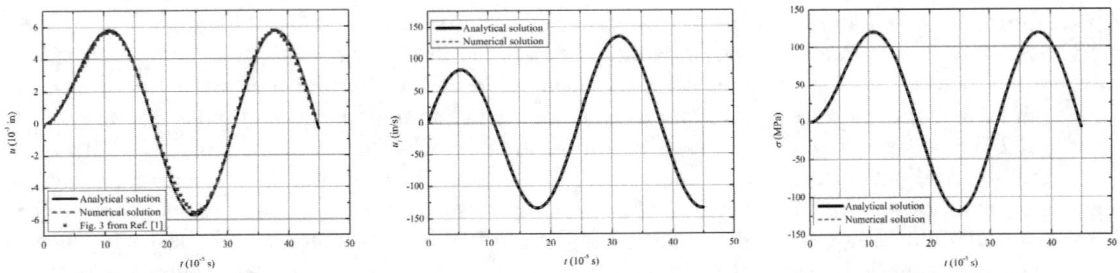

(a) Displacement (b) Velocity (c) Membrane stress

Figure 2. Evolutions of main quantities in the elastic response of spherical shell: Model 1, Loading 1.

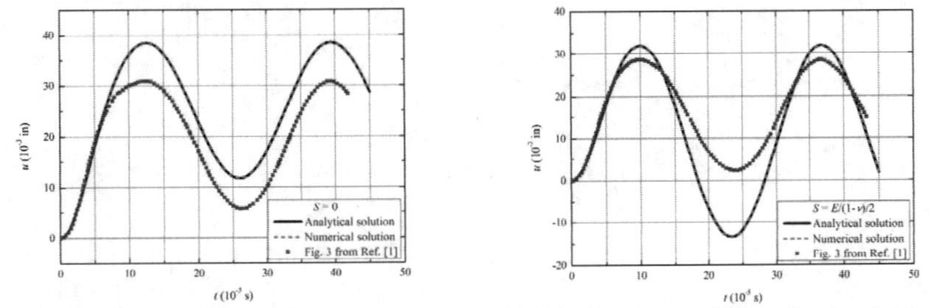

(a) Displacement of perfectly plastic shell (b) Displacement of linearly plastic shell

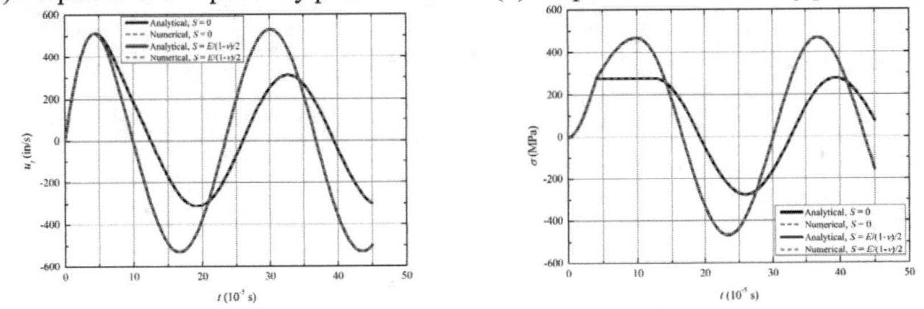

(c) Velocity (d) Membrane stress

Figure 3. Evolutions of main quantities in the elastic-plastic response of spherical shell: Model 2, Loading 2.

As $t < 6.4\times10^{-5}$ s, displacement curves are almost identical, and significant discrepancy occurs afterwards. Let us start to analyze by estimating the conditions of yielding. We obtain $t_A = 4.095\times10^{-5}$ s, $u_A = 13.36\times10^{-3}$ in, $\dot{u}_A = 509$ in/s, in good agreement with those from Ref. [1]. Next, the time of maximum displacements, t_B, is about 10×10^{-5} s for a linearly plastic material, and is even longer for a perfectly plastic material. Because of $t_A < T \leq t_B$, this example belongs the situation described in Subsection 3.1. As have been mentioned there, Baker's analytical solution is correct within $t \leq T$. Note that the numerical results in Figure 3 are only dependent on the equations of motion presented in Section 2, and completely independent of the steps followed in their analytical solution. Therefore, the numerical results in Figure 3 confirm that Baker's solution is not correct for $t > T$, while our new analytical solution is correct. At last, maximum displacements taken from Ref. [1] are 32.2×10^{-3} in and 28.7×10^{-3} in for the perfectly plastic and linearly plastic materials, while our results are 38.5×10^{-3} in and 31.9×10^{-3} in, with mean difference around 13%, much larger than the difference caused by computational precision or round-off error.

4.3. *Results of elastic-plastic response, considering radius variation and shell thinning*

Numerical solutions of Model 3 under Loading 2 are shown in Figure 4, with time step $\Delta t = 2\times10^{-8}$ s to achieve a better convergence. It is observed from Figure 4-(f) that the shell radius varies within 0.26%, thus the nonlinear effect from radius variation and shell thinning is negligible. In fact, results of Model 2 and Model 3 under Loading 2 are almost identical. As for the wrong conclusion drawn by Baker that the nonlinear effect is quite appreciable [1,8], it may be attributed to his incorrect analytical solutions. On this basis, we can infer that the Model 3 in this paper and the nonlinear theory from Ref. [1] are essentially the same as Model 2 due to small deformation. Therefore, the slight discrepancy of displacement curves in Figure 4-(b) and (c) may simply come from different numerical methods, computational precisions or round-off errors.

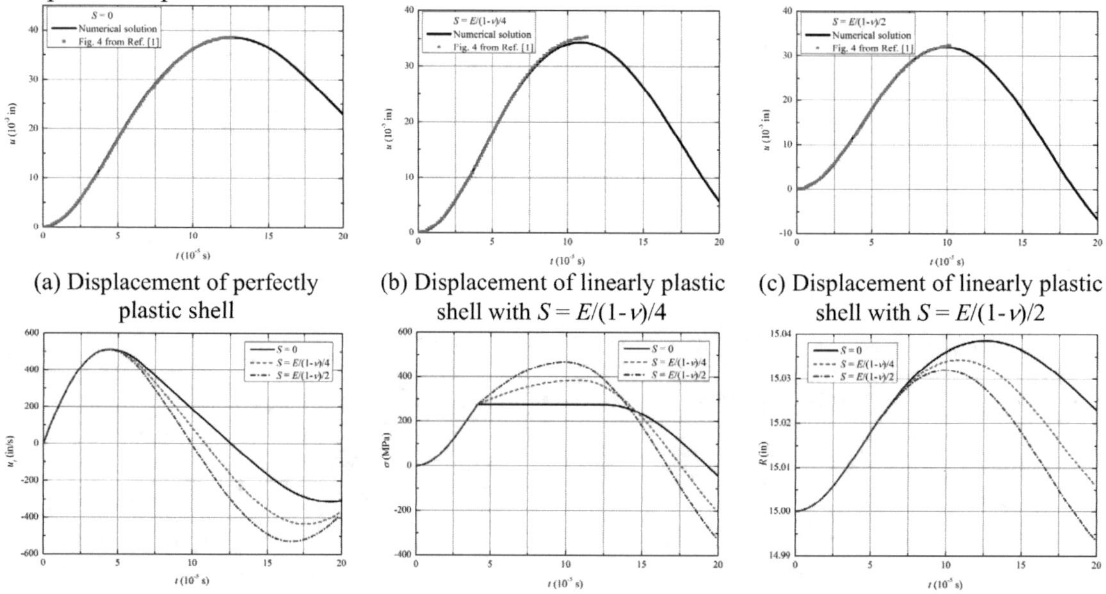

(a) Displacement of perfectly plastic shell (b) Displacement of linearly plastic shell with $S = E/(1-\nu)/4$ (c) Displacement of linearly plastic shell with $S = E/(1-\nu)/2$

(d) Velocity (e) Membrane stress (f) Radius

Figure 4. Evolutions of main quantities in the elastic-plastic response of spherical shell: Model 3, Loading 2.

5. Regularity studies

In this section, we apply the theory developed above to carry out some study about response regularities, mainly focusing on impulsive loading and loading duration.

5.1. *Regularity of impulsive loading*

It is noticed from above analytical derivation that in principle Laplace transform sets no restrictions on the form of the transient pressure $p(t)$. As a matter of fact, substitutions of other simplified pressure waveforms (such as rectangular, parabolic, and exponential) into equation (18) lead to the same formulae as in Refs. [3,11]. If the real loading duration T is much shorter than the period of the breathing mode T_e, or it is required to predict the dynamic behaviour of a spherical shell simply from loading impulse, then the transient pressure $p(t)$ can be regarded as an impulsive form,

$$p(t) = I\delta(t) \tag{31}$$

First consider the elastic response of the shell. The integral of equation (18) for impulsive $p(t)$ is

$$u(t) = \frac{I}{\rho_0 h_0 \omega_e} \sin \omega_e t \tag{32}$$

which can also be acquired by letting $T \to 0$ in the second formula of equation (19). Hence, maximum displacement of the elastic response is

$$u_{\text{max,elasticity}} = \frac{I}{\rho_0 h_0 \omega_e} \tag{33}$$

This is consistent with the results in Refs. [3,18], indicating that maximum displacement is proportional to impulse. Equation (33) is the basic impulse criterion in strength design of explosion containment vessels, which is appropriate in the case of $u_{\text{max, elasticity}} \leq R_0\varepsilon_Y$.

If $u_{\text{max, elastic}} > R_0\varepsilon_Y$, then plastic response occurs. Similar derivation yields

$$u_{\text{max,perfect plasticity}} = \frac{1}{2} R_0 \epsilon_Y + \frac{R_0 I^2}{4\rho_0 Y h_0^2} \tag{34}$$

$$u_{\text{max,linear plasticity}} = R_0 \epsilon_Y + \frac{R_0}{S}\left[\sqrt{Y^2 + \frac{S}{2\rho_0 h_0^2}\left[I^2 - \left(\rho_0 h_0 R_0 \omega_e \epsilon_Y\right)^2\right]} - Y\right] \tag{35}$$

As impulse becomes more intense, maximum displacements for perfectly plastic and linearly plastic materials can be approximated by,

$$u_{\text{max,perfect plasticity}} \approx \frac{R_0}{4\rho_0 Y_0 h_0^2} I^2, \qquad u_{\text{max,linear plasticity}} \approx \frac{R_0}{\sqrt{2\rho_0 S} h_0} I \tag{36}$$

The former is proportional to the square of impulse, while the latter is proportional to impulse.

Figure 5 displays maximum displacements for shell materials with different strain-hardenings under impulsive loading. Shell dimensions and material parameters are as follows: $R_0 = 39.4$ in, $h_0 = 0.197$ in, $\rho_0 = 0.283$ lb/in^3, $E = 3.05\times10^7$ psi, $\nu = 0.3$, $Y = 4.35\times10^4$ psi. Strain-hardening coefficient S takes various values, among which $S/E = 1.43 = 1/(1-\nu)$ stands for the special situation where the elastic and plastic responses are the same. If $I \leq \rho_0 h_0 R_0 \omega_e \varepsilon_Y$, the response of shell is always within the elastic regime, and different materials behave identically. As I becomes larger than $\rho_0 h_0 R_0 \omega_e \varepsilon_Y$, plasticity occurs and differences emerge. We can see from Figure 5 that the smaller strain-hardening coefficient S is, the larger maximum displacement u_{max} becomes, due to the fact that more deformation is needed to absorb the loading energy. Differences between the elastic response and the plastic response of small strain-hardening coefficient are quite remarkable. Explosion containment vessels are mainly made of steel, whose S/E is rather small, thus impulse criterion of elastic response is no longer suitable for vessels allowing certain plastic deformation.

5.2. Regularity of loading duration

Displacement curves of a spherical shell for perfectly plastic material subjected to the same impulse as Loading 2 (denoted as Impulse 2) with different loading durations are presented in Figure 6. Under this condition, we can estimate that $u_A = 13.33\times10^{-3}$ in, $I/\rho_0 h_0 R_0 \omega_e \varepsilon_Y = 2.33 > 1$, thus it may cause plastic deformation if loading duration is short enough. Denoting $T_0 = 6.65\times10^{-5}$ s, we find $u_{\text{max}} < u_A$ with $T = 10T_0$, as shown in Figure 6, meaning that the response is always restricted within the elastic range. Besides, we can further observe that displacement is vibrating in a harmonic mode centered at the origin with the period of the breathing mode T_e if motion time is longer, as indicated by the second formula of equation (19). If loading duration is reduced to about $3T_0$, however, plastic deformation commences. Furthermore, maximum displacement is found to get larger with decreasing loading duration and converges towards the extreme case of impulsive loading. Therefore, we can claim that it is an effective way to protect containment vessels from considerably deforming or even rupturing, if we can manage to prolong the loading process. By doing so, it is even possible to confine the shell motion within the elastic response.

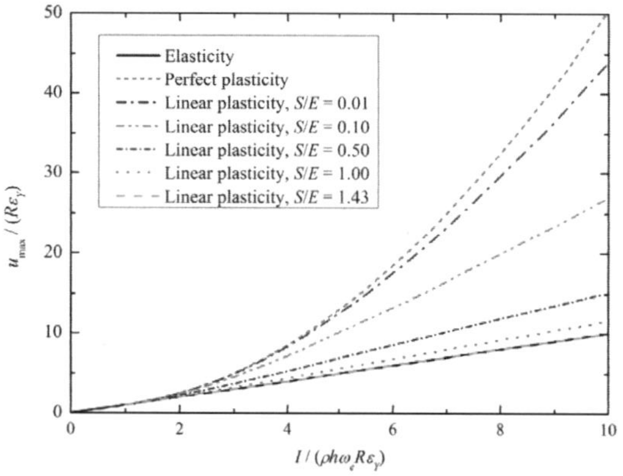

Figure 5. Maximum displacements of spherical shells with different strain-hardenings under impulsive loading.

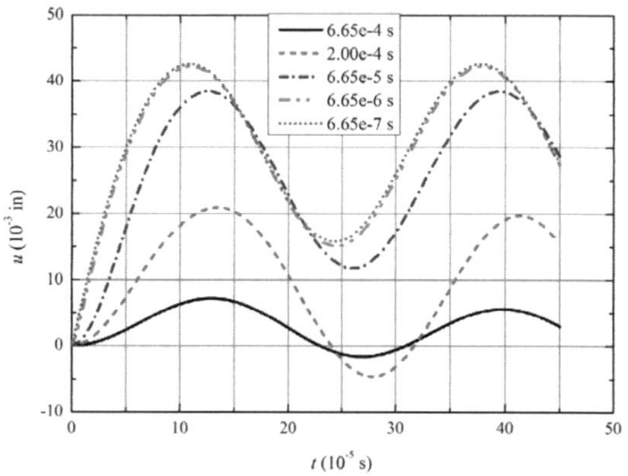

Figure 6. Displacements of a perfectly plastic spherical shell under different loading durations: Model 2, Impulse 2.

6. Summary

In this paper, we reinvestigate the elastic-plastic response of thin spherical shells to spherically symmetric internal blast loading, as a result of finding some mistakes in the classic literature of Baker [1]. We have developed three models that account for the elastic response, the elastic-plastic response neglecting radius variation and shell thinning, and the elastic-plastic response considering these nonlinear effects, separately. The main ideas of the analytical and numerical methods to solve the motions of equations are also presented, along with some important procedures and results by using Laplace transform techniques. On this basis, three examples are calculated to illustrate the differences between our models and Baker's [1,8]. In the end, our theory is applied to demonstrate some response regularities, including the case of impulsive loading and the effect of loading duration. Further advances may be made in two aspects:

(1) It is convenient to use the analytical and numerical methods presented in this paper to study the impacts of shell dimensions, loading waveforms, and material parameters on the shell response. Regularities obtained by doing so may help us to improve the design criteria and operation efficiency of containment vessels.

(2) Our theory and methods, in combination with numerical simulations, can be exploited to study the whole response of spherical shells subjected to real internal blast loading that is rather complicated. For example, the blast loading history is divided into three periods in Ref. [18], and each of them has its own interactive mechanisms. Here we mainly focus on the first period, i.e., the primary-shock period, while the latter two periods are also crucial for the safe use of explosion containment vessels. For example, we may gain some insights into mechanisms of strain growth from the shock-reflection period [9,16-18,23,24], thereby further optimizing the design and operation of containment vessels.

References

[1] Baker W E 1960 *J. Appl. Mech.* **27** 139-44
[2] White J J, Trott B D and Backofen J E 1977 *Physics in Tech.* **8** 94-100
[3] Zhao S D 1989 *Explosion and Shock Waves* **9** 89-96 (in Chinese)
[4] Cheng S 2014 *Research on Internal Blast Impulse Loading and Elastic-to-Plastic Response of Spherical Explosive Containment Vessel* (Xi'an: Northwest Institute of Nuclear Technology) pp 1-6 (in Chinese)
[5] Baker W E and Allen F J 1958 *Proc. of the Third U.S. National Congress of Applied Mechanics* (New York: ASME Press) p 79
[6] Baker W E 1961 *J. Acoust. Soc. Am.* **33** 1749–58
[7] Baker W E, Hu W C L and Jackson T R 1966 *J. Appl. Mech.* **33** 800-6

[8] Baker W E 1959 *Ballistic Research Laboratories Report* BRLM-1194 (Aberdeen Proving Ground)
[9] Buzukov A A 1976 *Fiz. Goreniya. Vzryva.* **12** 605-10
[10] Abakumov A I, Egunov V V, Ivanov A G, Uchaev A A, Tsypkin V I and Shitov A T 1984 *Zh. Prikl. Mekh. Tekh. Fiz.* **3** 127–30
[11] Hu B Y, Liu D M, Bai J S and Zhang M 1998 *J. Vib. Shock* **17** 19-23 (in Chinese)
[12] Hu B Y, Bai J S, Liu D M, Liu G Z, Wang D S and Zhang M 1998 *Press. Vessel Technol.* **17** 39-41 (in Chinese)
[13] Hu B Y, Bai J S, Zhang M and Cheng W 2001 *Eng. Mech.* **18** 136-9 (in Chinese)
[14] Hu B Y, Bai J S, Zhang M and Cheng W 2001 *Chin. J. of Appl. Mech.* **18** 91-5 (in Chinese)
[15] Duffey T A, Baker W E, Lewis B B and Greene J M 1993 *ASME Pressure Vessels and Piping Conf. (Denver)* vol 264 (New York: ASME Press) pp 85-91
[16] Duffey T A and Romero C 2003 *Int. J. Impact Eng.* **28** pp 967–83
[17] Dong Q, Li Q M and Zheng J Y 2010 *Int. J. Impact Eng.* **37** 196-206
[18] Dong Q, Li Q M and Zheng J Y 2010 *Int. J. Impact Eng.* **37** 349-358
[19] Chen S Y, Hu B Y, Gu Y and Dong Q 2010 *Acta Armamentarii* **31** 504-9 (in Chinese)
[20] Hu L S, Hu S Q and Wu W 2011 *J. Safety Sci. Technol.* **7** 151-154 (in Chinese)
[21] Wang D W, Zhang D Z, Li Y, Wang C L, Liu W X and Wang H 2012 *Acta Armamentarii* **33** 1493-7 (in Chinese)
[22] Liu W X, Tan S S, Jing J Y, Zhang X L, Zhu Y R, Liu G L and Zhong F P 2013 *Explosion and Shock Waves* 33 594-600 (in Chinese)
[23] Liu W X, Zhang Q M, Zhong F P, Cheng S, Zhang D Z and Li Y 2017 *Int. J. Impact Eng.* **104** 1-12
[24] Liu W X, Zhang D Z, Cheng S and et al 2017 *Explosion and Shock Waves* **37** 901-6 (in Chinese)
[25] Liu W X, Zhang D Z,Zhong F P, Cheng S and Zhang Q M 2018 *Explosion and Shock Waves* **38** 1045-50 (in Chinese)
[26] Yang G T 2012 *Dynamic Theory of Plasticity* (3rd Edition) (Beijing: Higher Eduacation Press) p33-6 (in Chinese)
[27] Huang K Z and Huang Y G 2013 *Advanced Solid Mechanics* (Beijing: Tsinghua University Press) p131-4 (in Chinese)
[28] Liang S M 2010 *Methods of Mathematical Physics* (4th Edition) (Beijing: Higher Eduacation Press) p90-102 (in Chinese)
[29] Guan Z and Lu J F 2005 *Fundamentals of Numerical Analysis* (2nd Edition) (Beijing: Higher Eduacation Press) p397-401 (in Chinese)

An analytical method for the spherical stress wave equation in linear hardening materials

Zhihuan Wang[1,2], Zhou Tian[2,3], Shufei Wang[2], Feng Chen[2], Zeng He[2] and Leiming Jia[2]

[1]Department of Engineering Physics, Tsinghua University, Beijing, 100084, China
[2]Northwest Institute of Nuclear Technology, Xi'an, 710024, China

[3]Corresponding author's e-mail: tianzhou@nint.ac.cn

Abstract. An analytical method for the spherical stress wave equation in linear hardening materials under impact loading is given in this paper. In this method, the elastic, unloading and plastically loading wave equations are solved, separately. In the elastic region, the solutions under different wave velocity ratio conditions are given. In the unloading region and the plastically loading region, the general solutions of the equations are given and the ordinary differential equations of the arbitrary functions are obtained. And then the exact solutions of the equations are given. The stress contour diagrams in the medium under typical loading conditions are given by the analytical method. The correctness of the analytical method is verified by comparing with the numerical results. On the stress contour diagrams, the discontinuities of the analytical results are clearer than that of the numerical results. A parametric study is conducted to show the effects of the parameters on the behaviour of stress distributions. This method can be useful for the rapid calculation of the elastic-plastic spherical stress waves in engineering and the theoretical analysing and law studying of the elastic-plastic spherical stress wave researches.

1. Introduction

Solving the spherical stress wave equation is one of the fundamental problems in impact dynamics. Wave equations can be solved by numerical methods or analytical methods. Analytical methods have some advantages, such as intuitionistic results, clear physical meanings, obvious laws and so on, which are important for the study of stress wave propagation laws and solid constitutive relations.

The analytical methods to solve wave equations include traveling wave method, variable separation approach, integral transform method and so on. Some analytical studies have been developed on the spherical stress wave equation in the past. The wave equation is relatively easy to solve when there are only elastic waves existing in the medium, and mature results have been obtained [1-3]. Many researchers also have studied the analytical solutions of the elastic-plastic spherical stress wave equation [4-11], but the effects of the unloading on the stress wave propagating has not been considered yet in the existing analytical solutions. In recent years, much attention has been paid to the influence of thermal field, electromagnetic field, rotation, initial stress and gravitation on the stress wave propagation and reflection [12-18]. However, researchers pay less attention to the influence of plasticity. Wang et al [19] developed a study on the theoretical calculation method for the spherical stress wave field in linear hardening materials under impact load, and obtained three different types of stress wave propagating images by analyzing the effects of the unloading. But the specific solving method for the wave equation was not given.

In this paper, we develop a study on the analytical method for the spherical stress wave equation using the travelling wave method on the basis of reference [19]. The elastic, unloading and plastically loading wave equations are solved, separately. In the elastic region, we study the effects of the ratio of the plastic wave velocity and the elastic wave velocity on the solutions. The formulas under different ratio conditions are given, separately. In the unloading region and the plastically loading region, we solve the equations analytically using the displacement potential method. The general solutions of the equations and the ordinary differential equations of the arbitrary functions in the general solutions are obtained. And then the exact solutions of the equations are given. The correctness of the analytical method is verified by comparing with the numerical results. Finally, a parametric study is conducted. The effects of the elastic modulus, the Poisson ratio, the yield strength and the peak loading on the stress waves are given, separately.

2. The analytical method

2.1. Physical model

The physical model in this paper refers to reference [19], as shown in Figure 1. Consider an infinite solid medium with a spherical cavity. The radius of the cavity is r_0. An impact loading acts on the cavity boundary uniformly. The pressure of the cavity boundary is p(t) at time t. The applied pressure p(t) has a jump from 0 to the peak value P_0 at time $t = 0$, and is released smoothly in $t > 0$ following the exponential form

$$p(t) = P_0 \exp\left(-\frac{t}{T_0}\right) \tag{1}$$

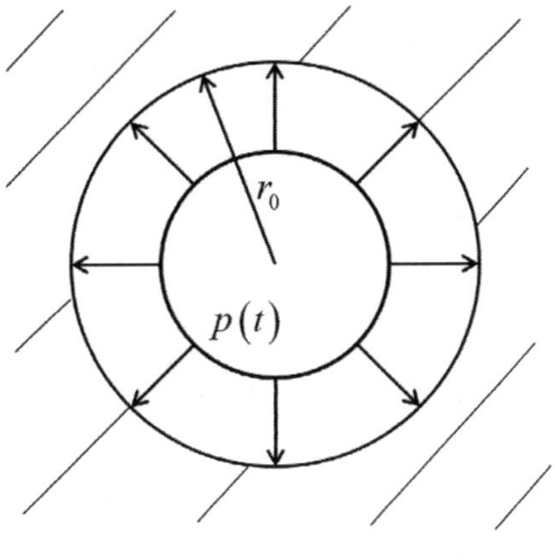

Figure 1. The schematic diagram of the model.

The constitutive relation is linear hardening model [20]

$$\sigma_r + 2\sigma_\theta = (3\lambda + 2\mu)(\varepsilon_r + 2\varepsilon_\theta) \tag{2}$$

$$\sigma_\theta - \sigma_r = 2\mu(\varepsilon_\theta - \varepsilon_r), \text{ elastic} \tag{3}$$

$$\sigma_\theta - \sigma_r = 2G_\mathrm{p}(\varepsilon_\theta - \varepsilon_r - \theta_0) + 2\mu\theta_0, \text{ plastically loading} \tag{4}$$

$$\sigma_\theta - \sigma_r = 2\mu(\varepsilon_\theta - \varepsilon_r) + 2(G_\mathrm{p} - \mu)(\theta_\mathrm{m} - \theta_0), \text{ unloading} \tag{5}$$

where σ_r and σ_θ are the radial stress and the tangential stress, respectively, ε_r and ε_θ are the radial strain and the tangential strain, respectively, λ and μ are the Lame constants, θ_0 is the yield deviatoric strain, $Y = 2\mu\theta_0$ is the yield strength, θ_m is the maximum deviatoric strain, G_p is the plastic shear modulus. Equation (2) shows that the volumetric deformation changes linearly with the hydrostatic pressure. Equation (3), (4) and (5) show that the deviatoric strain and deviatoric stress satisfy the linear hardening relation.

Under the assumption of small deformation, the spherical stress wave equation is [20]

$$\frac{\partial \sigma_r}{\partial r}+\frac{2(\sigma_r-\sigma_\theta)}{r}=\rho\frac{\partial^2 u}{\partial t^2} \tag{6}$$

where r and t are the radial and time coordinates, respectively, u is the displacement, ρ is the density. Equation of motion (6) shows the relation between the stress and the medium moving. The geometric equations are [20]

$$\varepsilon_r=\frac{\partial u}{\partial r} \tag{7}$$

$$\varepsilon_\theta=\frac{u}{r} \tag{8}$$

Equation (7) and (8) show the relations between the strain and the displacement. The initial and boundary conditions are

$$u|_{t=0}=0 \tag{9}$$

$$\sigma_r|_{r=r_0}=-P_0\exp\left(-\frac{t}{T_0}\right) \tag{10}$$

Reference [19] analyzed the propagation of the elastic-plastic spherical stress waves under the physical model above and obtained three different types of stress wave propagating images, as shown in Figure 2. We will solve the wave equation analytically in this section according to Figure 2.

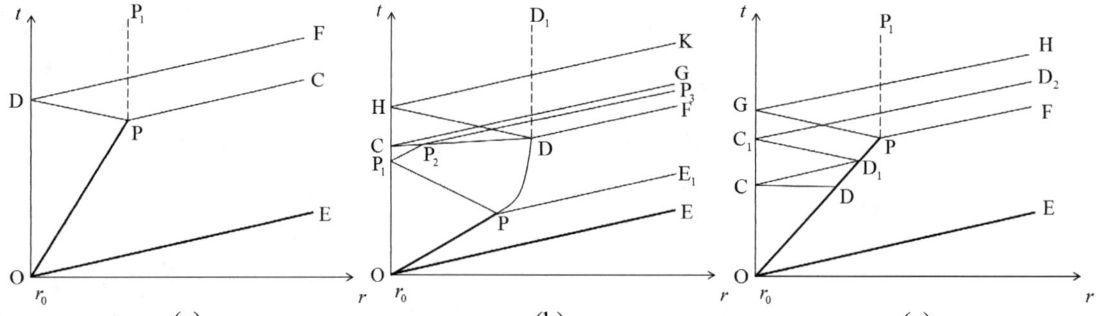

Figure 2. Schematic diagrams of the stress wave propagation (a) rapid unloading; (b) slow unloading; (c) middle unloading [19].

2.2. Elastic region

Substituting the elastic constitutive equations and the geometric equations into the wave equation, we can obtain the elastic wave equation

$$\frac{\partial^2 u}{\partial r^2}+\frac{2}{r}\frac{\partial u}{\partial r}-\frac{2u}{r^2}=\frac{1}{c^2}\frac{\partial^2 u}{\partial t^2} \tag{11}$$

where c is the elastic wave velocity

$$c=\left(\frac{\lambda+2\mu}{\rho}\right)^{1/2} \tag{12}$$

The plastic wave velocity

$$c_\mathrm{p}=\left(\frac{3\lambda+2\mu+4G_\mathrm{p}}{3\rho}\right)^{1/2} \tag{13}$$

Now we solve the elastic wave equation (11) taking the region EOPC in Figure 2(a) as an example. On the elastic wave surface OE, the displacement fits the continuity condition. On the plastic wave front OP, the stress fits the yield condition. That is,

$$u = 0,\quad r = r_0 + ct \tag{14}$$

$$\sigma_\theta - \sigma_r = Y,\quad r = r_0 + c_p t \tag{15}$$

Equations (11), (14) and (15) form the definite solution problem of the stress waves in region EOPC. Define the displacement potential function ϕ

$$u = \frac{\partial \phi}{\partial r} \tag{16}$$

Thus, the elastic wave equation (11) becomes

$$\frac{\partial^2 (r\phi)}{\partial r^2} = \frac{1}{c^2}\frac{\partial^2 (r\phi)}{\partial t^2} \tag{17}$$

There are only right-travelling waves in the elastic precursor, so the general solution of equation (17) is

$$\phi(r,t) = \frac{1}{r} f(\tau) \tag{18}$$

where f is an arbitrary function, and

$$\tau = t - \frac{r - r_0}{c} \tag{19}$$

Using the boundary condition (15), we can obtain the ordinary differential equation that f satisfies,

$$\frac{1}{(c-c_p)^2 (r_0 + c_p t)} \frac{\mathrm{d}^2 f\left[\left(1-\frac{c_p}{c}\right)t\right]}{\mathrm{d}t^2} + \frac{3}{(c-c_p)(r_0+c_p t)^2}\frac{\mathrm{d}f\left[\left(1-\frac{c_p}{c}\right)t\right]}{\mathrm{d}t} + \frac{3}{(r_0+c_p t)^3} f\left[\left(1-\frac{c_p}{c}\right)t\right] = -\frac{Y}{2\mu} \tag{20}$$

The existing solutions of equation (20) have not considered the effects of the wave velocity ratio. However, the solution of equation (20) has different forms in different wave velocity ratio conditions. When the ratio of the plastic wave velocity and the elastic wave velocity $c_p/c < \sqrt{3}/2$,

$$f(\tau) = \left[C_1 \sin\left(\omega \ln \frac{r_0 + a\tau}{r_0}\right) + C_2 \cos\left(\omega \ln\frac{r_0 + a\tau}{r_0}\right)\right]\left(\frac{r_0 + a\tau}{r_0}\right)^\alpha - \frac{(c-c_p)^2 (r_0 + a\tau)^3 Y}{6\mu c (c + c_p)} \tag{21}$$

where

$$a = \frac{c c_p}{c - c_p} \tag{22}$$

$$\omega = \frac{c}{2a}\left[12 - \left(\frac{a}{c} - 3\right)^2\right]^{1/2} \tag{23}$$

$$\alpha = \frac{1}{2} - \frac{3c}{2a} \tag{24}$$

Using the boundary condition (14), we can obtain that

$$C_1 = \left(\frac{1}{2\omega} - \frac{\alpha}{6\omega}\right)\frac{(c-c_p)^2 r_0^3 Y}{\mu c (c + c_p)} \tag{25}$$

$$C_2 = \frac{(c - c_p)^2 r_0^3 Y}{6\mu c(c + c_p)} \tag{26}$$

When the ratio of the plastic wave velocity and the elastic wave velocity $c_p/c > \sqrt{3}/2$,

$$f(\tau) = C_3\left(\frac{r_0 + a\tau}{r_0}\right)^{\beta_1} + C_4\left(\frac{r_0 + a\tau}{r_0}\right)^{\beta_2} - \frac{(c - c_p)^2 (r_0 + a\tau)^3 Y}{6\mu c(c + c_p)} \tag{27}$$

where

$$\beta_1 = \frac{1}{2} - \frac{3c}{2a} + \frac{c}{2a}\left[\left(\frac{a}{c} - 3\right)^2 - 12\right]^{1/2} \tag{28}$$

$$\beta_2 = \frac{1}{2} - \frac{3c}{2a} - \frac{c}{2a}\left[\left(\frac{a}{c} - 3\right)^2 - 12\right]^{1/2} \tag{29}$$

Using the boundary condition (14), we can obtain that

$$C_3 = \frac{(\beta_2 - 3)(c - c_p)^2 r_0^3 Y}{6(\beta_2 - \beta_1)\mu c(c + c_p)} \tag{30}$$

$$C_4 = \frac{(3 - \beta_1)(c - c_p)^2 r_0^3 Y}{6(\beta_2 - \beta_1)\mu c(c + c_p)} \tag{31}$$

When the ratio of the plastic wave velocity and the elastic wave velocity $c_p/c = \sqrt{3}/2$,

$$f(\tau) = C_5\left(\frac{r_0 + a\tau}{r_0}\right)^{2-\sqrt{3}} + C_6\left(\frac{r_0 + a\tau}{r_0}\right)^{2-\sqrt{3}} \ln\frac{r_0 + a\tau}{r_0} - \frac{(c - c_p)^2 (r_0 + a\tau)^3 Y}{6\mu c(c + c_p)} \tag{32}$$

Using the boundary condition (14), we can obtain that

$$C_5 = \frac{(c - c_p)^2 r_0^3 Y}{6\mu c(c + c_p)} \tag{33}$$

$$C_6 = \frac{(1 + \sqrt{3})(c - c_p)^2 r_0^3 Y}{6\mu c(c + c_p)} \tag{34}$$

The above gives the solutions of the definite solution problem under different wave velocity ratios. For most solid materials, the ratio $c_p/c < \sqrt{3}/2$. So we use formula (21) in the later discussion.

2.3. Unloading region

Substituting the unloading constitutive equations and the geometric equations into the wave equation, we can obtain the unloading wave equation

$$\frac{\partial^2 u}{\partial r^2} + \frac{2}{r}\frac{\partial u}{\partial r} - \frac{2u}{r^2} = \frac{1}{c^2}\frac{\partial^2 u}{\partial t^2} + \frac{c_p^2 - c^2}{c^2}\left[\frac{d\theta_m}{dr} + \frac{3(\theta_m - \theta_0)}{r}\right] \tag{35}$$

Now we solve the unloading wave equation (35) taking the region OPD in Figure 2(a) as an example. On the plastic wave surface OP, the displacement fits the continuity condition. On the boundary OD, the stress fits the boundary condition. That is,

$$u = u_1, \quad r = r_0 + c_p t \tag{36}$$

$$\sigma_r = -P_0 \exp\left(-\frac{t}{T_0}\right), \quad r = r_0 \tag{37}$$

Equations (35), (36) and (37) form the definite solution problem of the stress waves in region OPD, where u_1 is the displacement on the plastic wave surface OP which can be calculated by the formulas in section 2.2.

Decompose the displacement u [4]

$$u(r,t) = U(r,t) + g(r) \tag{38}$$

where U satisfies

$$\frac{\partial^2 U}{\partial r^2} + \frac{2}{r}\frac{\partial U}{\partial r} - \frac{2U}{r^2} = \frac{1}{c^2}\frac{\partial^2 U}{\partial t^2} \tag{39}$$

$$\frac{\partial U}{\partial r} + \frac{c_p}{c^2}\frac{\partial U}{\partial t} - \frac{U}{r} = \frac{c_p^2 - c^2}{c^2}\left\{\theta_0 + \left(\frac{r}{r_0}\right)^{\alpha-3}\left[\frac{(c-2c_p)Y}{4\mu\omega(c+c_p)}\sin\left(\omega\ln\frac{r}{r_0}\right) - \frac{c_p Y}{2\mu(c+c_p)}\cos\left(\omega\ln\frac{r}{r_0}\right)\right]\right\}, \quad r = r_0 + c_p t \tag{40}$$

$$(\lambda + 2\mu)\frac{\partial U}{\partial r} + \frac{2\lambda U}{r} = -P_0 \exp\left(-\frac{t}{T_0}\right), \quad r = r_0 \tag{41}$$

g satisfies

$$\frac{\mathrm{d}g}{\mathrm{d}r} - \frac{g}{r} = \frac{c_p^2 - c^2}{c^2}(\theta_m - \theta_0) \tag{42}$$

$$g(r_0) = 0 \tag{43}$$

U satisfies the elastic wave equation (39), and we use the displacement potential method to give the solution of U. Define the displacement potential function Φ

$$U = \frac{\partial \Phi}{\partial r} \tag{44}$$

The general solution of equation (39) is

$$\Phi(r,t) = \frac{1}{r}f_1(\tau) + \frac{1}{r}f_2(\tau_2) \tag{45}$$

where f_1 and f_2 are arbitrary functions, and

$$\tau_2 = t + \frac{r - r_0}{c} \tag{46}$$

Using the boundary conditions (40) and (41), we can obtain that

$$\begin{aligned}&\frac{1}{c(c-c_p)(r_0+c_p t)}\frac{\mathrm{d}^2}{\mathrm{d}t^2}f_1\left[\left(1-\frac{c_p}{c}\right)t\right] + \frac{3c-c_p}{c(c-c_p)(r_0+c_p t)^2}\frac{\mathrm{d}}{\mathrm{d}t}f_1\left[\left(1-\frac{c_p}{c}\right)t\right] + \frac{3}{(r_0+c_p t)^3}f_1\left[\left(1-\frac{c_p}{c}\right)t\right]\\ &+\frac{1}{c(c+c_p)(r_0+c_p t)}\frac{\mathrm{d}^2}{\mathrm{d}t^2}f_2\left[\left(1+\frac{c_p}{c}\right)t\right] - \frac{3c+c_p}{c(c+c_p)(r_0+c_p t)^2}\frac{\mathrm{d}}{\mathrm{d}t}f_2\left[\left(1+\frac{c_p}{c}\right)t\right] + \frac{3}{(r_0+c_p t)^3}f_2\left[\left(1+\frac{c_p}{c}\right)t\right]\\ &= \frac{c_p^2-c^2}{c^2}\left\{\theta_0 + \left(\frac{r_0+c_p t}{r_0}\right)^{\alpha-3}\left[\frac{(c-2c_p)Y}{4\mu\omega(c+c_p)}\sin\left(\omega\ln\frac{r_0+c_p t}{r_0}\right) - \frac{c_p Y}{2\mu(c+c_p)}\cos\left(\omega\ln\frac{r_0+c_p t}{r_0}\right)\right]\right\}\end{aligned} \tag{47}$$

$$\frac{\rho}{r_0}\frac{\mathrm{d}^2}{\mathrm{d}t^2}f_1(t) + \frac{4\mu}{r_0^2 c}\frac{\mathrm{d}}{\mathrm{d}t}f_1(t) + \frac{4\mu}{r_0^3}f_1(t) + \frac{\rho}{r_0}\frac{\mathrm{d}^2}{\mathrm{d}t^2}f_2(t) - \frac{4\mu}{r_0^2 c}\frac{\mathrm{d}}{\mathrm{d}t}f_2(t) + \frac{4\mu}{r_0^3}f_2(t) = -P_0\exp\left(-\frac{t}{T_0}\right) \tag{48}$$

The equations (47) and (48) are the ordinary differential equations of f_1 and f_2, and can be solved by a numerical method.

Solving the equations (42) and (43), we can obtain that

$$g(r)=\frac{c_{\mathrm{p}}^2-c^2}{c^2}r\left(\int_{r_0}^{r}\frac{\theta_{\mathrm{m}}(\hat{r})}{\hat{r}}\mathrm{d}\hat{r}-\theta_0\ln\frac{r}{r_0}\right) \tag{49}$$

where

$$\theta_{\mathrm{m}}=\theta_0+\frac{c^2}{c_{\mathrm{p}}^2}\left[\left(\frac{U}{r}-\frac{\partial U}{\partial r}\right)\Big|_{r=r_0+c_{\mathrm{p}}t}-\theta_0\right] \tag{50}$$

Equations (45) and (49) form the solutions of the definite solution problem in region OPD.

2.4. Plastically loading region

Substituting the plastically loading constitutive equations and the geometric equations into the wave equation, we can obtain the plastically loading wave equation

$$\frac{\partial^2 u}{\partial r^2}+\frac{2}{r}\frac{\partial u}{\partial r}-\frac{2u}{r^2}=\frac{1}{c_{\mathrm{p}}^2}\frac{\partial^2 u}{\partial t^2}+\frac{3}{r}\frac{c^2-c_{\mathrm{p}}^2}{c_{\mathrm{p}}^2}\theta_0 \tag{51}$$

Now we solve the plastically loading wave equation (51) taking the region OPP_1 in Figure 2(b) as an example. On the plastic wave surface OP, the displacement fits the continuity condition (36). On the boundary OP_1, the stress fits the boundary condition (37). Equations (51), (36) and (37) form the definite solution problem of the stress waves in region OPP_1.

Compared with the elastic equation, there is one more inhomogeneous item in equation (51). We can also solve it using the displacement potential method. Define the displacement potential function ϕ_1

$$u=\frac{\partial\phi_1}{\partial r} \tag{52}$$

The general solution of equation (51) is

$$\phi_1(r,t)=\frac{1}{r}f_3(\tau_3)+\frac{1}{r}f_4(\tau_4)+\frac{\left(c^2-c_{\mathrm{p}}^2\right)r^2\theta_0}{6c_{\mathrm{p}}^2}\left(3\ln\frac{r}{r_0}-\frac{5}{2}\right) \tag{53}$$

where f_3 and f_4 are arbitrary functions, and

$$\tau_3=t-\frac{r-r_0}{c_{\mathrm{p}}} \tag{54}$$

$$\tau_4=t+\frac{r-r_0}{c_{\mathrm{p}}} \tag{55}$$

Using the boundary conditions (36) and (37), we can obtain that

$$\begin{aligned}&-\frac{1}{c_{\mathrm{p}}(r_0+c_{\mathrm{p}}t)}\frac{\mathrm{d}}{\mathrm{d}t}f_3(0)-\frac{1}{(r_0+c_{\mathrm{p}}t)^2}f_3(0)+\frac{1}{2c_{\mathrm{p}}(r_0+c_{\mathrm{p}}t)}\frac{\mathrm{d}}{\mathrm{d}t}f_4(2t)-\frac{1}{(r_0+c_{\mathrm{p}}t)^2}f_4(2t)\\&+\frac{\left(c^2-c_{\mathrm{p}}^2\right)(r_0+c_{\mathrm{p}}t)\theta_0}{3c_{\mathrm{p}}^2}\left(3\ln\frac{r_0+c_{\mathrm{p}}t}{r_0}-1\right)=\frac{(c-c_{\mathrm{p}})(c+2c_{\mathrm{p}})(r_0+c_{\mathrm{p}}t)Y}{6\mu c(c+c_{\mathrm{p}})}+\left(\frac{r_0+c_{\mathrm{p}}t}{r_0}\right)^{\alpha-2}\\&\times\left[\frac{(c-c_{\mathrm{p}})\left(3c^2-2cc_{\mathrm{p}}-4c_{\mathrm{p}}^2\right)r_0Y}{12\omega\mu cc_{\mathrm{p}}(c+c_{\mathrm{p}})}\sin\left(\omega\ln\frac{r_0+c_{\mathrm{p}}t}{r_0}\right)-\frac{(c-c_{\mathrm{p}})(c+2c_{\mathrm{p}})r_0Y}{6\mu c(c+c_{\mathrm{p}})}\cos\left(\omega\ln\frac{r_0+c_{\mathrm{p}}t}{r_0}\right)\right]\end{aligned} \tag{56}$$

$$\frac{\rho}{r_0}\frac{d^2}{dt^2}f_3(t)+\frac{4G_p}{r_0^2 c_p}\frac{d}{dt}f_3(t)+\frac{4G_p}{r_0^3}f_3(t)+\frac{\rho}{r_0}\frac{d^2}{dt^2}f_4(t)-\frac{4G_p}{r_0^2 c_p}\frac{d}{dt}f_4(t)+\frac{4G_p}{r_0^3}f_4(t)$$
$$-\left(\lambda+\frac{2}{3}\mu\right)\frac{c^2-c_p^2}{c_p^2}\theta_0=-P_0\exp\left(-\frac{t}{T_0}\right) \tag{57}$$

The equations (56) and (57) are the ordinary differential equations of f_3 and f_4, and can be solved by a numerical method. And then we can give the solutions of the definite solution problem in region OPP_1.

For other regions in Figure 2, the methods to solve the wave equation are almost the same as the discussion above, and we can obtain the displacement. Then using the geometric relations and the constitutive relations, we can obtain the stress, strain and so on.

3. Results and discussion

In this section, we use AUTODYN to verify the solution in section 2, and use the solution to study the effects of the parameters on the stress waves.

3.1. Numerical verification and comparisons

We use AUTODYN to verify the solution. The parameters of the model are shown in Table 1. The attenuation index of the loading, T_0, is 0.01 ms, 0.2 ms and 0.03 ms, respectively, which are corresponding to the case (a), (b) and (c) in Figure 2, respectively [19]. Figure 3 shows the radial stress contours in the r-t plane of the three cases obtained by the analytical method in section 2 and AUTODYN, respectively. Define the radial stress error

$$e=\frac{\int_0^{t_1}\int_{r_0}^{r_1}\left|\sigma_r-\sigma_{r,\mathrm{N}}\right|\mathrm{d}r\mathrm{d}t}{\int_0^{t_1}\int_{r_0}^{r_1}\left|\sigma_{r,\mathrm{N}}\right|\mathrm{d}r\mathrm{d}t} \tag{58}$$

where σ_r and $\sigma_{r,\mathrm{N}}$ are the analytical result and the numerical result, separately. The e can reflect the average relative error between the analytical result and the numerical result in region $[r_0,r_1]\times[0,t_1]$. The errors of the three cases are shown in Table 2.

Table 1. Parameters of the model.

Density ρ/kg·m^3	Elastic modulus E/GPa	Poisson ratio υ	Yield strength Y/MPa	Plastic shear modulus G_p/MPa	Cavity radius r_0/m	Peak loading P_0/GPa
2630	50	0.16	40	12.5	0.2	0.2

Table 2. Relative errors.

Attenuation index	Calculating regions			Relative error
T_0/ms	r_0/m	r_1/m	t_1/m	e/%
0.01	0.2	0.6	0.2	6.39
0.2	0.2	1.4	0.6	5.25
0.03	0.2	0.6	0.2	3.18

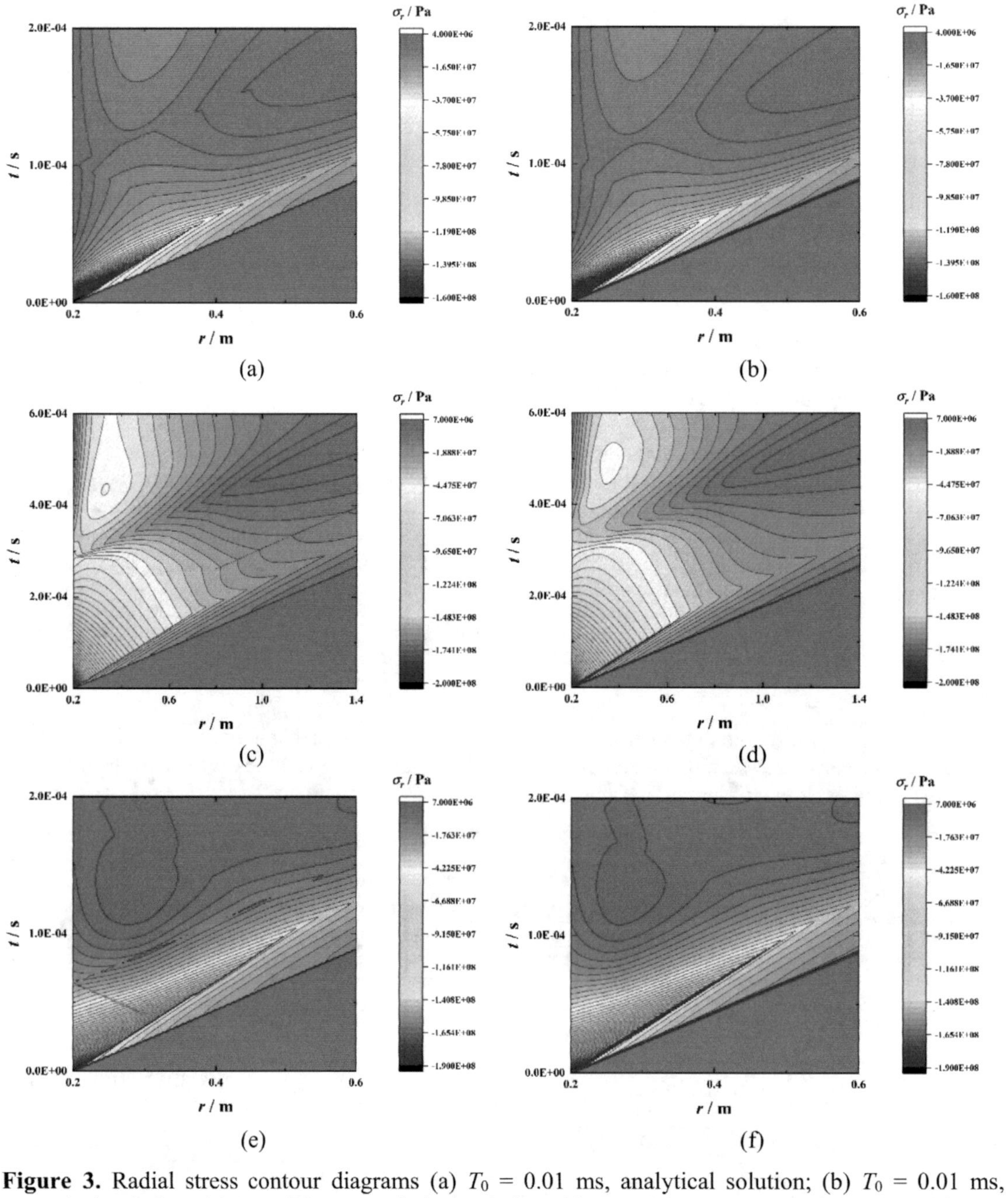

Figure 3. Radial stress contour diagrams (a) T_0 = 0.01 ms, analytical solution; (b) T_0 = 0.01 ms, numerical solution; (c) T_0 = 0.2 ms, analytical solution; (d) T_0 = 0.2 ms, numerical solution; (e) T_0 = 0.03 ms, analytical solution; (f) T_0 = 0.03 ms, numerical solution.

Figure 3 shows that the analytical results agree well with the numerical results in all the three cases. Table 2 shows that the average relative errors between the analytical result and the numerical result are all less than 10%. The results verify the correctness of the analytical method. The error of the analytical results is mainly caused by the simplification of the stress wave propagating [19]. In addition, there is a difference between the analytical method and the numerical method. As shown in Figure 3, the strong discontinuities of the analytical results are clearer than that of the numerical results. The reason is that the elastic and plastic stress wave surface are regarded as strong discontinuities in the analytical method. The variables are discontinuous across the wave surface. While there are no real strong discontinuities in the numerical method because of the grid division. The variables only vary sharply across the wave surface. In the same way, the weak discontinuities of the analytical results are clearer than that of the numerical results.

3.2. *Effects of the parameters*

We use the solution in section 2 to study the effects of the parameters on the stress waves. The initial parameters are the same as the parameters in section 3.1, as shown in Table 1, while the attenuation index of the loading $T_0 = 0.01$ ms. Keep other parameters unchanged and change the elastic modulus, the Poisson ratio, the yield strength and the peak loading, separately, then we can obtain the effects of the parameters mentioned above on the stress waves. Figure 4 shows the radial stress distributions along the radial coordinate at $t = 0.06$ ms for different elastic moduli, Poisson ratios, yield strengths and peak loadings, separately.

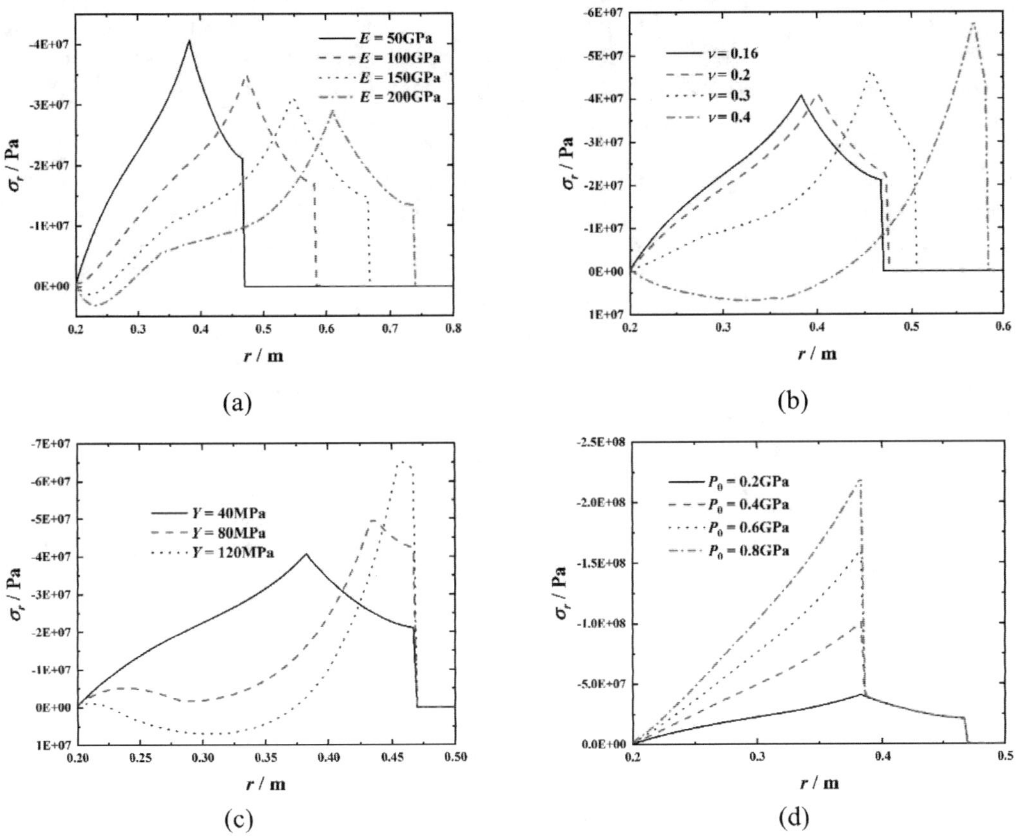

Figure 4. The radial stress distributions along the radial coordinate at $t = 0.06$ ms for different parameters (a) elastic modulus; (b) Poisson ratio; (c) yield strength; (d) peak loading.

Figure 4(a) shows that the wave speed increases with the increase of the elastic modulus, so the stress waves travel farther at the same time for larger elastic moduli. But the elastic modulus affects little on the stress wave form. Figure 4(b) shows that the wave speed and the peak value of the stress waves increase with the increase of the Poisson ratio and the stress wave width decreases with the increase of the Poisson ratio. Figure 4(c) shows that the peak value of the stress waves increases with the increase of the yield strength and the stress wave width decreases with the increase of the yield strength. However, when the yield strength is large enough, the plastic stress waves will disappear and there will be only elastic stress waves left. The stress waves will not change anymore with the increase of the yield strength. Figure 4(d) shows that the peak value of the stress waves increases with the increase of the peak loading. But there is no effect of the peak loading on the stress wave width and the elastic precursor.

4. Conclusion

In this paper, we give an analytical method for the spherical stress wave equation on the basis of reference [19]. The exact solutions of the elastic, unloading and plastically loading wave equations are obtained, separately. In the elastic region, we give the formulas under different wave velocity ratio conditions, separately. In the unloading region and the plastically loading region, we give the general solutions of the equations and the ordinary differential equations of the arbitrary functions in the general solutions. And then we give the exact solutions of the equations.

The stress wave equation under typical loading conditions is solved using the analytical method in this paper. The results are in good agreement with the numerical results. Because of the simplification of the stress wave propagating, there exist local errors between the analytical results and the numerical results. The discontinuities of the analytical results are clearer than that of the numerical results on the stress contour diagrams. A parametric study is conducted and the effects of the elastic modulus, the Poisson ratio, the yield strength and the peak loading on the stress waves are given, separately.

Compared with the previous methods, the analytical method in this paper takes the effects of the plasticity into account and give the solution of the stress waves on the whole *r-t* plane. And this method makes up for the deficiency of solving the unloading wave equation especially. This method can be useful for the rapid calculation of the elastic-plastic spherical stress waves in engineering and the theoretical analyzing and the law studying of the elastic-plastic spherical stress wave researches. In future, we will develop a further research on the law studying of the elastic-plastic spherical stress waves, including the propagating laws, the frequency analysis, the energy evolution and so on.

References

[1] Wlodarczyk E and Zielenkiewicz M 2009 *J. Theor. App. Mech.* **47** 127
[2] Li X L 2000 *Explo. Shock Waves* **20** 186 (in Chinese)
[3] Xiao J H 2001 *Oil Geophys. Prospect.* **36** 160 (in Chinese)
[4] Yang C Y 1970 *Int. J. Solids Struct.* **6** 757
[5] Morland L W 1971 *J. Mech. Phys. Solids* **19** 295
[6] Milne P C, Morland L W and Yeung W 1988 *J. Mech. Phys. Solids* **36** 15
[7] Milne P C and Morland L W 1988 *J. Mech. Phys. Solids* **36** 215
[8] Rapoport L, Katzir Z and Rubin M B 2011 *Wave Motion* **48** 441
[9] Santos T, Brezolin A, Rossi R and Rodriguez-Martinez J A 2020 *Acta Mech.* **231** 2381
[10] Xiao J H and Sun W T 1997 *Oil Geophys. Prospect.* **32** 809 (in Chinese)
[11] Chen S, Wu J and Zhang Z 2017 *J. Eng. Mech.* **143** 04017034
[12] Youssef H M and El-Bary A A 2009 *J. Therm. Stresses* **32** 1293
[13] Ezzat M, El-Bary A A and Ezzat S 2011 *Energ. Convers. Manage.* **52** 934
[14] Ezzat M A and El-Bary A A 2012 *Magnetohydrodynamics* **48** 587
[15] Ezzat M A and El-Bary A A 2016 *Int. J. Appl. Electrom.* **50** 549
[16] Abo-Dahab S M, Rida S Z, Mohamed R A and Kilany A A 2018 *Mechanics and Mechanical Engineering* **22** 313
[17] Abo-Dahab S M, Abd-Alla A M, Kilany A A and Elsagheer M 2018 *J. Micro. Tech.* **24** 3357
[18] Bayones F S, Abo-Dahab S M, Abd-Alla A M, Elhag S H, Kilany A A and Elsagheer M 2021 *Complexity* **2021** 5560900
[19] Wang Z H, Jia L M, He Z and Tian Z 2022 *Acta Phys. Sin.* **71** 018301 (in Chinese)
[20] Wang L L 2005 *Foundation of Stress Waves* (2nd Edition) (Beijing: National Defense Industry Press) p227-38 (in Chinese)

Development of Monte Carlo based treatment planning system for BNCT

Jiang Chen[1, 2], Yi-Chiao Teng[2, 3], Wan-Bing Zhong[2], Hai-Bo Yang[1], Qing Hong[1] and Yuan-Hao Liu[2, 4, 5, 6]

[1]Nanjing Vocational University of Industry Technology, Nanjing, Jiangsu Province, P.R. China; 2020101169@niit.edu.cn
[2]Neuboron Therapy System Ltd., Xiamen, Fujian Province, P.R. China
[3]National Tsing Hua University, Hsinchu, Taiwan 30013, ROC
[4]Nanjing University of Aeronautics and Astronautics, Nanjing, Jiangsu Province, P.R. China
[5]Neuboron Medtech Ltd., Nanjing, Jiangsu Province, P.R. China

[6]Corresponding author's e-mail: yhl.taiwan@gmail.com; liuyuanhao@neuboron.com

Abstract. A treatment planning system (TPS) plays a major role for a successful radiotherapy, directly influencing the calculated dose accuracy, and thus the treatment quality. Multifunctional Arithmetic for Neutron Transportation Analysis (NeuMANTA), a TPS especially designed for complete workflow for BNCT, which aims to provide high precision dose calculation with maximizing the efficacy. The dedicated MC engine COMPASS seamlessly embedded in NeuMANTA can be called for dose distribution calculation in background. Optimized cross-section library, parallel computing, and various variance reduction techniques are applied to speed up the particle transportations. A benchmark was performed between NeuMANTA and MCNP6, and results agreed well with no significant statistical difference, however, NeuMANTA needs only half of the calculation time. The accuracy of estimated doses, and complete treatment workflow are the main features in NeuMANTA.

1. Introduction

Compared to conventional radiation therapy modalities, boron neutron capture therapy (BNCT) is a unique type of radiation therapy that enables biological targeting of the cancer at cellular level [1]. In theory, it is based on the nuclear capture and fission reaction that occur when boron-10 is irradiated with neutron, as shown below. Accelerator-based neutron source is generalized with the reaction ^{7}Li (p, n) ^{7}Be. Recently, lots of development project for BNCT devices are being undertaken in many countries of the world, and BNCT facilities is expected to be established as standard therapy for cancer treatment in hospital [2-5].

$$^{10}\mathrm{B} + n \rightarrow {}^{4}\mathrm{He} + {}^{7}\mathrm{Li} + 2.31 MeV \quad (1)$$

To perform BNCT as a type of radiation therapy, a treatment planning system (TPS) works as the heart of the treatment process, which is used to compute the expected dose distribution and determine suitable irradiation parameters for the best of patient outcomes. In conventional radiation therapy such as photon beam, simplified analytical and model based on method are efficiently applied. However it is not feasible in BNCT, there are many aspects that are relevant for correctly describing the neutron transport, especially secondly produced particles and the mixture of components with different cross section in tissue complicate the radiobiology and make it difficult dose estimation during neutron irradiation. Therefore, BNCT related TPS always perform the radiation simulation based on MC algorithm. Some MC-based TPS used for BNCT are implemented to determine the accuracy. SERA [6] and NCTplan [7] were developed and have been successfully employed for research applications. It represents a significant advance in several areas for treatment planning, however, medical images of DICOM format are not supported, and the voxel size is very large due to the memory limitation, which my effect the dose accuracy. THORplan [8] developed by National Tsing Hua University produces input file for MCNP (Monte Carlo NParticle) [9] to calculate dose distribution, and the accuracy was verified by a Snyder phantom. Tsukuba Plan [10, 11] developed by a team including researchers from the University of Tsukuba, is another MC-based TPS [12] utilized PHITS [13] as MC engine. An original prototype software CARONTE [14] and BDTPS are both PET-based TPS, featured a pre-processed module for processing PET images for in vivo boron distribution [15]. NeuCure [16] is a TPS designed especially for BNCT, which developed by Sumitomo Heavy Industries, must be used in combination with another radiotherapy treatment planning product RayStation for image processing.

Table 1. Features comparison between different MC-based TPS.

	Developer	**Image format**	**Image fusion**	**Voxel size(cm^3)**	**MC engine**	**Cross Section**
NeuMANTA	Neuboron	DICOM	PET/CT PET/MR	1 pixel to 1	MCNP/ PHITS/ COMPASS	ENDF/B-VIII.0
SERA	INEEL/MSU	QSH	NO	1	Self-embedded	ENDF/B-VI
NCTPlan	Havard/MIT	TIFF	NO	1	MCNP 4b	ENDF/B-VI
JCDS	JAEA	DICOM	NO	1	PHITS	ENDF/B-VI
THORPlan	Tsing Hua University	DICOM	PET/CT	8 pixels to 1 homogenized	MCNP	ENDF/B-VI
TsukubaPlan	Tsukuba University	DICOM	NO	2 pixels to 1 homogenized	PHITS	ENDF/B-VII.1
CARONTE	Genova University/ Pisa University	DICOM	PET/CT	0.125	MCNP 4a	ENDF/B-VI
NeuCure	Sumitomo	DICOM	NO	0.003	PHITS	JENDL 4.0

Accuracy and efficiency are two main considerations for a TPS. Although developing MC-based TPS with general purpose Monte Carlo engine is a fast way, it is not an efficient algorithm for BNCT. In NeuMANTA, an optimized MC engine COMPASS is seamlessly integrated to greatly speed up the calculation process. Table 1 listed the main feathers and comparison between NeuMANTA and other MC-based TPS.

2. Monte Carlo dose engine in NeuMANTA

The architecture of NeuMANTA was designed with Object-oriented method, and it was implemented with C++ language. It contains three main modules, Plan, COMPASS, and Dose. Plan module is responsible for treatment plan generation, COMPASS module is a MC engine, and Dose module is used to combine and visualize the physical and biological dose calculated by MC engine. BNCT absorbed dose consists boron dose, neutron dose and photon dose. Boron dose is from $^{10}B(n,\alpha)^{7}Li$ reaction. Neutron dose is major from $^{1}H(n,n)p$ and $^{14}N(n,p)^{14}C$ reactions. Photon dose comes from beam port and $^{1}H(n,\gamma)^{2}H$, $^{16}O(n,\gamma)^{17}O$, and $^{12}C(n,\gamma)^{13}C$ reactions. The main GUI is shown as Figure 1, and the whole workflow is illustrated in Figure 2.

Journal of Physics: Conference Series **2313** (2022) 012012 doi:10.1088/1742-6596/2313/1/012012

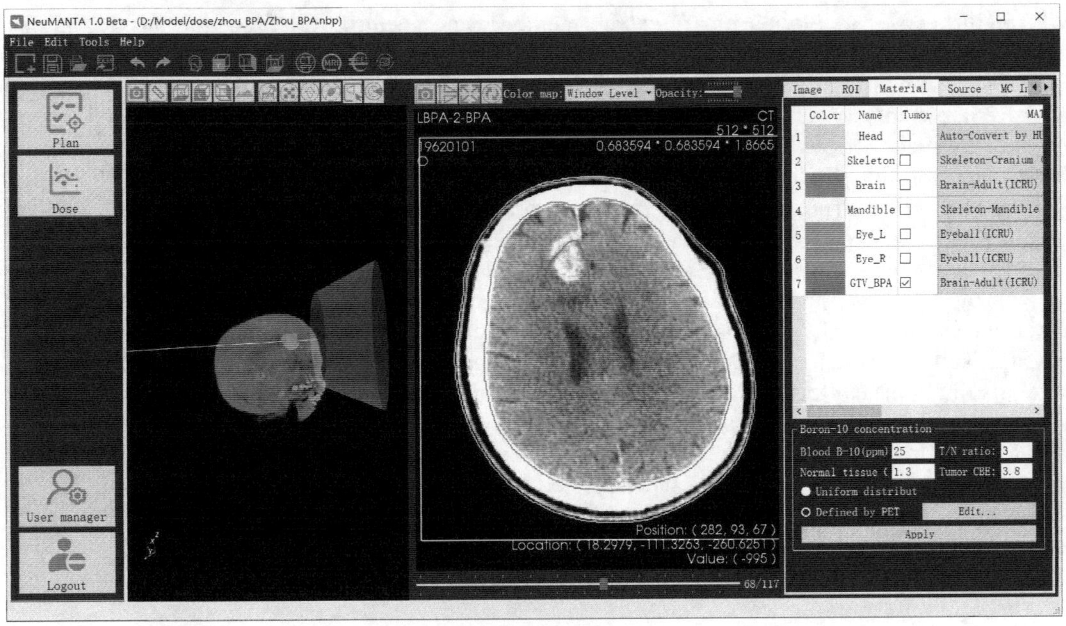

Figure 1. Main GUI of NeuMANTA.

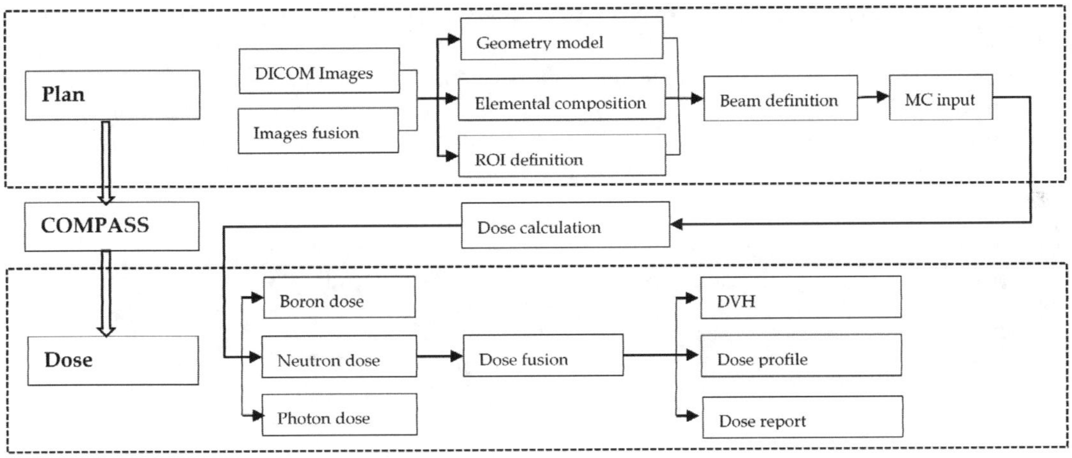

Figure 2. Workflow of NeuMANTA.

Dose calculation with MC code is really complicated and time consuming. COMPASS (COMpact PArticle Simulation System) is a Monte Carlo engine especially developed for BNCT dose calculation with maximizing efficiency. For dose estimation, all nuclide component should be calculated, and the absorbed dose $D(E)$ from the reaction of neutron and each atom is given by

$$D(E)=\phi(E)\cdot\sum_{i}\left(\rho_i\sigma_i(E)H_i(E)\right) \tag{2}$$

where $\phi(E)$ is flux of neutron, ρ_i is particle number density of ith nuclide in cross section library, $\sigma_i(E)$ is the cross section of this nuclide, and $H_i(E)$ is the heating number of this nuclide.

In order to speed up the calculation, a simplified lattice was applied to construct the geometry model as particle simulation in lattice was more efficient than general purposed Monte Carlo code. Also, the continuous neutron energies were divided into multiple groups, a hash table was maintained for index by

COMPASS instead of traditional binary search method. Since neutron cross section has a complex structure, a well-organized HDF5 (Hierarchical Data Format) cross section database (ENDF/B7.1 version) was adapted to simulate random walk of neutron. In the process of simulating particle motion, three dose components were tallied for each lattice, including and boron dose, neutron dose and photon dose. All of these changes are applied to accelerate calculation speed.

In addition, for full performance of the computing power of multi-core CPUs, the MPI parallel library and OpenMP instruction are used to achieve process-level and thread-level parallelism. Users can set the number of processes and threads according to the specific equipment conditions, so that the computer performance can be fully utilized for dose calculation.

In terms of source definition, COMPASS can currently define multiple sources that are independent of spatial angle energy. The spatial distribution can be defined by Cartesian coordinates, cylindrical coordinates, and spherical coordinates. The definition of each component of the coordinate can be a discrete distribution. Uniform distribution, Tubular distribution, etc.; the distribution of particle emission direction can be isotropic distribution, unidirectional emission, or directly define the distribution of azimuth angle and polar angle, etc. If it is a correlation source, such as the correlation between energy and spatial location, it can be defined by decomposing into multiple sources.

The dose distribution of multiple models was calculated using COMPASS and compared with the calculation results of MCNP to verify the calculation accuracy and efficiency of COMPASS.

3. Result

Since MCNP, now in the version MCNP6, developed at Los Alamos National Laboratories, is one of the most commonly used general purpose MC engine, and often it is used as the dose calculation standard in BNCT. COMPASS and MCNP have been compared and verified in a lot of examples, one of which is the classic Snyder model.

Snyder model includes skin, skull, brain, and tumors located in the brain. The size of the entire model is 15.2cm*18.4cm*20.4cm, and the number of grids in the model is 152*184*102, with a total of 2852736 grids. The proportion of boron concentration in the tumor area is set to 60 ppm, and the proportion of boron concentration in other materials (such as other parts of the brain, scalp, etc.) is set to 20 ppm.

Figure 3. Heat map of neutron dose distribution in different frames (the upper line is MCNP and lower line is COMPASS).

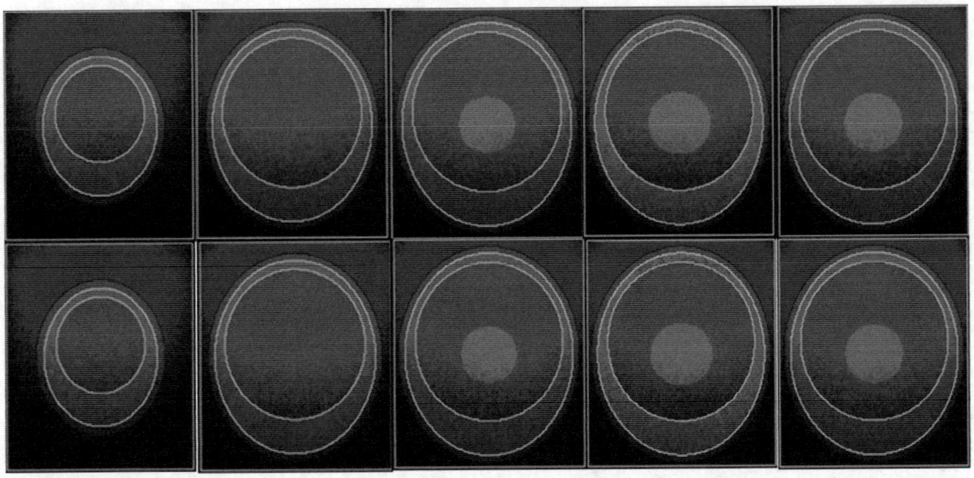

(MeV/g)

Figure 4. Heat map of photon dose distribution in different frames (the upper line is MCNP and lower line is COMPASS).

The source used in the calculation is the neutron source at the exit of a certain proton accelerator. The energy distribution and angular distribution of the neutron source depend on the spatial distribution of the exit. Using COMPASS and MCNP to simulate 50 million particles in each code, calculate the boron dose, neutron dose and photon dose of each grid.

Figure 3 and Figure 4 show the neutron dose distribution heat map and photon dose distribution heat map of MCNP and COMPASS at different heights (from left to right, corresponding to 20th frame, 40th frame, 50th frame, 55th frame, 60th frame, different frame numbers correspond to different heights in the Z direction)

Table 2 shows the tally of the dose of COMPASS and MCNP and their relative deviations. The results show that the difference between them is within 1%, which is within the acceptable error range.

Table 2. Comparison of mean Dose for COMPASS and MCNP.

Dose	material	COMPASS result(MeV/g)	MCNP result(MeV/g)	Relative error(%)
Boron dose	tumour	3.201E-8	3.202E-8	0.03
	brain	1.674E-8	1.676E-8	0.12
	skeleton	1.530E-8	1.534E-8	0.26
	skin	1.651E-8	1.653E-8	0.12
Neutron dose	tumour	3.317E-8	3.318E-8	0.03
	brain	1.875E-8	1.878E-8	0.16
	skeleton	1.865E-8	1.870E-8	0.27
	skin	2.020E-8	2.022E-8	0.10
Photon dose	tumour	2.058E-8	2.061E-8	0.14
	brain	2.064E-8	2.066E-8	0.10
	skeleton	1.983E-8	1.985E-8	0.10
	skin	1.961E-8	1.960E-8	0.05

The computational efficiency of the Monte Carlo program is quantified by the FOM (Figure of Merit) value. The FOM value is defined as follows:

$$FOM = \frac{1}{R^2 T} \tag{3}$$

where R is the relative standard deviation of a certain statistical tally, and T is the calculation time of the program. It can be seen from the above formula that the smaller the relative standard deviation and the shorter the calculation time, the larger the FOM value and the higher the calculation efficiency of the Monte Carlo program.

In order to avoid the impact of parallel efficiency, the single-core operation time of COMPASS and MCNP is compared here. The single-core operation time of COMPASS is 6856s, and the single-core operation time of MCNP is 13979s. The following Table 3 shows the average relative standard deviation of the dose result in the tumor area and the calculated FOM value. It can be seen that the relative standard deviation of the COMPASS result is almost the same with that of MCNP, but since the COMPASS calculation time is shorter, the FOM value of COMPASS is about twice the FOM value of MCNP, which proves that the calculation efficiency of COMPASS is higher than that of MCNP.

Table 3. Comparison of FOM for COMPASS and MCNP.

dose	MCNP result error	MCNP FOM value(s^{-1})	COMPASS result error	COMPASS FOM value(s^{-1})
Boron dose	0.0419521	0.03909	0.0441155	0.07495
Neutron dose	0.0419611	0.03908	0.0441177	0.07494
Photon dose	0.0816028	0.01033	0.07743	0.02433

The parallel efficiency of the COMPASS program has also been tested. The test results show that on the i9-7940X CPU, the 14-core parallel computing speed of COMPASS is 12.4 times the single-core computing speed, the parallel efficiency exceeds 88%, which proves that COMPASS can fully utilize the performance of high-performance computers to achieve the purpose of reducing computing time.

4. Discussion

BNCT is based on the direct use of neutron, and it means that treatment planning and dosimetry are strongly based on MC. General purpose MC code such as MCNP, PHITS are widely used for treatment planning system. However, the comparison of results obtained by MCNP and COMPASS shows that MC-based simulation is time consuming. Different kinds of the strategies should be developed to enhance the MC efficiency in application for BNCT.

5. Conclusion

The TPS NeuMANTA designed for BNCT provides unique features. The integrated MC engine COMPASS can provide significant performance speed enhancements, and users can now attain calculation speeds up to 2 times faster than general purpose MC engine MCNP. With NeuMANTA, users can achieve robust accuracy combined with speed and performance for BNCT treatment.

Acknowledgments

This research was funded by Scientific Research Fund of NIIT, Grant No. YK21-01-06, and the Science and Technology Major Project of Xiamen Municipal Bureau of Science and Technology, Grant No. 3502720201031.

References

[1] Porra L, Seppälä T, Wendland L, Revitzer H, Joensuu H, Eide P, Koivunoro H, Smick N, Smick T, Tenhunen M. 2021 Accelerator-based boron neutron capture therapy facility at the Helsinki University Hospital. Acta Oncol. Sep 25: 1-5.

[2] Decab C, Mecab C, Mb A, et al. 2020 Status of low-energy accelerator-based BNCT worldwide and in Argentina. Applied Radiation and Isotopes, 166.

[3] Chanana A D, Capala J, Chadha M, Coderre J A, Diaz A Z, Elowitz E H, et al. 1999 Boron Neutron Capture Therapy for Glioblastoma Multiforme: Interim Results from the Phase I/II Dose escalation Studies. Neurosurgery 44, 1182-1193.

[4] Hideghety K, Sauerwein W, Haselsberger K, Grochulla F, Fankhauser H, Moss R, Huiskamp R, Gabel D, de Vries M. 1999 Postoperative Treatment of Glioblastoma with BNCT at the Petten Irradiation Facility (EORTC protocol 11, 961). Strahlenther Onkol 175 Suppl 2, 111-114.

[5] Joensuu H, Kankaanranta L, Seppala T, Auterinen I, Kallio M, Kulvik M, et al. 2003 Boron Neutron Capture Therapy of Brain Tumors: Clinical Trials at the Finnish Facility using Boronophenylalanine. J. Neurooncol. 62, 123-134.

[6] Nigg D W, Wemple C A, Wessol D E, et al. 1999 SERA -- An advanced treatment planning system for neutron therapy and BNCT. Trans.am.nus.soc, 80.

[7] González S J, G Cruz, Iii W, et al. 2002 NCTPlan, the new PC version of MacNCTPlan: Improvements and verification of a BNCT treatment planning system.

[8] Lin T Y, Liu Y. 2011 Development and verification of THORplan—A BNCT treatment planning system for THOR. Applied Radiation & Isotopes Including Data Instrumentation & Methods for Use in Agriculture Industry & Medicine, 69(12): 1878-1881.

[9] Goorley T, James M, Booth T, Brown F, Bull J, Cox L J, et al. 2012 Initial MCNP6 release overview MCNP6 version 0.1. Nucl Technol. 180: 298–315.

[10] Kumada H, Matsumura A, Sakurai H, Sakae T, Yoshioka M, Kobayashi H, et al. 2014 Project for the development of the linac based NCT facility in University of Tsukuba. Appl Radiat Isot; 88: 211–5.

[11] Kumada H, Takada K, Yamanashi K, Sakae T, Matsumura A, Sakurai H. 2015 Verification of nuclear data for the Tsukuba plan, a newly developed treatment planning system for boron neutron capture therapy. Appl Radiat Isot 106: 111–5.

[12] Takada K, Kumada H, Liem P H, et al. 2016 Development of Monte Carlo based real-time treatment planning system with fast calculation algorithm for boron neutron capture therapy. Physica Medica.

[13] Sato T, Niita K, Matsuda N, Hashimoto S, Iwamoto Y, Noda S, et al. 2013 Particle and heavy ion transport code system PHITS, version 2.52. Nucl Sci Technol, 50: 913–23.

[14] Cerullo N, Daquino G 1998 "CARONTE", A Treatment Planning System Based on Real Macroscopic Boron Distribution and MCNP-4A Code. Frontiers in Neutron Capture Therapy, Proc. 8th Int. Symp. Neutron Capture Therapy, New York, vol. I, pp. 225–230.

[15] Cerullo N, D Aquino G G, Muzi L 2006 An overview on the developments and improvements of a treatment planning system for BNCT. IEEE Transactions on Nuclear Science, 53(3): 1333-1338.

[16] Hu N, Tanaka H, Kakino R, Yoshikawa S, Miyao M, Akita K, Isohashi K, Aihara T, Nihei K, Ono K. 2021 Evaluation of a treatment planning system developed for clinical boron neutron capture therapy and validation against an independent Monte Carlo dose calculation system. Radiat Oncol. Dec 24; 16(1): 243.

Journal of Physics: Conference Series **2313** (2022) 012013 doi:10.1088/1742-6596/2313/1/012013

Synchronization time in two coupled cluster networks based on discontinuous map

Yunfeng Yang[1, 2] and Xue Bai[1]

[1]College of Science, Xi'an University of Science and Technology, Xi'an 710054, China

[2]Corresponding author's e-mail: yangyunfeng_1978@126.com

Abstract. The synchronization time in the coupled discontinuous maps is investigated. The results show that the synchronization time of the coupled discontinuous systems exhibits a non-monotonic behaviour as the coupling strength increases. Moreover, a coexistence attractor, which consists of a period state and synchronization one, is found, and it exhibits a riddle basin character. The initial conditions of coupled systems, which is close to the basin boundary of period attractor, can lead to a long quasiperiodic transient, and the trajectory jumps from one region to another one in the phase space. Finally, the non-monotonic behaviour of the synchronization time of the coupled discontinuous systems is also checked in other types of discontinuous maps.

1. Introduction

Synchronization is one of the most frequently observed collective dynamics in many physical and biological systems [1-4]. In recent years, a number of theoretical methods have been presented to determine the conditions of synchronization, such as, the methods of master stability functions [5-7] and eigenvalue analysis [8]. However, these previous studies pay much more attention to the coupling strengths and topologies coupled units than the synchronization time given that a network synchronizes.

In recent years, the synchronization time has been investigated with increasing interest [9-12]. We note that these studies are main based on the continuous systems. As is known to all, in the real networks the node shows discontinuous dynamical properties, and the discontinuity should be considered. There exist a plenty of discontinuous systems, for example, models of nerve cells [13], relay control systems [14], electrical systems [15], synchronization of physical and biological systems [16] and so on. We have been studied the synchronization transition process of coupled piecewise map lattices, and found a new period attractor which exists [17]. Lai [18] has been found a long irregular transient with a negative largest Lyapunov exponent in discontinuous map systems. In this Letter, the synchronization time in two coupled cluster networks based on discontinuous map is investigated, and the relation between the synchronization time and the coupling strength is discovered.

2. Model

The first paragraph after a heading is not indented (Bodytext style). A typically clustered structure network consists of a number of sparsely connected sub-networks, or clusters, where the connectivity within each individual cluster is dense. The results in Ref. [17] have suggested the global coupled discontinuous systems can form two synchronization clusters with different sizes, and it means that the global coupled discontinuous system can be treated as two coupled one-dimensional systems. Based on these factors, the system employed in this letter is described by the following equation:

$$\begin{aligned} x_n &= (1-\varepsilon+\varepsilon r) f(x_n) + \varepsilon(1-r) f(y_n), \\ y_n &= (1-\varepsilon r) f(y_n) + \varepsilon r f(x_n). \end{aligned} \qquad (1)$$

Where n is the discrete time index, $r = N_1 / N$ denotes the fraction of the total population N that synchronizes into state x, and ε gives the strength of coupling between x_n and y_n. $f(x)$ or $f(y)$ is the local one-site map, and it is as follows:

$$f(x) = \begin{cases} \lambda\mu - \sqrt{\mu - x} & x \le \mu, \\ \lambda x & x > \mu. \end{cases} \qquad (2)$$

A single map exhibits chaotic behavior under the conditions $\mu > 0$ and $0 < \lambda < 1$. In the following simulation, the parameters are set as μ=0.2, μ=0.07, respectively. λ is taken as 0.7. The network site is $N = 100$.

3. Results and discussions

Assuming that the coupled system reaches complete synchronization in the end, the synchronization time t for the system (1) for different coupling strength ε is calculated. Firstly, the r is set as 0.5, and it means that the two clusters are in the same sizes. The complete synchronization state is described by the synchronization error $\delta = |x_n - y_n|$. When the synchronization error $\delta < 10^{-8}$, the complete synchronization state is obtained. The synchronization time t of the coupled discontinuous maps as a function of the coupling strength ε with different μ is plotted, and is shown in Figure 1. The synchronization time t appears a large spike near ε=0.57 in Figure 1(a) and two large spikes near ε=0.68 and ε=0.82 in Figure 1(b), respectively. It demonstrates that the synchronization time t exhibits a non-monotonic behavior as the coupling strength ε increases in the coupled discontinuous maps.

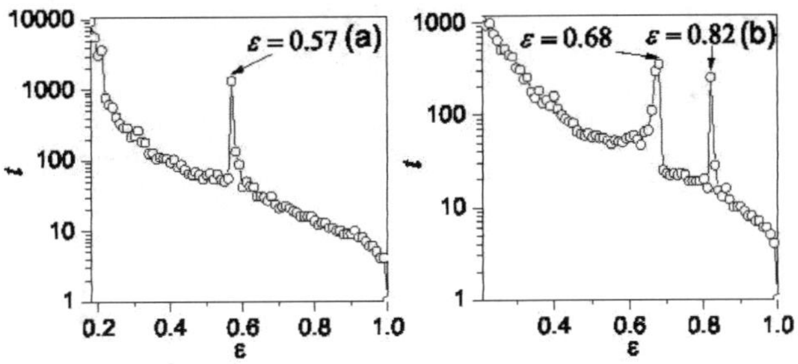

Figure 1. The synchronization time t of the coupled discontinuous maps is as a function of the coupling strength ε. The network site is N=100, $\lambda = 0.7$ and $r = 0.5$. (a) $\mu = 0.07$, (b) $\mu = 0.2$.

Figure 2 (a) and (b) show a typical quasi-periodic transient in this system, the black and red dot correspond to the x_n and y_n, respectively. The green dash line denotes the discontinuous boundary of isolated map. To our surprise, we find a quasi-periodic transient dynamic. Figure 2(a) and (b) clearly show that the iterations trajectories gradually approach the basin boundaries of isolated map with the time evaluating. Once the iterations trajectories reach discontinuous boundaries, the coupled system reaches global synchronization. For example, see the Figure 2(a), the iterations trajectories of x_n (black dot) reach discontinuous boundaries at $n = 1310$, then, the system reaches the complete synchronization afterward (at $n = 1330$). The quasiperiodic transient state is also plotted in the phase space (see Figure 2 (c) and (d)). The phase space can be divided into the individual zones with different dynamical features by the discontinuous boundary. The transient trajectories repeatedly jump from one region to anther region in the phase space, and form a typical quasi-periodic transient.

Based on the mutual interactions of the local nonlinearity and the spatial coupling, the typical quasi-periodic transient dynamics can be analyzed. The coupled dynamics consists of two independent stages: the local dynamics and the coupling dynamics. The former is the application of the one-dimensional isolated map $f(x)$ to every node, and the latter couples the dynamics by means of a weighted sum over a neighbourhood ε . The map function Eq. (1) can be reduced as follows:

$$\begin{aligned} x_n &= f(x_n) - \varepsilon(1-r)(f(x_n) - f(y_n)), \\ y_n &= f(y_n) + \varepsilon r(f(x_n) - f(y_n)). \end{aligned} \tag{3}$$

The first term on the right side of Eq. (3) is the isolated map and the second term is the coupling perturbation dynamics. While the former tends to constraint the map to the predefined trajectory, the latter prevents this tendency by providing a perturbation. If the perturbation equals zero, the system reaches global synchronization. If the perturbation equals to isolated map dynamics, the coupled system can form a stable period attractor, see Ref. [17]. Even a small difference between perturbation and isolated map dynamics can lead to its stability of the period attractor vanishing and a quasi-periodic states transient appearing. Once the quasi-periodic states transient reaches discontinuous boundaries, the dynamics of the isolated map could mutate, then the balance between the perturbation and the isolated map dynamic will be disrupted, the quasi-periodic states transient vanish and the coupled system reaches the global synchronization.

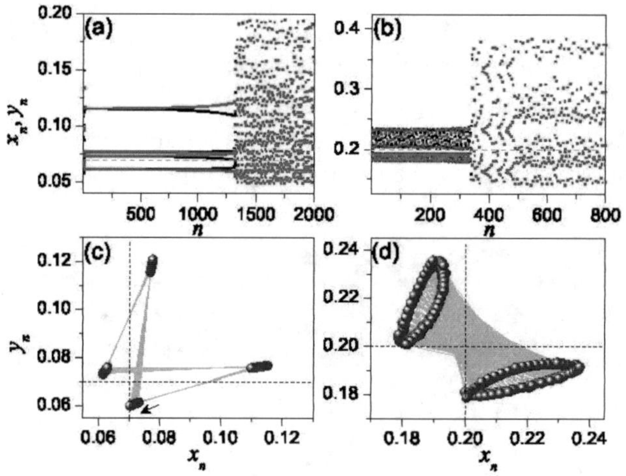

Figure 2. The iterations trajectories of the coupled system at the spike of the synchronization time and the corresponding transient trajectories are plotted in the phase space. (a) The iterations trajectories of the coupled systems with $\lambda = 0.7, r = 0.5, \mu = 0.07$ and $\varepsilon = 0.57$, (b) The iterations trajectories of the coupled systems with $\lambda = 0.7, r = 0.5, \mu = 0.2$ and $\varepsilon = 0.68$,(c) Figure 2(a) corresponding the transient trajectory in the phase space, (d) Figure 2(b) corresponding the transient trajectory in the phase space.

In order to demonstrate that the non-monotonic behavior of the synchronization time can be induced by a quasiperiodic state transient, the riddle basin of the coupled system is plotted in Figure 3. It shows a coexistence attractor which consists of a period states and synchronization states, which has a riddle basin character. The blue dots denote the period attractor, and the red regions indicate the synchronization states. When the two initial conditions, which are close to the boundary between period and synchronization basin, have tiny difference (see the green dot in Figure 3(a) and (b)), and the synchronization error is plotted as shown in Figure 3(c) and (d). The results show that the initial conditions $[x, y]$ in blue regions rapidly forms a period attractor, see the blue dot in Figure 3(c) and (d). However, the initial conditions $[x, y]$ in red regions, which close to red regions (the green dot in Figure 3(a) and (b)), can form a long quasi-periodic state transient.

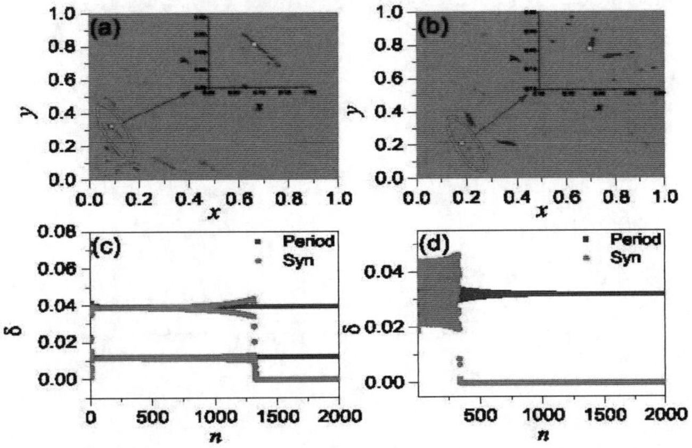

Figure 3. (a,b) The riddle basin for coupled piecewise map systems, blue indicates that the initial conditions $[x, y]$ lead to the period attractor, while red indicates that the initial conditions $[x, y]$ lead the system to complete synchronization. (c,d) The synchronization error with tiny difference in initial conditions.

It should be noted that there are two large spikes near ε=0.68 and ε=0.82 in the $t-\varepsilon$ plateau of Figure 1(b), we only analyze the first spike (ε=0.68), the result shows the transient trajectory forms two circles in the phase space. Can the long quasi-periodic transient states be observed in the second spike (ε=0.82)? To address this question, the iterations trajectories and the quasi-periodic transient state of the coupled systems with the coupling strength ε=0.82 in the phase space are also plotted as shown in Figure 4.

We have set the size of clusters $r = 0.5$ in the above discussion. However, for clustered networks, the size of clus-ters is not always the same number. In practice, many real networks have different size of clusters. In order to demonstrate the non-monotonic behavior of the synchronization time can also find in the clustered networks with the different size of clusters, we now change r to verify it. We change $r = 0.1$ and the other parameter same as Figure 1(b). The synchronization time t of the coupled discontinuous maps as a function of the coupling strength ε is shown in Figure 5.

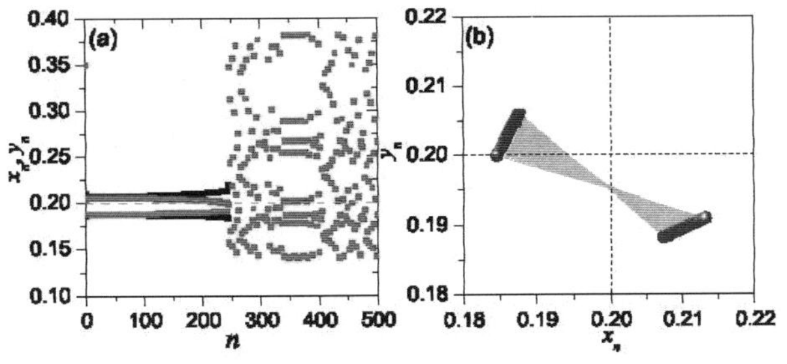

Figure 4. The iterations trajectories and the corresponding transient trajectories of the coupled system at the second spike of the synchronization time in Figure 1(b) are plotted in the phase space with $\lambda = 0.7, r = 0.5, \mu = 0.2$ and ε=0.82 . (a) The iterations trajectories of the coupled systems, (b) Figure 4(a) corresponding the transient trajectory in the phase space.

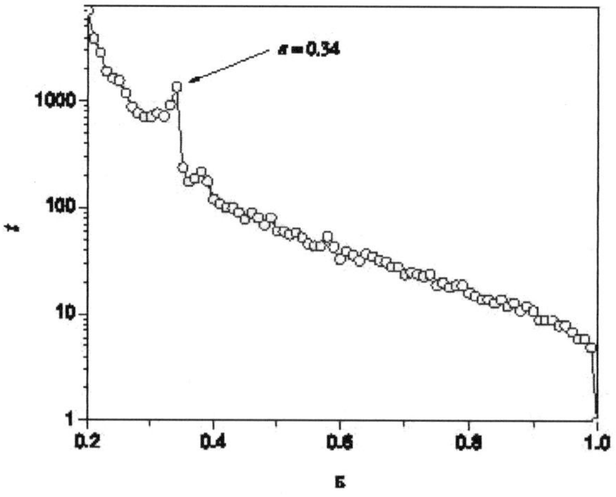

Figure 5. The synchronization time t of the coupled discontinuous maps is as a function of the coupling strength ε with asymmetric clustered networks $r=0.1$, the other parameters are $\lambda=0.7, r=0.5$ and $\mu=0.2$.

Can the non-monotonic behavior for the synchronization time be observed in other discontinuous maps, ie., the map function is changed. In order to address this problem, the Eq. (2) is replaced by a different map function, and the relation between the synchronization time and the coupling strength is explored. The new map function is as follow

$$x_n = f_m(x_n) = k_m x_n + b_m \bmod 1, \tag{4}$$

with $m=1,2,3,4$ and

$$k_1 = \frac{y_b - y_a - \mu}{x_b - x_a}, b_1 = y_a, \qquad x \in [x_a, x_b];$$

$$k_2 = \frac{y_c - y_a + \mu}{x_g - x_b}, b_2 = y_A - \mu - k_2 x_b, \qquad x \in [x_b, x_g];$$

$$k_3 = \text{constant}, b_3 = \text{constant}, \qquad x \in [x_g, x_f];$$

$$k_4 = \text{constant}, b_4 = \text{constant}, \qquad x \in [x_f, x_d];$$

where μ is the control parameter. The other parameters are selected as $x_a=0$, y_a=0.203921, x_b=0.107663, x_g=0.350000, y_c=0.460000, x_f=0.497121, x_d=1, y_d=y_A, k_3=3.07055, b_3=−0.530165, k_4=0.405507, and b_4=−0.201586.Calculating the synchronization time for the system (1) with the local dynamics (4), the synchronization time t of the coupled discontinuous maps as a function of the coupling strength e is plotted in Figure 6. It is seen that the synchronization time exhibits the same phenomenon as shown in Figure 1 and Figure 5. This result is in accordance with expectations, the non-monotonic behavior of synchronization time is a com phenomenon in the cluster network with discontinuous systems.

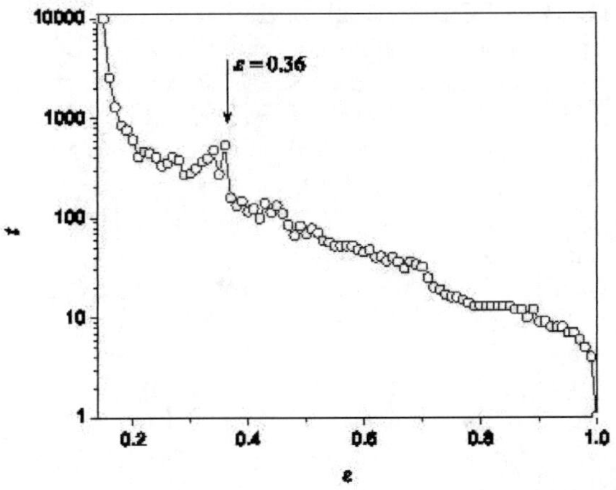

Figure 6. The non-monotonic behavior for synchronization time be observed in other clustered discontinuous maps (4), $r=0.2, N=100$ and the control parameter is $\mu=0.064$.

4. Conclusions

In summary, the relation between the synchronization time and the coupling strength is investigated in the cluster network with a discontinuous system. The non-monotonic behavior is induced by a long quasi-periodic transient in the cluster network with a discontinuous system. Based on the detailed analysis, these long quasi-periodic transient actually has certain structure, which can be illustrated by the transient trajectories. The mutual interactions of the local dynamics and the spatial coupling generate a coexistence attractor, which consists of a period states and synchronization one. Its basin has a riddle basin character. The initial condition closing to the boundary between period and synchronization basin, results in a long quasiperiodic transient. The trajectories of this quasi-periodic transient jump from one region to another one in the phase space. The relation between the synchronization time and the coupling strength is exhibited with the different parameters and other map function. The results show that the synchronization time shows a same phenomenon, ie., a non-monotonic behavior with the coupling strength increasing. It suggested that the non-monotonic behavior of the synchronization time may be a common phenomenon in the coupled discontinuous system. The findings will enrich our knowledge about the synchronization dynamics in coupled discontinuous systems.

References

[1] Arenas A, Diaz-Guilera A, Kurths J, Moreno Y and Zhou C 2008 "Synchronization in Complex Networks," Physics Reports, Vol. 469, no. 3, pp. 93-153.
[2] Pikovsky A, Rosenblum M and Kurths J, Synchronization 2001 A universal concept in nonlinear sciences, (Cambridge Univ. Press,Cambridge, UK).
[3] Strogatz S H 2004 Sync: The emerging science of spontaneous order, (Penguin Books, London, UK).
[4] Kanter I, Kinzel W and Kanter E 2002 "Secure exchange neuralnetworks," Europhys Letters, Vol. 57, no. 1, pp. 141-147.
[5] Pecora L M and Carroll T L 1998 "Master Stability Functions for Synchronized Coupled Systems," Physical Review Letters, Vol. 80, no. 10, pp. 2109-2112.
[6] Huang L, Chen Q, Lai Y C and Pecora L M 2009 "Generic behavior of master-stability functions in coupled nonlinear dynamical systems," Physical Review E, Vol. 80, 036204.
[7] Huang L, Lai Y C, Park K, Wang X and Lai C 2007 "Synchronization in complex clustered

networks," Frontiers of Physics, Vol. 2, no. 4, pp. 446-459.
[8] Yang J, Hu G and Xiao J 1998 "Chaos synchronization in coupled chaotic oscillators with multiple positive lyapunov exponents," Physical review letters, Vol. 80, no.3, pp. 496-499.
[9] Grabow C, Grosskinsky S and Timme M 2011 "Speed of complex network synchronization," The European Physical Journal B, Vol. 84, no. 4, pp. 613-626.
[10] Timme M, Wolf F and Geisel T 2004 "Topological Speed Limits to Network Synchronization," Physical review letters,Vol. 92, no. 7, Article ID 074101.
[11] Motter A E, Zhou C S and Kurths J 2005 "Enhancing complex-network synchronization," Europhysics Letters, Vol. 69, no. 3, pp. 334-340.
[12] Timme M, Geisel T and Wolf F 2006 "Speed of synchronization in complex networks of neural oscillators: Analytic results based on Random Matrix Theory," Chaos, Vol. 16, Article ID 015108.
[13] Ibarz B, Casado J M and Sanjuan M A F 2011 "Map-based models in neuronal dynamics," Physics Reports, Vol. 501, no. 1-2, pp.1-74.
[14] Zhusubaliyev Z T and Mosekilde E 2003 Bifurcation and Chaos in Piecewise-Smooth Dynamical Systems. (World Scientific, Singapore).
[15] Gala S, Ragusa M A, Sawano Y, Tanaka H 2014 Uniqueness criterion of weak solutions for the dissipative quasi-geostrophic equations in Orlicz–Morrey spaces, Applicable Analysis 93(2), 356-368.
[16] J. Leonel Rocha, Sandra M. Aleixo and Acilina Caneco 2014 Synchronization in Richards' Chaotic Systems, Journal of Applied Nonlinear Dynamics, 3, 2, 115-130.
[17] Yang K L and Wang C J 2015 "Coexistence of attractors and effects of noise on coupled piecewise maps," Nonlinear Dynamics, Vol. 79, no. 1, pp. 377-385.
[18] Zou H, Guan S and Lai C H 2009 "Dynamical formation of stable irregular transients in discontinuous map systems," Physical Review E, Vol. 80, Article ID 046214.

Journal of Physics: Conference Series **2313** (2022) 012014 doi:10.1088/1742-6596/2313/1/012014

A novel voltage protection method for multi-cell lithium-ion battery protection IC

Zhijian Chen[1, 3], Xinyi Wu[1], Bin Li[1], Xiaoling Lin[2], Rui Xiao[1], Guangyin Feng[1] and Yanjie Wang[1]

[1] School of Microelectronics, South China University of Technology Guangzhou, Guangdong 510640, China
[2] Science and Technology on Reliability Physics and Application Technology of Electronic Component Laboratory, Guangzhou, 510610, China

[3] Corresponding author's e-mail: chenzhijian@scut.edu.cn

Abstract. Lithium batteries have been widely used in portable electronic devices and other electric products. To ensure safety and long life, a lithium battery needs to be equipped with a protection chip. A novel voltage protection method for three-cell lithium-ion battery protection IC (Integrated Circuit) is proposed in this paper. A new voltage protection circuit structure and a three-cell lithium battery voltage sampling circuit are presented to improve the circuit performance of the chip and reduce the dynamic power consumption and chip area of the system. With a new time-division multiplexing circuit structure, the area and power consumption of the chip are greatly reduced. Moreover, this solution can be extended to any number of lithium battery packs connected in series for voltage protection. Testing results show that the overcharge detection accuracy can achieve better than ±15mV, the over-discharge detection accuracy can achieve ±7.5mV, the power consumption of the voltage protection module is 1.7μA, and the area is 300μm×500μm^2.

1. Introduction

Lithium batteries have the advantages of low cost, high capacity, and pollution-free, which are widely used in portable electronic devices and large electric products. Nevertheless, lithium batteries have potential safety hazards such as overcharging, over-discharging, and short-circuiting. Consequently, each lithium battery needs to be equipped with a protection chip to ensure the safety and life of the lithium battery.

Voltage protection is the core function of the lithium-ion battery protection IC (Integrated Circuit), including overcharge, over-discharge, and 0V charging protection [1]. The current research on lithium-ion battery protection ICs usually uses multiple comparators to compare different voltage functions. This structure will double the number of comparators in multi-cell lithium battery protection applications, seriously affecting the chip power consumption and area [2-4]. This paper uses GF (Global Foundries) 0.18-um BCD (Bipolar CMOS DMOS) process to design a new voltage protection circuit architecture and a multi-cell lithium battery sampling circuit. Only a single comparator is used to achieve multiple voltage detection of three lithium batteries. The power consumption and area of the chip are significantly reduced. Tape-out test results show that the voltage overcharge detection accuracy can reach ±15mV, the over-discharge detection accuracy can reach ±7.5mV, the power consumption of the voltage protection module is 1.7μA and total area is 300μm×500μm^2.

Journal of Physics: Conference Series **2313** (2022) 012014 doi:10.1088/1742-6596/2313/1/012014

2. Circuit design

2.1. Structure of voltage protection circuit

The structure of the new lithium battery voltage protection circuit is shown in Figure 1. The circuit consists of a lithium battery voltage sampling circuit, a reference voltage circuit, a strobe circuit, a voltage comparator, nine latches, two delay circuits, and an output logic circuit.

The lithium battery voltage sampling circuit samples the voltage of the three lithium batteries in real-time at different periods and sends the sampled voltage to the positive port of the comparator for comparison and detection. The reference voltage circuit is used to generate a reference voltage for overcharge, over-discharge, and 0V detection. The strobe gates different reference voltages to the negative port of the comparator in different periods and is used to implement the time-division multiplexing of the voltage protection function. The hysteresis logic control circuit is used to control the reference voltage hysteresis in the detection to prevent factors such as voltage jitter from affecting the detection. A total of 9 latches in the circuit respectively latch the three voltage states of the three lithium batteries output by the comparator. The overcharge and over-discharge delay circuits are responsible for delaying the detection result. When the detection result is high during the delay time, the delay circuit outputs effectively; the output logic and the 0V charge control circuit perform logical output and prohibit 0V Logic calculation control of charging.

This architecture is designed by time-division multiplexing to perform different functional tests under other clock pulses. Only one comparator is used to complete the detection and protection of 9 voltage states of overcharge, over-discharge, and 0V detection of three lithium batteries.

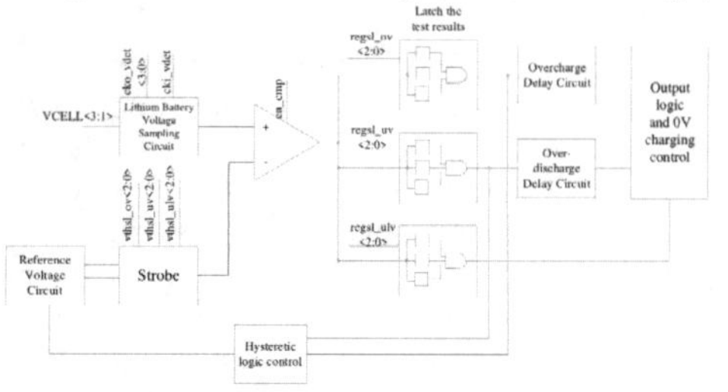

Figure 1. Novel lithium battery voltage protection circuit architecture.

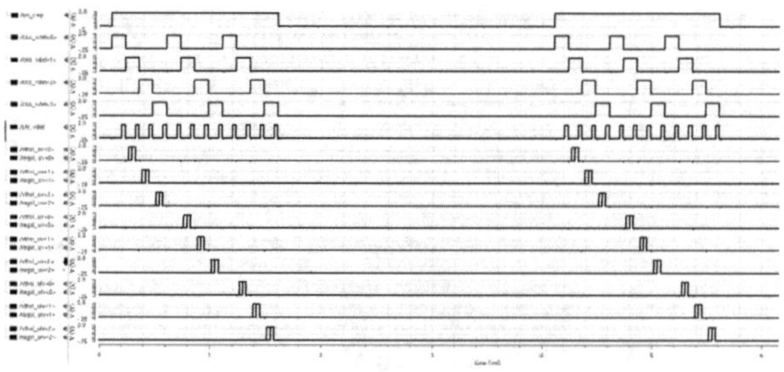

Figure 2. Voltage detection clock.

The sequential clock of voltage detection is shown in Figure 2. The en_cmp is the working clock of the comparator, covering all other clock pulses to ensure that the comparator is normally turned on during detection. The clocks cko_vdet<3:0> and cki_vdet are used to sample the voltages of the three lithium batteries, where cko_vdet<1:3> samples the voltages of the first to third lithium batteries respectively. In the en_comp clock pulse, each cko_vdet signal has three pulses, which are respectively used for the overcharge, over-discharge, and 0V battery detection of the lithium battery.

In the first 75% of each lithium battery sampling pulse, there is an equal phase reference voltage gating clock: vthsl_ov<2:0>, vthsl_uv<2:0>, vthsl_ulv<2:0>; they are used for overcharge respectively, over-discharge and reference voltage gating for 0V battery detection. In one work cycle of the comparator, three voltage state detections of three lithium batteries are completed under 9 clock pulses of vthsl. The clocks regsl_ov<2:0>, regsl_uv<2:0>, and regsl_ulv<2:0> are the latch clocks of the detection results, which are located in the second half of the strobe clock vthsl pulse.

With the reasonable cooperation of the above 24 clocks, this architecture realizes the time-division multiplexing of the voltage protection module. Using a comparator loop, within a period of 4ms, the module completes the voltage detection function of all lithium batteries in 37.5% of the time, which greatly reduces the dynamic power consumption of the system, improves the utilization rate of the circuit, and reduces the layout area of the module.

2.2. Multi-cell lithium battery voltage sampling circuit

For a multi-cell lithium-ion battery protection IC, the voltage of the previous lithium battery will be directly superimposed on the latter battery, resulting in a larger voltage value to be detected. If the voltage at both ends of the lithium battery is detected directly, multiple comparators are required to perform multiple detections and comparisons [5]. To cooperate with the new voltage protection circuit architecture and reduce the power consumption of the system, it is necessary to sample the voltage of each lithium battery, obtain the real-time voltage of each lithium battery, and then transmit it to the comparator for detection and comparison in a specific period. A low-power, high-precision three-cell voltage sampling circuit for lithium batteries is proposed in this paper.

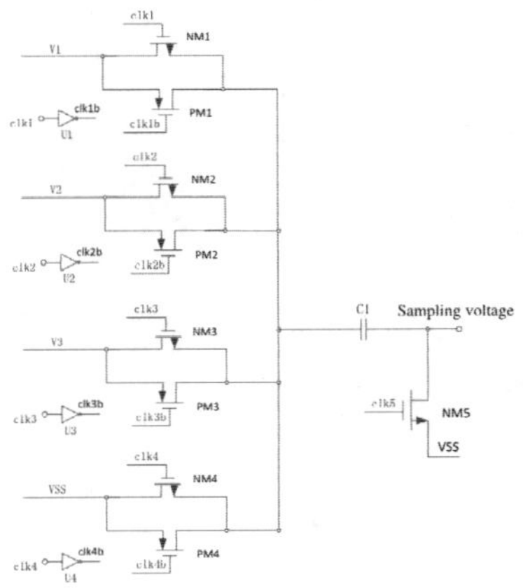

Figure 3. Lithium battery voltage sampling circuit.

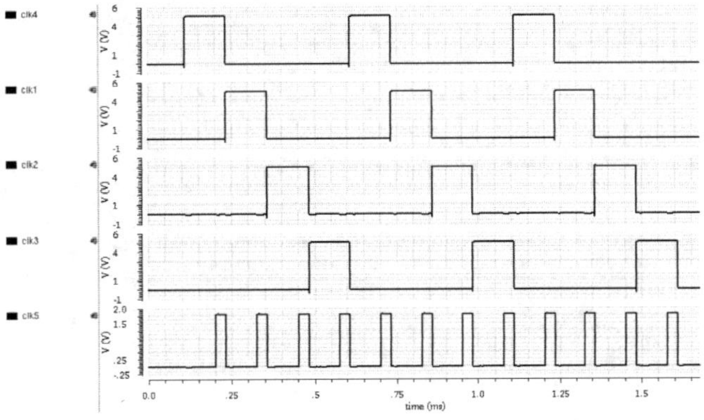

Figure 4. Lithium battery voltage sampling clock.

The voltage sampling circuit for three lithium batteries is shown in Figure 3. The voltage sampling circuit is opened under the control of four non-overlapping clock pulses to sample the three lithium batteries' voltage. The sampling clock is shown in Figure 4.

The voltage sampling circuit is mainly composed of four mux, a sampling capacitor C1 and a reset NMOS NM5 [6-7]. The clock CLK1~CLK3 respectively control the first to third lithium battery positive terminal voltage V1~V3, the clock CLK4 gating VSS=0V, the clock CLK5 controls the reset switch NM5, CLK4 and CLK5 play the role of sampling reset. The period of the clock CLK1~CLK4 is 500μs and the duty cycle is 25%; the period of the clock CLK5 is 125μs and the duty cycle is 25%. The pulses of clock CLK4, CLK1~3 generate in sequence, and the pulse of clock CLK5 is located at the last 25% of each pulse of clock CLK1~CLK4.

When the sampling circuit starts to work, the first phase of the clock CLK4 pulse arrives, and the positive port of the capacitor C1 gates the 0V voltage. The CLK5 clock pulse arrives during the last 25% of the CLK4 pulse, and NM3 is turned on, the negative port of the capacitor is grounded, the voltage drop of C1 is 0V and the stored charge Q1=0. The output sampling voltage is 0V, and the sampling circuit is reset.

In the second phase, the clock pulse of CLK1 arrives, and the positive port of the capacitor C1 gates the positive port voltage V1 of the first lithium battery. Since the voltage drop of the capacitor is 0 at the previous moment, the output sampling voltage is V1 at this time. After 25% of the CLK1 clock pulse, the CLK5 clock arrives, and NM5 is turned on to form a charging circuit for the capacitor C1. The capacitor charging Q1=V1×C1, the voltage drop of the capacitor rises to V1, and the output sampling voltage is 0. Therefore, during the first 75% of the pulse of the clock CLK1, the sampled output voltage is the voltage of the first lithium battery.

The third phase clock CLK2 arrives, and the positive port of the capacitor C1 gates the positive terminal voltage V2 of the second lithium battery. Since the voltage drop of the capacitor was V1 at the previous moment, the output sampling voltage is V2-V1 at this time. When the clock CLK5 pulse comes, the output sampling voltage is 0, the capacitor charging Q1=V2×C1, and the voltage drop of the capacitor is V2. During the first 75% of the clock CLK2 pulse, the sampled output voltage is the voltage of the second lithium battery. In the same way, it can be obtained that during the first 75% of the clock CLK3 pulse, the sampled output voltage is the third lithium battery voltage V3-V2. Finally, when the clock CLK4 pulse arrives again, the capacitor C1 is completely discharged to 0V, and the reset is completed. The system starts sampling in the first to third lithium batteries.

At CLK4, both ends of the capacitor are grounded and discharged; when CLK1~3 overlap with CLK5, one port of the capacitor is grounded and the other port is connected to the power supply for charging. According to the charging formula of the capacitor:

$$Q = CV\left(1 - e^{-\frac{t}{RC}}\right) \tag{1}$$

The value of the charge stored on the capacitor C1 can be obtained:

$$\begin{cases} Q_{CLK4} = 0 \\ Q_{CLK1,5} = V_1 C_1 \\ Q_{CLK2,5} = V_2 C_1 \\ Q_{CLK3,5} = V_3 C_1 \end{cases} \tag{2}$$

Q_{CLK} indicates the charge stored in a capacitor in a clock. The voltage across the capacitor will not change instantaneously, the formula for sampling voltage can be obtained as:

$$\begin{cases} V_{CLK4} = 0 \\ V_{CLK1_{0\sim75\%}} = V_1 \\ V_{CLK2_{0\sim75\%}} = V_2 - V_1 \\ V_{CLK3_{0\sim75\%}} = V_3 - V_2 \end{cases} \tag{3}$$

V_{CLK} represents the sampling voltage within a certain clock time. It can also be seen from the formula that the voltage sampling circuit samples the voltages of three lithium batteries under the control of four clock pulses.

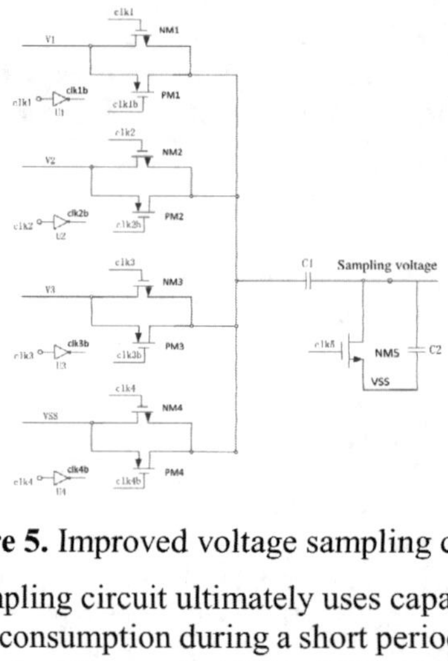

Figure 5. Improved voltage sampling circuit.

The lithium battery voltage sampling circuit ultimately uses capacitors to sample the lithium battery voltage and only generates current consumption during a short period when the capacitor is charged and discharged, and the capacitor charging time accounts for 25% of the time in one cycle. The charging voltage is the voltage of a single-cell lithium battery, and the static power consumption at the rest of the time is approximately equal to zero [8-9]. The average power consumption in one cycle of the sampling circuit is:

$$P = \frac{0.125CV^2}{t} \tag{4}$$

In formula (4), C is the capacitance value of the sampling capacitor, V is the voltage of a single-cell lithium battery, and t is the charging time.

Figure 5 is an improved lithium battery voltage sampling circuit. By adding capacitor C2, the voltage of each lithium battery can be sampled by voltage division, which is convenient for low-voltage domain detection and comparison in the comparator without generating additional voltage divider current.

The standard working voltage of a lithium battery is usually 2.8V~4.2V. The sampling accuracy of the voltage sampling circuit directly affects the detection accuracy of the chip overcharge and over-discharge protection. The PVT simulation results of the sampling circuit in Figure 5 after 1/4 voltage division are shown in Figure 6, including three process angles of tt, ss, and ff; four temperatures of -20℃, 0℃, 27℃, and 85℃.

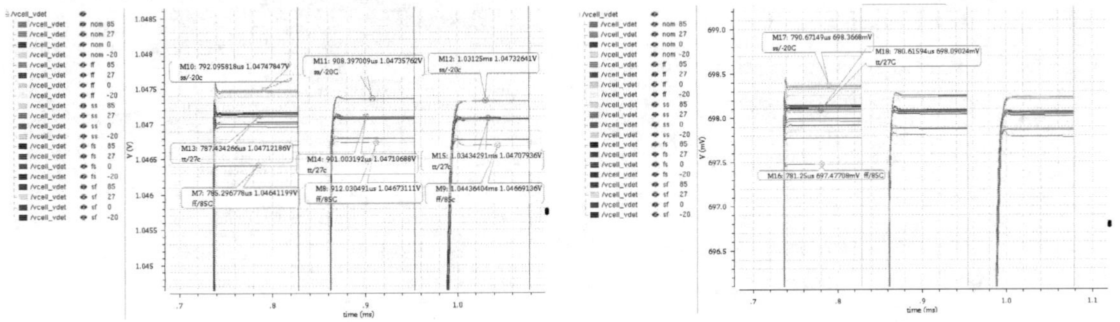

(a) 1/4 sampling result when lithium battery voltage is 4.2V

(b) 1/4 sampling result when lithium battery voltage is 2.8V

Figure 6. PVT simulation results of sampled partial voltage under different voltages of three lithium batteries.

The simulation results are summarized in Table 1. When the lithium battery voltage is 4.2V, the typical value of the 1/4 sampling result is 1.0471V, the maximum sampling error of the lithium battery is -14.4mV, and the minimum sampling error is -10mV. When the lithium battery voltage is 2.8V, the typical sampling value is 698.09mV, the sampling error is -7.64mV, the maximum error is -10.08mV, and the minimum is -6.52mV.

Table 1. PVT simulation results of the voltage sampling circuit.

Battery Voltage	1/4 Sampling			Sampling error		
	Min (mV)	Typical(mV)	Max (mV)	Min (mV)	Typical (mV)	Max (mV)
4.2V	1046.4	1047.1	1047.5	-10	-11.6	-14.4
2.8V	697.48	698.09	698.37	-6.52	-7.64	-10.08

3. Test results

The photomicrograph of the lithium-ion battery protection IC based on the new voltage protection circuit architecture is shown in Figure 7. The overall area of the chip is 1029μm×1205μm^2, of which the area of the voltage protection module is 300μm×500μm^2. The total static power consumption of the whole chip is 11μA (including LDO, oscillator, voltage, current and temperature protection module, etc.). The power consumption of the voltage protection module is 1.7μA.

In voltage protection, the detection accuracy of overcharge and over-discharge is an important indicator to measure the performance of the sampling circuit, which is affected by the accuracy of the lithium battery voltage sampling circuit, the input offset voltage of the comparator, and the accuracy of the reference voltage. In this paper, the voltage detection function of 4~5 lithium-ion battery protection ICs in the above scheme is tested, and the accuracy of the overcharge and over-discharge of each chip is tested in the application scenario of three lithium batteries. To verify the configurability of the chip positions and ensure the reliability of the test results, the accuracy test of each chip can be set three detection threshold voltages. The test system is shown in Figure 8, and the test plan is shown in Table 2.

Figure 7. Microscope photo of lithium battery protection chip.

Figure 8. Test platform.

Table 2. Lithium battery protection chip voltage detection threshold voltage plan.

	Overcharge	Overdischarge
Test threshold voltage1 （V）	4.6	3.2
Test threshold voltage2 （V）	4.2	2.8
Test threshold voltage3 （V）	3.34	1.6

During the test, the overcharge and over-discharge detection and delay functions work normally. The accuracy test results are shown in Figure 9. Under the test conditions covering all thresholds, the maximum detection accuracy of overcharge is ±15mV; the maximum detection accuracy of over-discharge is ±7.5mV. Compared with the simulation results, the error is within an acceptable range.

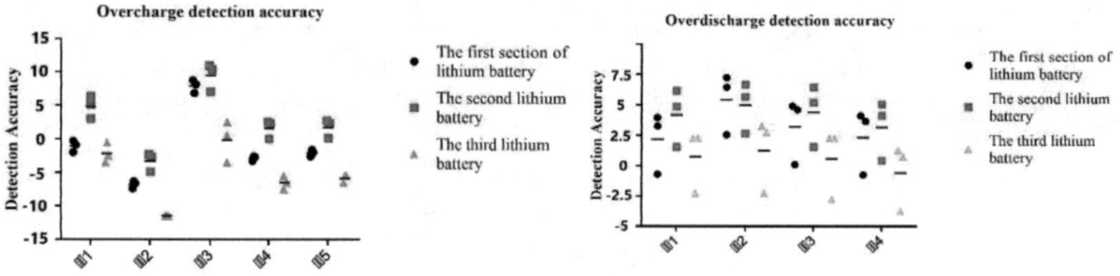

Figure 9. Overcharge and over-discharge detection accuracy test results.

Table 3. Performance summary of lithium battery protection chips.

	[10]	[11]	[12]	[13]	This work
Detection accuracy of over-charge (mV)	±17	±30	±25	±2	±15
Detection accuracy of over-discharge (mV)	±34	±30	±80	±2	±7.5
Current (μA)	3	-	>2.16	45	1.7

This work is compared to earlier reported lithium battery protection chips in Table 3. Nevertheless, [13] used 12-bit ADC to improve the performance of accuracy at the cost of power. The structure proposed in this paper has better detection accuracy performance and can be oriented to more complex application scenarios. Aiming at design difficulties such as high voltage generated by multiple lithium batteries in series, the new three-cell lithium battery voltage protection circuit architecture and three-cell lithium battery voltage sampling circuit proposed in this paper provide a good solution.

4. Conclusion

This paper introduces a new structure of three-cell lithium battery voltage protection and a three-cell lithium battery voltage sampling circuit. Through the new circuit structure design of time-division multiplexing, the area and power consumption of the chip are significantly reduced. This solution can be extended to any number of lithium battery packs connected in series for voltage protection. The tape-out test results show that the overcharge detection accuracy can reach ±15mV, the over-discharge detection accuracy can reach ±7.5mV, the power consumption of the voltage protection module is 1.7μA, and the area is $300\mu m \times 500\mu m^2$.

Acknowledgment

Supported by the Project of Science and Technology on Reliability Physics and Application Technology of Electronic Component Laboratory under granted No. 6142806200502 and the Fundamental Research Funds for the Central Universities under granted No. 2020ZYGXZR067

References

[1] Wang Y, Han Z, Shi J 2014 One cell Lithium-ion battery protection IC [C]//2014 9th *IEEE Conference on Industrial Electronics and Applications*. IEEE, 2014: 1427-1432.

[2] Lee T J and Lin W R 2014 VCT Protection IC for Li-Ion Battery [J]. *Springer International Publishing Switzerland*, 2014: 649-657.

[3] Alamgir M 2017 Lithium has transformed vehicle technology: how trends in Li-ion battery technology have developed for vehicle electrification[J]. *IEEE Electrification Magazine*, 5(1): 43-52.

[4] Shen W J, Li H X 2017 A sensitivity based group-wise parameter identification algorithm for the electric model of Li-ion Battery[J]. *IEEE Access*, 5: 4377-4387.

[5] Son J, Du Y 2019 Model-Based Stochastic Fault Detection and Diagnosis of Lithium-Ion Batteries[J]. *Processes*, 7(1): 38.

[6] Zhang F 2017 Modeling and Control of a Modular Battery Management System for Lithium-Ion Battery Packs[D]. University of Colorado at Boulder.

[7] Hao Yunqiang, Yi Dongbai, Zhang Xiaowei, Yu Wenxin, Xi Jianxiong, He Lenian 2021 A Power Management IC Used for Monitoring and Protection of Li-Ion Battery Packs[J]. *Journal of Sensors*.

[8] Anandh B A, Ganesh A Shankar, Sakthivel R, Kumar D, Mahesh Kumar E 2021 PremSmart battery management system using LiFePO4 battery for offline UPS[C]. *Journal of Physics: Conference Series*, v 2040, n 1, October 25, 2021, International Conference on Physics and Energy 2021.

[9] Michael Sujatmiko Reza Pernandito, Abuzairi Tomy, Rizkinia Mia, Kurniawan Taufiq Alif 2019 Design of overcharging protection and passive balancing circuits using dioda for lithium-ion battery management system[C]. *2019 16th International Conference on Quality in Research*, July 2019.

[10] Lijun Xiong 2015 Design of high-performance lithium-ion battery charge-discharge protection chip [D]. *Guizhou University*.

[11] Siliang Chen 2019 Research and Design of Single-cell Lithium Battery Charge and Discharge Protection Circuit[D]. *Chongqing University of Posts and Telecommunications*.

[12] Dandan Hu 2018 Research on the Key Technology of Multi-cell Lithium Battery Protection Chip [D]. *North China University of Technology*.

[13] Wang Tong, Zhao Ye, Chen Jie 2018 A battery monitoring IC with an isolated communication interface for electric vehicles[J]. *IEICE Electronics Express*, v15, n12.

Construction and properties analysis of Z-A-Z graphene nanoribbons transistors

Hanxi Wang[1, 3], Yuanzhi Ding[2], Guojun Li[1] and Yuxuan Song[1]

[1]School of Microelectronics, Hefei University of Technology, Hefei, 230601, China
[2]School of mechanical and electrical engineering, Anhui Jianzhu University, Hefei, 230009, China

[3]Corresponding author's e-mail: 2020213561@mail.hfut.edu.cn

Abstract. Based on the first-principles theory combining density-functional theory and non-equilibrium Green's function, A zigzag-armchair-zigzag (Z-A-Z) graphene nanoribbon transistor model was constructed using the quantumATK tool, and Stone-Wales (SW) defects were introduced. The transmission spectrum, electrical conductivity, gate potential and current-voltage (I-V) characteristics of Z-A-Z graphene nanoribbons were analyzed. By constructing Z-A-Z graphene nanoribbons with SW defects, the effect of defects on their transport properties was explored. The research results can provide a reference for the research on the characteristics of Z-A-Z graphene nanoribbons.

1. Introduction

Graphene is two-dimensional carbon atomic layer with hexagonal structure, in which the atoms are connected together by SP2 hybridization. Because of unique physical, chemical and electrical properties, it has very important research value and application prospects [1-3]. Graphene has attracted extensive attention in the field of molecular electronics research due to its excellent electrical and mechanical properties [4]. Graphene nanoribbons are also ideal materials for the preparation of nanoscale field effect tubes [5], and the simulation of their electron transport properties is of great significance.

According to the difference of edge shape, graphene nanoribbons (GNRs) can be divided into zigzag graphene nanoribbons (zGNRs) [6] and armchair graphene nanoribbons (aGNRs) [7], in which zGNRs are metallic, while aGNRs are semiconducting [8]. The properties of graphene can be improved by chemical modification and defects introduction.

There are many different kinds of defects in graphene nanoribbons, such as five-seven defects, mechanical deformation and other topological defects, which have a great impact on its electronic structure and transport properties. Among them, Stone-Wales (SW) defects are one of the important types of topological defects.

SW defects are crystallographic defects that commonly occur in graphene and can strongly change electronic, chemical, and mechanical properties. It is named after Anthony Stone and David Wales of Cambridge University [9], and Peter Thrower described similar defects in a paper on graphite defects [10]. In single-layer graphene, by rotating a carbon-carbon bond between two adjacent carbon atoms 90°, the four hexagons around it will become two heptagons and two pentagons, so that SW defects in graphene nanoribbons are formed.

Most of the current literature studies the effect of defects on the transport properties of Z-type or A-type graphene nanoribbons [11-13]. In practical applications, zigzag-armchair-zigzag (Z-A-Z) nanoribbons model has more practical significance.

Muhammad H. R. gave a model of Z-A-Z nanoribbons [14], but did not give a property analysis of the model. Amretashis S. explored the effect of SW defects on silicon, germanium, and phosphorus nanotubes [15], but not involving graphene nanoribbons. As a typical two-dimensional material, SW defects have an important impact on Z-A-Z graphene nanoribbon transistors, and there are few literatures on this aspect. This paper presents the construction and property analysis of Z-A-Z graphene nanoribbon transistor model, and explores the effect of SW defects on its electrical transport properties.

2. Construction of graphene nanoribbons and SW defects

2.1. Construction of Z-A-Z graphene nanoribbons

Here a special kind of graphene nanoribbon transistor is introduced whose metallic gate is separated from the ribbon by a dielectric. This transistor design was inspired by the work presented in Ref. [8]. QuantumATK developed by Atomistix company is used as the construction, calculation and analysis platform.

The main steps in the construction of Z-A-Z-type graphene nanoribbons are described as follows. The central region (armchair nanoribbons) and two electrodes (zigzag nanoribbons) are first constructed, then, combine the three components together to build a complete device.

One armchair nanoribbon and two zigzag nanoribbons are created. The former has an atomic width of 4 and is repeated 5 times along the C axis. The latter has an atomic width of 6 and is repeated 4 times along the C axis. From this, three components that make up Z-A-Z-type graphene nanoribbons are obtained (Figure 1a, b).

Connect the three components in Z-A-Z pattern. Rotate the armchair nanoribbons 30° along X axis and connect the assemblies end-to-end. The resulting structure does not have electrodes, so new electrodes need to be added. Among them, the left and right electrodes are added in "Device Tools" parameter, and the system will automatically select the appropriate electrode length, so just keep the default parameters unchanged.

Dielectric and metallic gates are added to the structure using "Spatial Regions" parameter. Two spatial regions are inserted into the structure, corresponding to the dielectric and metallic gates, respectively. Set the first space region as the dielectric region, and set the permittivity to 4.0, use fractional coordinates to define the boundary of the dielectric region, where the specific parameters are set as: a, 0.05-0.358 Å, b, 0-1 Å, c, 0.2679-0.7323 Å. If geometric coordinates are used, the parameters are defined as: a, 0.5-3.58 Å, b, 0-27 Å, c, 10.4593-26.4558 Å.

The second space region is added then, setting it as metallic gate and setting the gate potential to -1V. The size of the gate is also defined using fractional coordinates. The specific parameters are set as: a, 0-0.05 Å, b, 0-1 Å, c, 0.2679-0.7323 Å. If geometric coordinates are used, the parameters are defined as: a, 0-0.5 Å, b, 0-27 Å, c, 10.4593-26.4558 Å. In this paper, geometric coordinates are used in the following calculation. The resulting structure is shown in Figure 1c.

2.2. Construction of SW defects graphene nanoribbons

Firstly, a common nanoribbon structure is introduced. In order to simplify and speed up the calculation process, a very narrow band is chosen. Its chirality index is (2, 2), corresponding to a zigzag graphene nanoribbon, 8 carbon atoms wide, and other default parameters remain unchanged. Then, the resulting nanoribbon structure is repeated 12 times along Z axis, and two adjacent carbon atoms are selected and rotated 90° along X axis to introduce simple SW defects in the graphene nanoribbons (Figure 1d).

Two script blocks "TransmissionSpectrum" and "IVCurve" are added on the script writer in ATK. When setting parameters, the parameters in "TransmissionSpectrum" keep the default values, the bias voltage in "IVCurve" is set to V1=1V, and the energy range is set to E0=-1eV, E1=1eV, so as to facilitate the calculation of transmission spectrum and I-V characteristic.

Journal of Physics: Conference Series **2313** (2022) 012015 doi:10.1088/1742-6596/2313/1/012015

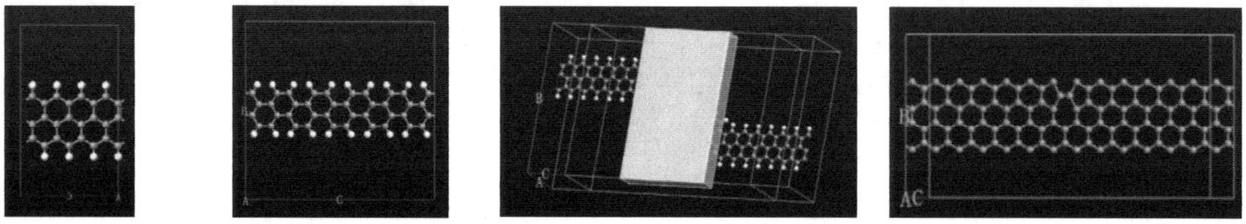

(a) (b) (c) (d)

Figure 1. (a) Zigzag nanoribbon, (b) armchair nanoribbon, (c) Final structure of Z-A-Z graphene nanoribbon transistor, (d) SW defects graphene nanoribbon.

3. Properties analysis of Z-A-Z graphene nanoribbons

3.1. Transmission spectrum and electrostatic potential

The calculation results can be viewed through main window. The resulting shows that this transmission spectrum has lower values in the energy range [−0.5, 1.5] eV, corresponding to another energy window within the band-gap of the central semiconducting armchair edge. The valence band edge in the central region is shifted towards the electrode Fermi level because the gate potential is -1V (Figure 2a).

In the corresponding contour plot, the effect of the gate potential on the electrostatic differential potential is visualized (Figure 2b).

3.2. Influence analysis of gate voltage and conductivity

The effect of gate potential is explored. When calculating the transmission spectrum, the gate voltage is -1V. Next, change the gate voltage from -2 to 2V and calculate the transmission spectrum corresponding to each gate voltage.

For each specific gate voltage, a new self-consistent calculation should be performed to obtain the corresponding transmission spectrum.

After calculation, it is found that the transmission spectrum of each specific gate voltage is very similar to the gate voltage of -1 V, and the main difference among the different spectra is the shift of the Fermi level at the edge of the valence and conduction band. In fact, this is also the main function of the metallic gate, which can be used to tune the position of the band edge relative to the Fermi level, thereby controlling the electrical properties of the device.

It can also be seen that all transmission spectra can be calculated without self-consistent cycles, just by shifting the Fermi level and using linear response theory. This also provides another efficient way to calculate the transmission spectrum at all gate voltages. Assuming that the effect of the gate potential is a simple shift in the relative position of the Fermi level, the function is calculated based on the linear response (Figure 2c).

The temperature distribution of this graph is from 0K to 2000K, and step size is 100. As can be seen graphically, the linear response process reproduces the main trend of a fully self-consistent calculation.

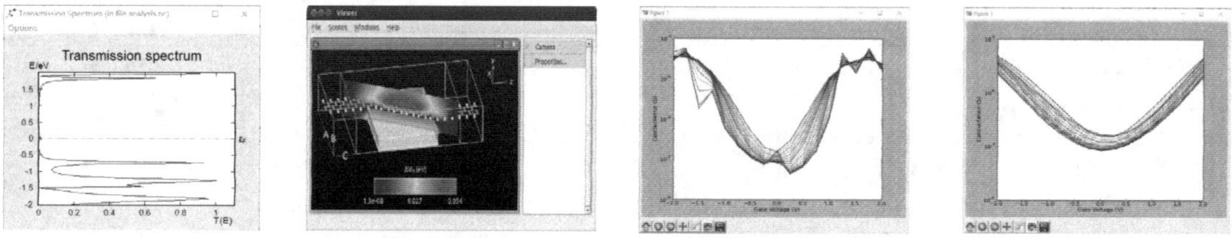

(a) (b) (c) (d)

Figure 2. (a) Simulation of transmission spectrum, (b) Simulation of contour plot, (c) Linear response calculation results, (d) Gate potential function.

The conductivity of the structure is next calculated as a function of gate potential at different electrode temperatures.

The functional relationship between conductivity and gate potential is shown (Figure 2d). Different curves are for different electron temperatures, i.e. 0, 100, 200, ..., 2000K. The minimum gate potential for this conductivity is about 0.25V. At this time, the Fermi level is located in the gap between the valence band and the conduction band.

Next step, use the transmission spectrum to calculate the device conductivity and the electrodes temperature. At a finite electrode temperature, the conductivity is determined by the transmission coefficient of the energy window near the Fermi level [16]. The conductivity at zero bias is given by the following equations.

$$T_L T_R T_L = T_R = 0 \tag{1}$$

$$\sigma(T_L) = \frac{2e^2}{h} \int T(E) f'(\eta(E)) \frac{dE}{k_b T_R} \tag{2}$$

In equation (2), the derivative of the Fermi function is:

$$f'(\eta(E)) = f'\left(\frac{E - E_F^L}{k_b T_R}\right) \tag{3}$$

Where T_L and T_R are temperature of the left and right electrodes, E the energy level of the electron, h Planck's constant, σ conductivity, *T(E)* represents the projection coefficient of the system at the energy of E, $f(\eta(E))$ is Fermi function.

In semiconductors, electrical conductivity is usually determined by hot electrons or holes propagating within the conduction or valence band, so-called thermionic emission. Hot electrons are located in the energy range of the tail of the Fermi function. Therefore, in order to accurately calculate the conductivity of thermionic emission, the energy range of the transmission spectrum needs to be properly sampled. The transmission spectrum plot above shows that both the valence and conduction band edges in the central region are in the energy range of the transmission spectrum (Figure 2a). Therefore, from this data, we are able to accurately determine the temperature-dependent conductivity (Figure 3a).

Temperature dependent conductivity for this device is typical for a field effect transistor. The initial decrease in conductivity as a function of temperature is related to a large conductivity at the Fermi level caused by electron tunneling. At larger temperatures, the usual increase of conductivity with temperature starts dominating.

3.3. *I-V characteristic analysis*

3.3.1. Comparison of SW defects and perfect nanoribbons. After calculation incorporating SW defects, Figure 3b shows the transmission spectrum and corresponding I-V points for all biases within the set range. Each point on the I-V curve corresponds to a transmission spectrum. In order to calculate the current, it is necessary to specify the temperature of the electrons in the electrodes, and if the electron temperature changes, the current is to be recalculated. The I-V points are connected by spline interpolation, and the distances among the interpolation points are determined by the step size. The figure shows the dI/dV curve obtained by differential spline interpolation. For systems with anti-symmetry or z-direction mirror symmetry, the I-V curve will be symmetric in the applied bias. For such systems, performing an I-V sweep in one bias direction can obtain a complete I-V curve. If "Symmetrize" parameter in the figure is checked, positive bias points will be copied, resulting in negative bias points as well.

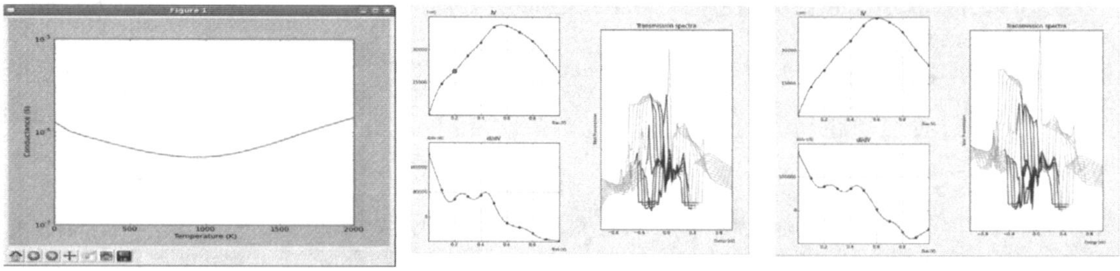

(a) (b) (c)

Figure 3. (a) Conductivity as a function of temperature, (b) Simulation of SW defects, (c) Simulation of perfect graphene nanoribbons.

The same calculation process is performed for perfect graphene nanoribbons (Figure 3c). For the convenience of comparison, the width of the nanoribbon is set to be the same as that of SW defects graphene nanoribbons. By comparison, it is found that the characteristic curves of both SW defects and perfect graphene nanoribbons have the same trend. For the I-V characteristic curve, the former is always slightly lower than the latter, while for the dI/dV-V curve, there is a clear enhancement of the former. This shows that the existence of SW defects plays a key role in regulating the electrical transport of graphene nanoribbons. Under the same bias voltage, the current of graphene nanoribbons with SW defects becomes smaller, but does not hinder the transport of electrons, and significant gain in conductivity occurs. The figure also shows that the current of the two forms of nanoribbons continuously increases from 0 to 0.6V and decreases from 0.6 to 1V.

3.3.2. I-V characteristic curves of different bias voltages. The I-V characteristic curve was plotted by varying the external bias voltage in the range of 0-2V. The applied gate voltage for this calculation is -1V, and the gate voltage value remains unchanged throughout the calculation process. The variation of the bias voltage appears as the variation of the source-drain electrode voltage.

Figure 4a shows the electrostatic differential potential at a gate voltage of -1V and a bias voltage of 1V, i.e. -0.5V for the left electrode and +0.5V for the right electrode.

From the transmission spectrum at each bias voltage, I-V curves for the -1V gate potential are calculated (Figure 4b).

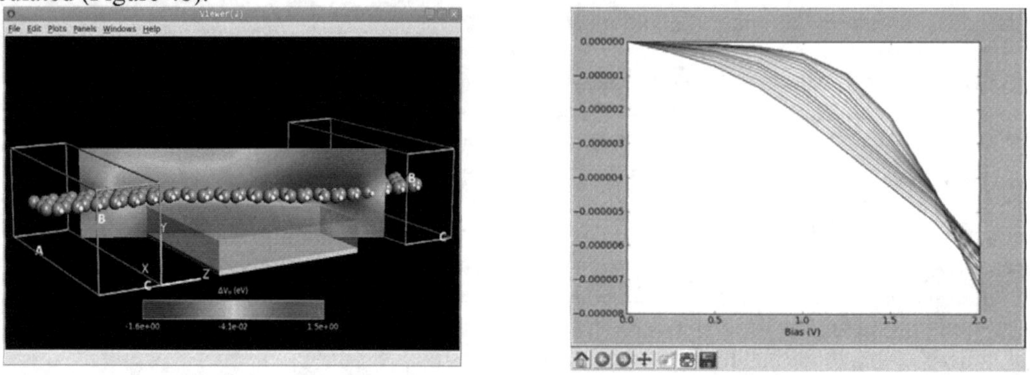

(a) (b)

Figure 4. (a) Electrostatic potential, (b) I–V curve (gate potential -1 V).

This curve shows the self-consistent current-voltage characteristics of Z-A-Z graphene nanoribbons when the gate voltage is -1V. It can be seen that as the temperature increases, the current increases. As the bias voltage approaches 2V, the currents tend to be consistent.

3.3.3. I–V characteristics of linear response. Linear response results are compressed into self-consistent response results. This self-consistency is quite good for low bias voltages, but at high bias voltages the linear response current begins to deviate from the self-consistent current (Figure 5a).

For a gate voltage of -1V and an electrode temperature of 300K, the blue curve is the self-consistent response, and the green curve is the linear response (Figure 5b).

The above calculation method is based on the extended Huckel formalism theory, and the calculation amount is large and time-consuming. An alternative is to invoke the linear response approximation in and use non-self-consistent calculations. This calculation method is much faster than that of self-consistent.

This linear response calculation way shows that the linear response works well at relatively small electrode bias voltages, but for smaller gate voltages, the two calculations tend to agree, suggesting that the linear response current is a good first approximation.

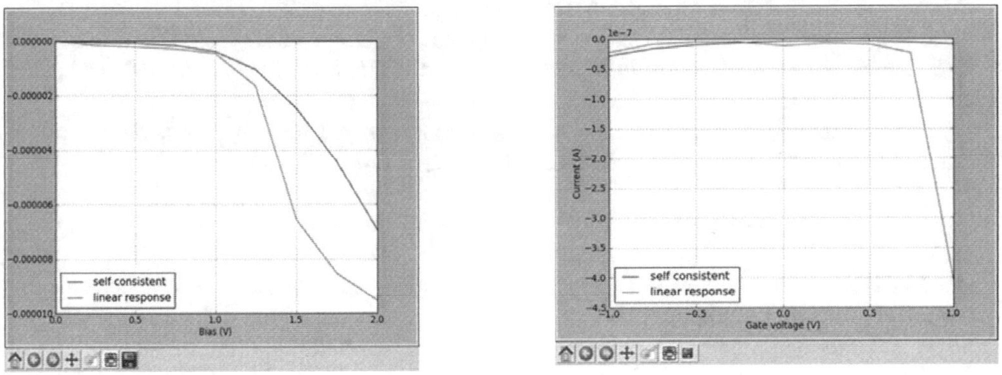

(a) (b)

Figure 5. (a) Linear and self-consistent response results, (b) Self-consistent and linear response current.

3.4. Discussion of results

The design and construction of the Z-A-Z graphene nanoribbon model have some key potential advantages. Due to the small contact area, there is usually a large contact resistance between the metal electrode and the molecule, so it is usually difficult to find a reliable method to make the electrode contact with the molecular device unit. This difficulty can be solved by using Z-A-Z graphene nanoribbons. Because devices based on this structure can be connected to external circuits only through the zGNRs, used as extensions of the metal electrodes to make contact with the semiconductor GNRs and to make the metal-semiconductor interface maintains minimal contact resistance. The metallic zGNRs at the poles of the model acted as a transition. On the other hand, graphene nanoribbons with SW defect can induce more additional conductive channels, which can significantly enhance the electrical conductivity of graphene nanoribbons, making their currents smaller than that of perfect graphene nanoribbons.

4. Conclusion

Using the first-principles method combining density functional theory and non-equilibrium Green's function, Z-A-Z graphene nanoribbons are selected as the research object, Z-A-Z graphene nanoribbons are constructed, and SW defects are introduced, which are calculated and analyzed by ATK software. The characteristic parameters such as transmission spectrum and IV curve of this defect nanoribbon are obtained. These provide an effective way for the regulation and design of the electrical transport properties of graphene molecular devices, and provide a design idea for constructing ideal molecular devices through defect design.

Acknowledgement
This work has been financially supporting by Innovation Training Program of National College Student (Grant No. G202110359082).

References

[1] Nnovoselov K S, Gelm A K and Morozov S V 2004 Electric field effect in atomically thin carbon films. *Science*, 306(5696:666-9), DOI: 10.1126/science.1102896.
[2] Rodgers P 2010 Nanomechanics: Welcome to the quantum ground state. *Nature Nanotechnology*, 5(4): 245, DOI: 10.1038/nnano.2010.70.
[3] Young-Woo S, Marvin L and Steven G L 2006 Half-metallic Graphene Nanoribbons. *Nature*, vol. 444, no. 7117, 347-49.
[4] Ahmed S and Yi J 2017 Two-dimensional transition mental dichalcogenides and their charge carrier mobilities in field-effect transistors. *Nano-Microletters*, 9(4): 50-52.
[5] Azad N and James D M 2009 Compact Physics-based Circuit Models for Graphene Nanoribbon Interconnects. *IEEE Transactions on Electron Devices*, vol. 56, no. 9, 1822-33, DOI: 10.1109/TED.2009.2026122.
[6] An Y and Yang Z 2011 Abnormal electronic transport and negative differential resistance of graphene nanorib-bons with defects. *Applied Physics Letters*, 99(19): 192102, DOI: 10.1016/j.reactfunctpolym.2014.03.006.
[7] Son Y W, Cohen M L and Louie S G 2006 Half-metallic graphene nanorib-bons. Nature, 444(7117): 347-349, DOI: 10.1038/nature05180.
[8] Liu Y, Xia C J and Zhang B Q 2018 Effect of chemical doping on the electronic transport properties of tailoring graphene nanoribbons. *Chinese Physics Letters*, 35(6): 067101.
[9] Stone A J and Wales D J 1986 Theoretical studies of icosahedral c60 and some related species. *Chemical Physics Letters*, 128(5-6): 501-03.
[10] Thrower P 1969 The study of defects in graphite by transmission electron microscopy. *Chemistry and Physics of Carbon*, 5: 217-320.
[11] Wind S, Appenzeller J J and Martel R 2002 Vertical scaling of carbon nanotube field-effect transistors using top gate electrodes. *Appl Phys Lett*, 80(20): 3817-19.
[12] Horiuchi S, Gotou T and Fujiwara M 2004 Single graphene sheet detected in a carbon nanofilm. *Applied Physics Letters*, 84(13): 2403-05.
[13] An Y P and Yang Z Q 2011 Abnormal electronic transport and negative differential resistance of graphene nanoribbons with defects, *Applied Physics Letters*, 99(19): 192102, DOI: 10.1063/1.3660228.
[14] Muhammad H R, Ants K and Toomas R 2020 Simulations of Graphene Nanoribbon Field Effect Transistor for the Detection of Propane and Butane Gases: A First Principles Study. *Nanomaterials,* 1-17, DOI: 10.3390/nano10010098.
[15] Amretashis S 2016 Atomistic simulation of transport properties of nongraphitic armchair nanotubes and effect of Stone Wales defects. *Simulation of Semiconductor Processes and Devices*. pp: 97-101.
[16] Ravinder K and Derick E 2015 Anatomizing the Impact of High Dielectric Gate Materials on the Charge Transport in Graphene Field Effect Transistors. *Recent Advances In Nano Science And Technology,* 1933-39.

Journal of Physics: Conference Series **2313** (2022) 012016 doi:10.1088/1742-6596/2313/1/012016

Cascade integration of nonlinear phenomena exhibited by monometallic nanoparticles

Soledad D. Estrada-León

Centro de Estudios Superiores del Sureste, Mérida, 97000, Mexico

Corresponding author's e-mail: soledad.estrada@cess.edu.mx; soledad_estrada@hotmail.com

Abstract. In this work were numerically analyzed the third-order nonlinear optical effects exhibited by combinations of uncoupled monometallic silver, gold and platinum nanoparticles. The nanosystems were considered to be homogeneous and surrounded by a dielectric matrix. The design of the nanostructures was evaluated as a collective behavior resulting from the specific order of the set in trilayer tandem form. The characteristic Localized Surface Plasmon Resonance phenomena in the yellow, green and ultraviolet absorption bands for the Ag, Au and Pt; respectively, were considered. Different permutations of the three kind of monometallic elements selected in sequence were studied by using the wave equation and the finite difference method. Immediate indications useful for exploring nonlinear nanodevices and for building up nanostructured platforms based on nanoscale functions can be anticipated from this work.

1. Introduction

Combinatorial concepts in the research of material science represent a widely used tool as a strategy to develop advanced compounds and evaluating active materials [1]. For studying chemistry and physics of materials, there have been many efforts to describe the resulting effects derived by the integration of diverse items in a set [2]. Also the complexity associated with the same elements in different order as a permutation has been extensively studied in different areas of research [3]. Particularly, in nonlinear optics, the collective behavior emerging from the influence of anisotropy makes possible the design of multi-functions that can be remarkable for signal processing [4], encryption [5] or decision making operations [6].

The structure and morphology at the nanoscale are responsible for fascinating phenomena like the Localized Surface Plasmon Resonance (LSPR), quantum confinement in discrete states and the sharp-selective optical response exhibited by metallic nanoparticles [6]. Nonlinear optical properties can be tailored by nanoparticle size [7], entropy [8], mechanical stress [9], plasmonics [10], morphology [11], distribution [12] or structure [13]. Permutation of processes in nonlinear optical systems dependent on LSPR can be tailored to derive in a dramatic increase in propagation modes with potential applications for developing all-optical switching devices [14]. Remarkably, noble metal nanoparticles have been pointed out with key factors of tandem processes and good candidates for designing optical platforms with ultrafast response in different regions of the electromagnetic spectrum. Gold and silver nanoparticles present their LSPR in the visible region while platinum displays LSPR for more powerful photons in the UV [15].

Considerable attention in the dynamical systems for building up all-optical functions has been undertaken by synthetizing randomized nanocomposites [16]. However, controlled approaches with specific distributions including nanoparticles promote a significant transformation with an enhancement of their nonlinear optical response [17]. Furthermore advantages in the proper selection of surrounding media encapsulating metallic nanoparticles for engineering particular nonlinear optical parameters have been reported [18]. A reinforcement of multi-photonic interactions for absorption and refraction by different combinations of nanoparticles involved in tandem processes seems to be attractive as basic strategies to yield complex nonlinearities and it is the main topic of this work.

2. Materials and methods

For simplicity of the analysis, steady-state samples with no evolution of physical and chemical properties by aging or optical damage under optical irradiation were considered. Ion-implanted samples were selected with well-controlled homogeneous distribution of monometallic nanoparticles. Individual SiO_2 substrates separately bombarded by Ag, Au and Pt ions were analyzed. The samples can be described by nanoparticles with 7.5 nm average size and nanoparticle density close to 4×10^4 particles/μm^2. The estimated optical path for the ion-implanted distribution was approximated to 250 nm in all cases and the geometric length of the SiO_2 was 1 mm. In order to explore the magnitude of the third-order optical nonlinearities exhibited by the combination of the different samples, a multi-wave mixing experiment with two incident beams was numerically analyzed. The trajectory, irradiance and phase wavefront of the interacting beams can be studied by the wave equation [19] and the finite difference method [20]. The incident beams in the samples were linearly polarized at 532 nm wavelength. The incident irradiance of each beam was 250 MW/cm^2 with a diameter of the incident beam equal to 1.5 mm focused along the samples and nanosecond pulses with 5 ns full width half maximum. Here, a monochromatic 532 nm light source is considered. A degenerated two-wave mixing can be obtained by the superposition of two optical coherent beams with the same wavelength. It results in a modulation of optical irradiance by a fringe pattern that derives in a change of the polarization of the incident beams by an induced birefringence that is governed by the optical Kerr effect.

The Mie theory and the Fresnel-Kirchhoff diffraction integral were considered to guarantee that the distribution of electric fields involved in a mixing of two coherent waves with spatially uncoupled and independent monometallic nanoparticles [21],

$$U(x,y,z)=\frac{1}{i\lambda}\iint_S U(x_0,y_0)\left[\frac{\cos(\vec{n},r)-\cos(\vec{n},r')}{2}\right]\frac{e^{i\kappa r}}{r}ds \tag{1}$$

where λ represents the optical wavelength and $\kappa = 2\pi/\lambda$ corresponds to the wavevector. It was considered $U(x_0, y_0)$ for describing the amplitude distribution at the position S with a normal direction $\vec{n}$ at z = 0, r and r' are the vectors at the plane z = 0 and z; respectively.

The propagation of the travelling waves and their transmittance was calculated by considering the vectorial form of the wave equation [19]:

$$\nabla^2 E_\pm = -\frac{n_\pm^2\omega^2}{c^2}E_\pm \tag{2}$$

with ω as the angular optical frequency, c as the parameter describing the speed of light, and n is the index of refraction dependent on irradiance $n=n_0+n_2I$, with n_0 the refractive index at low irradiance, n_2 the nonlinear refractive index and I the optical irradiance. The real part of the third-order nonlinear optical susceptibility $\chi^{(3)}$ can be represented by the optical Kerr effect responsible for nonlinear refraction, mathematically described by n_2. A negative sign in the nonlinear refraction index, n_2, corresponds to a self-defocusing effect; while a positive n_2 gives origin to a self-focusing phenomenon. The imaginary part of the $\chi^{(3)}$ is associated with nonlinear optical absorption that is manifested as a two-photon absorption process or an induced transparency dependent on incident irradiance.

The optical irradiance, *I(L)*, transmitted in samples with nonlinear optical absorption coefficient β through a distance L, can be described as a function on incident optical irradiance I_0 as:

$$I(L) = \frac{I_o \exp(-\alpha_o L)}{1+\beta I_o L_{eff}}, \tag{3}$$

the magnitude of the optical absorption coefficient is with α_0, and the effective length L_{eff} is expressed as:

$$L_{eff} = \frac{\left(1-\exp\left(-\alpha_0 L\right)\right)}{\alpha_0} \tag{4}$$

The magnitude of the third order optical susceptibility $|\chi^{(3)}|$, β, and n_2 employed in the numerical simulations are presented in Table 1. The magnitudes for the nonlinear refractive index and nonlinear absorption coefficient for comparative samples studied by a two-wave mixing method have been previously reported [22-24]. The data in Table 1 correspond to experimental results reported in the cited references mentioned within the rows.

Table 1. Magnitude of the nonlinear parameters exhibited by the monometallic samples experimentally measured at 532 nm in comparative ion-implanted samples.

Sample	n_2 [m^2/W]	β [m/W]	$\lvert\chi^{(3)}\rvert$ [esu]	Reference
Au	-2.8×10^{-15}	-6.43×10^{-9}	2.2×10^{-9}	[22]
Ag	2.2×10^{-17}	5×10^{-11}	1.75×10^{-9}	[23]
Pt	1.9×10^{-15}	8.28×10^{-8}	3.17×10^{-9}	[24]

3. Results and discussion

In Figure 1(a) are presented the numerical data for the UV-vis spectra exhibited by the monometallic samples. Mie scattering formula in calculations have been employed in order to properly describe the absorbance spectra considering the quasi-spherical size of the nanoparticles studied. Rayleigh scattering can be used for describing the optical behavior of quasi-spherical nanoparticles. However, for simplicity the electrostatic approximation related to Mie theory also fits with the optical absorbance exhibited by plasmonic nanoparticles, and it was used for presenting the data plotted in Figure 1(a). In Figure 1(b) are represented three cases of different kind of permutations of the nanoparticles in sequence. Figure 1(b) illustrates a scheme with the yellow, blue and red fields as the LSPR for the Au, Pt and Ag nanoparticles. The propagation vectors of the incident beams are k1 and k2; while the angle between the parallel plane of incident linear polarizations of the electric field vectors E1 and E2 is cero. Evidently, according to Eq. (1), different excitations are expected for the three different arrangements integrated by dissimilar samples.

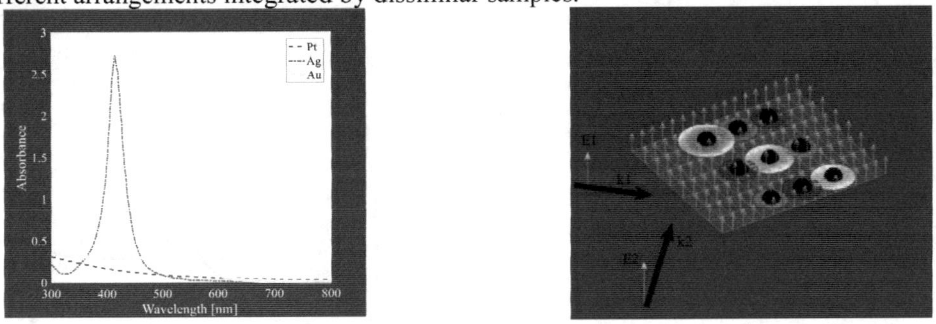

(a) (b)

Figure 1. (a) UV-vis spectra of the monometallic nanosystems studied. **(b)** Schematic representation of three different systems of nanoparticles with the expected field of their characteristic LSPR.

From Eq. (2) and the parameters shown in Table 1 was estimated the nonlinear refraction and absorption in the propagation of a two-wave mixing with a total irradiance of 500 MW/cm^2 and an irradiance relation 1:1 for the pump and probe beams. In Figure 2(a) are plotted the transmitted irradiance as a function of the incident optical intensity. Figure 2(b) illustrates the evolution of the nonlinear optical response of the triple-stage samples in their possible permutations.

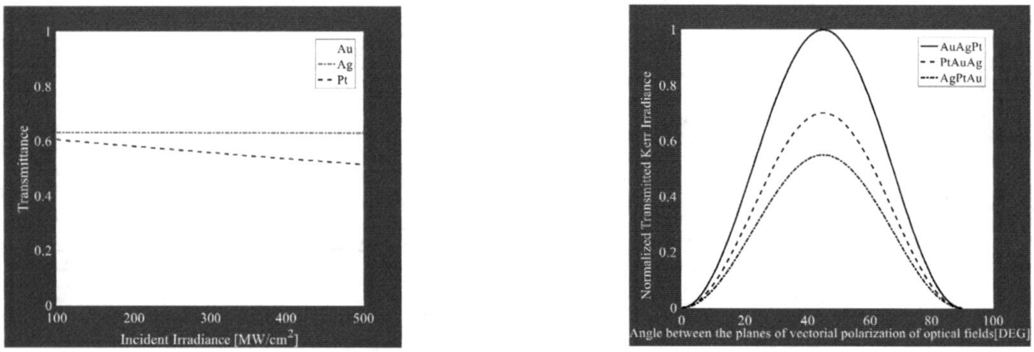

(a) (b)

Figure 2(a). Transmittance of a single-beam mode. **(b)** Normalized transmittance for a two-wave mixing of the sample.

From the data plotted in Figure 2(b), it can be deduced that the major contribution for obtaining a Kerr transmittance is given by the Au and the sequence of the three different nanoparticles presents a higher transmitted irradiance when the Au receives first the incident light. On the other hand, the maximum quenching of the transmitted irradiance is related to the nonlinear optical absorption associated with the Pt, it is clear that a stronger reduction of the transmitted irradiance can be obtained when the Pt receives first the incident light in the sequence of the three kind of nanoparticles.

For further describe the change in the nonlinear response of the samples, it is considered the variation in the refractive index [25]:

$$n_2 \approx \frac{4}{3} N m^4 \left[\frac{1}{\hbar \Delta} \right]^3 \frac{T_1}{T_2} \frac{7.91 \times 10^2}{n_0 c} \tag{5}$$

The Planck constant is $\hbar$, the detuning is Δ, and the atomic dipole moment is m. The numerical representation of the two-level atoms per unit volume in the nanostructures is N, while the magnitude describing the evolution of the two-level system from a dynamic aspect is $1/T_2$ with a relation of the rate of polarization loss for the off-diagonal matrix elements. On the other hand, the population loss through processes of radiative and non-radiative energy of the upper quantum level of excitation is $1/T_1$. In Figure 3 is shown a normalized variation of n_2 with no significant influence of multi-photonic absorption in the studied case for a population change in the excited nanoparticles by high irradiance in the three stage nanoparticles integrated by Pt, Ag and Au systems.

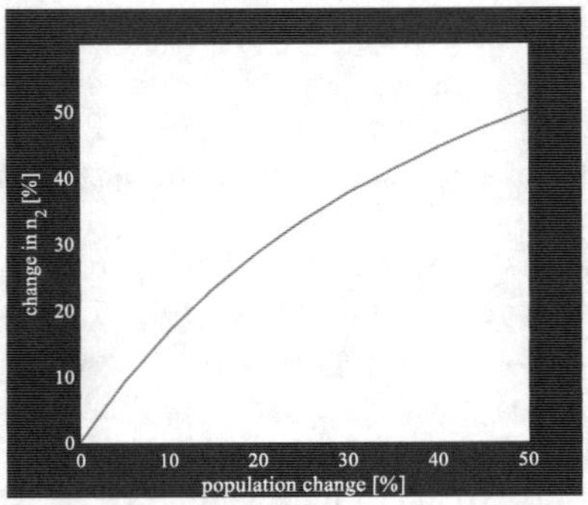

Figure 3. Calculated change in the nonlinear refractive index vs. population change in the monometallic systems studied.

Experimental verification of changes in nonlinear refractive index derived by the multi-metallic nanoparticles [26] or changes in ordered nanocomposites [27] have been previously reported and those findings are in good agreement with the results presented in this report.

4. Conclusions

The potential tuning and tailoring of nonlinear effects by a strategic distribution of nanoparticles is highlighted in this report. The combination of different third-order optical nonlinearities was analyzed to better present fundamental concepts related to the origin of unique multi-photonic effects exhibited by ultrafast photonic systems assisted by a contrast of nanoscale elements. Experimental data provided by multi-wave mixing measurements were considered for conducting numerical simulations in miniaturized samples. Uncoupled optical phenomena were initially contemplated to describe the importance in the selection of particular combinations of nanoobjects with different plasmonic response in 2D arrays. In order to understand the root of enhancement or inhibition of the electronic mechanisms responsible for the nonlinear refraction and nonlinear absorption, the influence of the vectorial effects dependent on optical irradiance was calculated in monometallic nanoparticles in integrated configurations. A sharp selectable dynamics in the performance of nonlinear optical properties can be suitable for building up functional all-optical devices and advanced quantum functions.

Acknowledgments

The author kindly acknowledges the financial support from the Graduate Study Program in Mathematics at the Escuela Superior Normal de Yucatán and Instituto Politécnico Nacional.

References

[1] Ludwig, A. 2019 Discovery of new materials using combinatorial synthesis and high-throughput characterization of thin-film materials libraries combined with computational methods *npj Comput Mater* **5** 70.
[2] Jiang, J., Chen, M. and Fan, J.A. 2021 Deep neural networks for the evaluation and design of photonic devices *Nat Rev Mater* **6**, 679-700.
[3] Tao, H., Wu, T., Aldeghi, M. et al. Nanoparticle synthesis assisted by machine learning Nat Rev Mater 6, 701–716 (2021).
[4] Bradshaw D. S., Forbes K. A. and Andrews D. L. 2019 Off-Resonance Control and All-Optical Switching: Expanded Dimensions in Nonlinear Optics *Appl. Sci.* **9** 4252.
[5] Yabin Zhang, Xiong Zhang, Mingguang Shan, Zhi Zhong, Bin Liu, Lei Yu, and Lei Liu, 2022 Asymmetric double-image encryption via wavelength multiplexing *Appl. Opt.* **61** 1248-1253.
[6] Chen W. 2022 Spatial nonlinear optics for securing information *Light Sci Appl* **7** 11.
[7] Karimarji, S., Sabouri, S.G. and Khorsandi, A. 2022 Nonlinear optical properties of size-

variable silver nanoparticles dispersed in water and potassium bromide Opt Quant Electron 54, 133.

[8] Meiqi Hao, Yuhua Wang, Ding Chen, Dulan Wu, Peiling Li, Haijun Zhang 2022 Ion implantation synthesis of long-term stable high-entropy metallic glass nanoparticles J Alloys Compd 906, 164303.

[9] Gutiérrez-Menchaca, J, Garay-Tapia, A M, Torres-Torres, D, Arizmendi-Morquecho A. M., Leyva-Porras C., Torres-Torres C., Oliver A. 2021 Enhanced fracture toughness of silica glass by ion-implanted platinum nanoparticles Fatigue Fract Eng Mater Struct 44, 1423–1438.

[10] Shahid Iqbal, Masoud Shabaninezhad, Mohammad Hatshan, Prashanta M. Niraula, Abubaker Abuhagr, Hasna Alali, Ramakrishna Guda, and Asghar Kayani 2018 Ion-implanted silver nanoparticles for metal-enhanced fluorescence AIP Adv 8, 095217.

[11] Yu-xi Zhanga and Yu-hua Wan 2017 Nonlinear optical properties of metal nanoparticles: a review. RSC Adv 7, 45129-45144.

[12] Panoiu N. C., Sha W. E. I., Lei D.Y. and Li G.-C. 2018 Nonlinear optics in plasmonic nanostructures J. Opt. 20 083001

[13] Gavrilenko V. I., Murzina T. V., Mizutani G. 2012 Nonlinear Optics of Nanostructures Phys Res Int 2012, 648758.

[14] Qi Wang, Zi-Han Ren, Wan-Ming Zhao, Lei Wang, Xin Yan, Ai-song Zhu, Feng-mei Qiu and Ke-Ke Zhang 2022 Research advances on surface plasmon resonance biosensors *Nanoscale* **14** 564-591.

[15] Whiting E. B., Goldflam M. D., Kang L., Sinclair M. B., Musick K. M., Campbell S. D., Burckel D. B. and Werner D. H. 2022 Broadband Asymmetric Transmission of Linearly Polarized Mid-Infrared Light Based on Quasi-3D Metamaterials *Adv Func Mat* 2109659.

[16] Torres-Torres D., Trejo-Valdez M., Castañeda L., Torres-Torres C., Tamayo-Rivera L., Fernández-Hernández R. C., Reyes-Esqueda J. A., Muñoz-Saldaña J., Rangel-Rojo R., and Oliver A. 2010 Inhibition of the two-photon absorption response exhibited by a bilayer TiO2 film with embedded Au nanoparticles *Opt. Express* **18** 16406-16417.

[17] Shangjr Gwo, Chun-Yuan Wang, Hung-Ying Chen, Meng-Hsien Lin, Liuyang Sun, Xiaoqin Li, Wei-Liang Chen, Yu-Ming Chang, and Hyeyoung Ahn 2016 Plasmonic Metasurfaces for Nonlinear Optics and Quantitative SERS *ACS Photonics* **3** 1371-1384.

[18] Trejo-Valdez M., Torres-Martínez R., Peréa-López N., Santiago-Jacinto P. and Torres-Torres C. 2010 Contribution of the two-photon absorption to the third order nonlinearity of Au nanoparticles embedded in TiO2 films and in ethanol suspension *J. Phys. Chem. C* **114** 10108-10113.

[19] Robert W B 1992 *Nonlinear Optics* vol 1, ed Academic Press p 456

[20] Reddy, J. N. 2006 *An Introduction to the Finite Element Method* (Third ed.). McGraw-Hill.

[21] Eugene H 2017 *Optics* vol 5, ed Pearson, p 532-535

[22] Torres-Torres C., López-Suárez A., Can-Uc B., Rangel-Rojo R., Tamayo-Rivera L. and Oliver A. 2015 Collective optical Kerr effect exhibited by an integrated configuration of silicon quantum dots and gold nanoparticles embedded in ion-implanted silica *Nanotechnol* **26** 295701.

[23] Torres-Torres C., Tamayo-Rivera L., Rangel-Rojo R., Torres-Martínez R., Silva-Pereyra H. G., Reyes-Esqueda J. A., Rodríguez-Fernández L., Crespo-Sosa A., Cheang-Wong J. C. and Oliver A. 2011 Ultrafast optical phase modulation with metallic nanoparticles in ion-implanted bilayer silica *Nanotechnol* **22** 355710.

[24] Torres-Torres C., Bornacelli J., Can-Uc B., Silva-Pereyra H. G., Rodríguez-Fernández L., Avalos-Borja M., Labrada-Delgado G. J., Cheang-Wong J. C., Rangel-Rojo R. and Oliver A. 2018 Coexistence of two-photon absorption and saturable absorption in ion-implanted platinum nanoparticles in silica plates *J. Opt. Soc. Am. B* **35** 1295-1300.

[25] Hurtado-Aviles E. A., Vila M., Vilatela J. J., Martines-Arano H., Bornacelli J., García-Merino J.

A., Cervantes-Sodi F. and C. Torres-Torres 2022 Structured light using carbon nanostructures driven by Kerr nonlinearities and a magnetic field *Phys Chem Chem Phys* **24** 1081.

[26] Niv A. Unique Nonlinear Optical Response from a Deep Subwavelength Bi-Metal Dimer, in OSA Nonlinear Optics 2021, R. Boyd, C. Conti, D. Christodoulides, and P. Rakich, eds., OSA Technical Digest (Optica Publishing Group, 2021),

[27] Michieli N., Balasa I. G., Kalinic B., Cesca T. and Mattei G. 2020 Optimal geometry for plasmonic sensing with non-interacting Au nanodisk arrays *Nanoscale Adv* **2** 3304-3315.

Anticorrosive behavior assessment of electrodeposited zinc salicylate coating on carbon steel

Ronghui Yuan[1], Yan Li[2, 4], Xiangyu Ye[1], Furong Liu[1], Qiping Tian[3] and Qiwang Wu[1]

[1]Zhejiang Light Industrial Products Inspection and Research Institute, Hangzhou 310018, China;
[2]Hangzhou Maysun Biotechnology Co., Ltd, Hangzhou 310030, China;
[3]Zhejiang Institute of Mechanical & Electrical Engineering Co., Ltd, Hangzhou 310051, China.

[4]Corresponding author's e-mail: liyan9664@163.com

Abstract. The study focuses on the evaluation of a novel composite coating obtained from zinc salicylate (ZnSal). Ion distributions of salicylic acid and zinc(II) were surveyed at various acidities, and pH 9.5 was selected for the direct current electrodeposition of ZnSal/K_2SO_4 on carbon steel substrate. The surface morpholgy of the ZnSal-steel coating was revealed by scanning electron microscopy, and compact deposited films with dendrite structures (specific surface area 247 m^2 g^{-1}, type-H3 BET hysteresis loop at relative pressures from 0.4 to 1.0) could be formed within 25-min plating time. NSST was conducted, and the test duration was 120 h. Preliminary assays were employed to assess its anticorrosive performance in salt spray, and the corrosion rate of the steel plates exhibited a nonlinear relationship with the exposure time. The results showed that the ZnSal-steel had a low corrosion rate <0.11 μm h^{-1}, thereby increasing the potential of ZnSal composites on actual carbon steel for anticorrosive coatings.

1. Introduction

Recently, the superiority of zinc electrodeposition is essential due to its practical applications, including protection and modification [1−3]. General scenarios and mutual importance of steel corrosion resistance are applied to building work [4,5] and ship transportation [6,7]. The standard electrode potential of zinc is −0.76 V. Galvanised layer on iron and steel protects them from chemicals [8−11]. Zinc coatings in the field of corrosion resistances have been popularized in production lines [12−14] and zinc-steel is a well-practised system because of its enhanced mechanical properties, including functional alloy deposit, highly porous morphology and anticorrosive coating [15−18].

Zinc salicylate (ZnSal) is regarded as an alternative antifouling coating that is employed to prevent fouling [19]. Moreover, high-quality steel should conquer corrosion [20,21], erosion [22] and fouling [23]. However, the slow deposition rate and large amounts of sediment restrict the good corrosion resistance of carbon steel as well as the difficult morphology control, expensive cost and incompact deposition parameters, leading to poor anticorrosion effects on steel. The influences of complex selection and pH value on zinc-containing electrodeposition are rarely reported, particularly for the ZnSal electrodeposition on carbon steel (ZnSal-steel).

The aim of this work is to develop an effective anticorrosive method on carbon steel substrate. Herein the ligand structure is an important factor, and ZnSal (salicylate ligand has a large π bond) mixed with

K_2SO_4 is used as the supporting electrolyte. The ion distributions of salicylic acid and Zn(II) were specifically investigated, and the specific surface area (*S*) of ZnSal was determined using BET theory. The morphology of as-prepared ZnSal and ZnSal-steel was characterized by scanning electron microscopy (SEM) and transmission electron microscopy (TEM) separately. The anticorrosive performance of electrodeposited ZnSal coatings was evaluated by neutral salt spray test (NSST).

2. Materials and procedure

2.1. Reagents and instruments

Carbon steel (steel grade: 20, R_m: 430 MPa, $R_{P0.2}$: 245 MPa, elongation: 21%) was purchased from Runlu Steel (China). The mass fraction of chemical compositions for the used carbon steel was listed hereafter: C (0.22%), Si (0.18%), Mn (0.37%), Fe (>96%). Salicylic acid (purity ≥99%), zinc sulphate (purity ≥99%), potassium sulphate, potassium hydroxide, hydrochloric acid, ethanol and sodium chloride were purchased from Leyan Agentia (China). Deionized water was prepared with a Dura Pro 12FV water purifier (China).

Apparatuses including a Zygo white light interferometric profiler (China), a JEM-Flash transmission electron microscope (Japan), a Hitachi SU3500 scanning electron microscope (Japan), an ASAP 2020Plus automatic surface and porosity analyzer (USA), an AOTSI-90H salt spraying machine (China), a KST CS1008 potentiostat (China) and a Mettler SevenMulti acidometer (Switzerland) were used.

2.2. Preparation of ZnSal and electrodeposition for ZnSal-steel coatings

The following steps were taken to fabricate ZnSal. First, 64 mL $ZnSO_4$ (3.0 mol L^{-1}) and 100 mL KOH (1.5 mol L^{-1}) solutions were poured into a beaker and stirred to acquire $Zn(OH)_x$. Next, 110 mL salicylic acid (1.0 mol L^{-1}) and 50 mL $Zn(OH)_x$ were placed on a stirrer. The liquid acidity was regulated to 9.5 using dilute KOH or HCl. The water-bath temperature was maintained at 70 °C for 2 h to acquire the ZnSal.

The mirror-polished carbon steel was subjected to 5-min ultrasonic treatment in ethanol [24]. The mode was constant current (−140 mA, 8 cm^{-2}) with carbon steel in 50 g L^{-1} supporting electrolyte (ZnSal-K_2SO_4). The rotating speed was 700 rpm and the plating time was 25 min. The deposition potential was monitored in the whole electroplating process.

2.3. Material characterisaions

TEM image of ZnSal was observed at 200 kV, while morphologies of ZnSal-steel were observed by SEM whose accelerating voltage was 15 kV. The *S* was scrutinized from N_2 using an automatic volumetric apparatus. The range of corrosion pits was measured by white light interference profilometer.

NSST was conducted to simulate a coastal atmospheric environment (corrosion tests in artificial atmospheres), drawing on the experience of ISO 9227:2017 [25]. The steel plates were constantly sprayed with sodium chloride (55 g L^{-1}, pH = 7.5) for 120 h in the salt-spray box. The fog deposition rate of the solution was 0.03 mL cm^{-2} h^{-1}, and the plates were lightly rinsed with cool water. After drying at 90 °C for 20 min, the rusty stain and corrosion degree were analysed. Each datum was the average of three parallel experiments.

3. Results and discussion

3.1. Ion distribution state of salicylic acid and zinc at different pH values

The dissociation equilibrium of salicylic acid ($C_7H_6O_3$) exists in an aqueous solution (pKa_1 = 3), and $C_7H_5O_3^-$ will always dissociate to $C_7H_4O_3^{2-}$ (pKa_2 = 12.38). Figure 1 displays the dissociation constants of ligand, and Figure 2 displays the distribution fraction at different acidities. The molecular form is available with pH <5, whereas the compound form exists with pH <2. With the increase in pH value to 5, $C_7H_6O_3$ gradually decreases to 1/100. The second form, $C_7H_5O_3^-$, has the widest distribution in the pH range of 1–14, and its maximum occurs at pH 5–11. $C_7H_4O_3^{2-}$ plays the main role after pH >14, and coexists with $C_7H_5O_3^-$ when pH is >10.

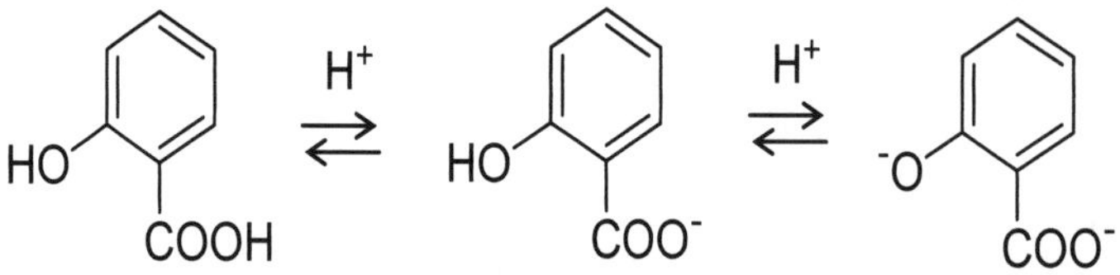

Figure 1. Dissociation equilibrium of salicylic acid in water.

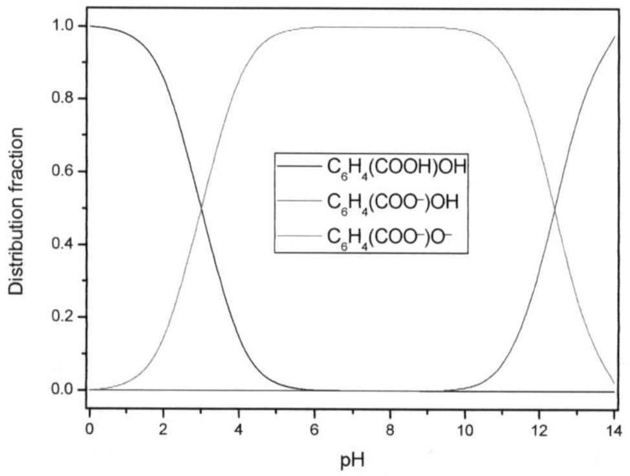

Figure 2. Distribution fraction of salicylic acid from pH 0–14.

Zinc ions in water take five forms under different pH conditions. Since $Zn(OH)_2$ is an amphoteric compound, it forms oxyhydrogen complexes in partial alkaline condition, while acts as Zn^{2+} in an acidic environment based on molecular formulas and conditional stability constants ($\log K_1 = 4.4$, $\log\beta_2 = 10.1$, $\log\beta_3 = 14.2$, $\log\beta_4 = 15.5$). The distribution state of zinc is plotted using numerical calculations, demonstrating the form of neutral and acidic solutions when pH is <8, of which the ratio is >96.4% when pH is within the interval of 0 to 8. The transient state $Zn(OH)^+$ has a ratio <10%. In addition, $Zn(OH)_3^-$ distributes between pH of 8.3 and 14, and increases after pH 8.6, which reaches the ratio of 92.6% at pH 11.3. $Zn(OH)_4^{2-}$ originates at pH >11. The maximum of amphoteric $Zn(OH)_2$ is 65.7% at pH 9.4. Therefore, the electrodeposition of ZnSal over the steel surface was evaluated at pH 9.5.

3.2. Morphology of ZnSal and ZnSal-steel

Zinc-containing coatings are applied to control corrosion according to isolating metal from an aggressive environment. Figure 3*A* shows the morphology of ZnSal revealed by TEM, which was an aggregation of nanocrystallines. The surface of carbon steel was flat before modification (Figure 4*A*), but the variation occurred on the steel surface after 25 min electrodeposition (Figure 3*B*). A compact film with a dendrite structure was provided where the ZnSal deposited tightly together. Based on the N_2 adsorption-desorption isotherm (Figure 3*C*), the BET-*S* of ZnSal was 247 $m^2\ g^{-1}$, and the pore capacity was detected to be 0.35 $cm^3\ g^{-1}$. At relative pressures from 0.4 to 1.0, the curve was a type-H3 hysteresis loop.

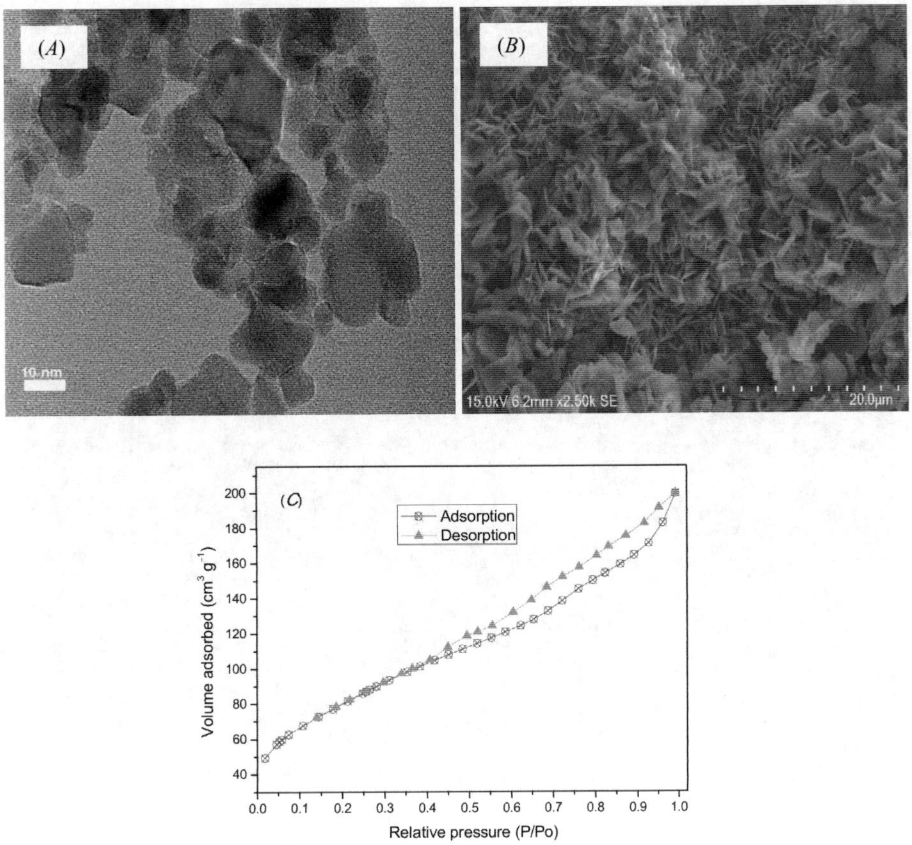

Figure 3. TEM image of ZnSal (*A*), SEM image of ZnSal-steel (*B*), and N_2 adsorption-desorption isotherm of ZnSal (*C*).

3.3. Estimation of anticorrosive property for ZnSal-steel

The comparative test of anticorrosive property for carbon steel and ZnSal-steel was performed by means of NSST (Figure 4). The top view of performed specimens was proven by a SLR camera as inserted in Figure 4*B* and 4*D* on the upper right. The specimen appearance indicated that all surfaces of the carbon steel corroded. Salt spray contributed to an increase in oxidation characteristics. Oxygen reacted with carbon steel, resulting in rust formation on the steel surface after 120 h. ZnSal-steel plates exhibited lower corrosion on their surfaces, where a slight amount of corrosion was observed in the composite coatings.

The anticorrosive behavior varied according to the test duration of NSST. The corrosion of steel plates was continuously caused by salt spray, and the corrosion rate decreased with time. It might be due to the oxidation films (acted as protective coatings) that limited further corrosion (Figure 5). The corrosion degree of steel surfaces primarily relied on the chemical medium and exposure time. A lower corrosion rate of ZnSal-steel (<0.11 $\mu m\ h^{-1}$) was observed compared with that of nascent carbon steel (<0.72 $\mu m\ h^{-1}$) whose increasing tendencies were slowed down. The average corrosion rate of the specimens kept track of the corrosive category, showing a nonlinear relationship with the exposure time.

Figure 4. Appearance of carbon steel (*A*,*B*) and ZnSal-steel coatings (*C*,*D*) for NSST (*A*,*C*: initial state; *B*,*D*: after 120 h).

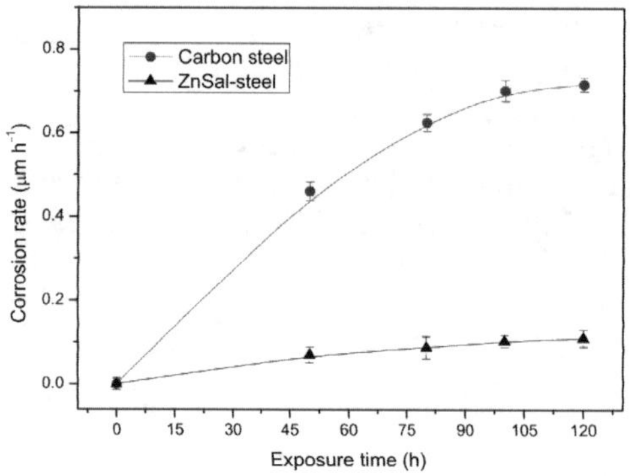

Figure 5. Correlation of exposure time and corrosion rate for carbon steel and ZnSal-steel coatings.

4. Conclusion

In summary, the ZnSal-steel coatings on carbon steel substrate were successfully prepared using a direct current electrodeposition of ZnSal/K_2SO_4 media. Various distribution states of zinc(II) and salicylic acid were surveyed from pH 0–14, and the optimum proportion of $C_6H_4(COO^-)OH$ and $Zn(OH)_x$ complexes were obtained at pH 9.5. The N_2 adsorption-desorption curve of ZnSal was a type-H3 hysteresis loop. The deposition efficiency was dramatically improved due to the specific structure of salicylate ligand, and the compact deposited film with dendrite structures was formed within 25 min. The corrosion resistance of steel plates in neutral salt spray and the corrosion degree demonstrated nonlinear relations with the exposure time. Furthermore, the NSST results indicated that ZnSal-steel had a much lower corrosion rate (*ca.* 0.1 μm h^{-1}) than that of nascent carbon steel (*ca.* 0.7 μm h^{-1}). This work provides helpful suggestions of composite coatings for developing rapid anticorrosive methods.

Acknowledgement
Thanks for the support from Eyas Incubation Project of Zhejiang (No. CY2022003).

References

[1] Kim J W, Lee J Y and Park S M 2004 Effects of organic additives on zinc electrodeposition at iron electrodes studied by EQCM and in situ STM *Langmuir* **20** 459
[2] Burlyaev D V, Tinaeva A E, Tinaeva K E and Kozaderov O A 2020 Electrodeposition of zinc–nickel coatings from glycine-containing ammonium-chloride electrolyte *Protection of metals and physical chemistry of surfaces* **56** 552
[3] Liao D H, Chen Z, Guo Z C and Lu L F 2011 Research status and development trend of inert anode materials for zinc electrodeposition *Electroplating & Finishing* **30** 50
[4] Aljoboury A I, Mourad A, Alawar A, Zour M A and Abuzeid O A 2010 Stress corrosion cracking of stainless steels recommended for building brine recirculation pumps *Engineering Failure Analysis* **17** 1337
[5] Marra E, Zirkelbach D and Künzel H M 2015 Prediction of steel corrosion in porous building materials by means of a new hygrothermal model *Energy Procedia* **78** 1299
[6] Li Q, Li C, Ping L and Xue L L 2007 Behavior of microbiological influenced corrosion of the ship plate steel in marine environment *Key Engineering Materials* **348** 25
[7] Abdel-Samad A, Soud Y and Zaki M 2014 Influence of paint on steel corrosion for marine applications *Journal of Surface Engineered Materials and Advanced Technology* **4** 189
[8] Wasekar N P, Jyothirmayi A, Hebalkar N and Sundararajan G 2015 Influence of pulsed current on the aqueous corrosion resistance of electrodeposited zinc *Surface and Coatings Technology* **272** 373
[9] Ramanauskas R, Juškėnas R, Kaliničenko A, Garfias-Mesias L F 2004 Microstructure and corrosion resistance of electrodeposited zinc alloy coatings *Journal of Solid State Electrochemistry* **8** 416
[10] Liu S, Sun H Y, Sun L J and Fan H J 2012 Effects of pH and Cl^- concentration on corrosion behavior of the galvanized steel in simulated rust layer solution *Corrosion Science* **65** 520
[11] Jiang J Y 2014 Research on galvanized layer uniformity detecting technology of transmission angle steel tower materials *Applied Mechanics and Materials* **596** 56
[12] Arman S Y, Ramezanzadeh B, Farghadani S, Mehdipour M and Rajabi A 2013 Application of the electrochemical noise to investigate the corrosion resistance of an epoxy zinc-rich coating loaded with lamellar aluminum and micaceous iron oxide particles *Corrosion Science* **77** 118
[13] Yang F, Liu T, Li J Y, Qiu S H and Zhao H C 2018 Anticorrosive behavior of a zinc-rich epoxy coating containing sulfonated polyaniline in 3.5% NaCl solution *RSC Advances* **8** 13237
[14] Long Z L, Zhou Y C and Yang C Q 2003 Chromate conversion coating treatments for electrodeposited zinc-iron alloy coatings from an acidic sulphate bath *Transactions of the IMF* **81** 148
[15] Meng W Q, Li F, Evans D G and Duan X 2004 Photocatalytic activity of highly porous zinc ferrite prepared from a zinc-iron(III)-sulfate layered double hydroxide precursor *Journal of Porous Materials* **11** 97
[16] Yuan R H, He H H, Huang J F and Su C 2019 Characterization of interfacial behaviour of *ortho*-positioned diaryl disulfide on silver substrates by surface-enhanced Raman scattering and electroreduction *Materials Chemistry and Physics* **235** 121726
[17] Ataxca C G, Martinez D, Aguilar G G, Chavez-Rosales R and Cruz R O 2021 Electrochemical evaluation of three zinc-rich paints on carbon steel substrates exposed to 3.5% NaCl *ECS Transactions* **101** 343

[18] Zhang H, Wen X and Wang Y X 2007 Synthesis and characterization of sulfate and dodecylbenzenesulfonate intercalated zinc–iron layered double hydroxides by one-step coprecipitation route *Journal of Solid State Chemistry* **180** 1636

[19] Bellotti N and Romagnoli R 2014 Assessment of zinc salicylate as antifouling product for marine coatings *Industrial & Engineering Chemistry Research* **53** 14559

[20] Che X H, Yuan R H, Tang L C, Xiong Y D, Wu M F and Zhu M H 2021 Study of catecholate-functionalized zinc complex on carbon steel and its application to corrosion resistance *Journal of Physics: Conference Series* **1978** 012009

[21] Permeh S, Lau K and Duncan M 2020 Degradation of coatings for steel in environments susceptible to corrosion associated with fouling *Structure and Infrastructure Engineering* **16** 1186

[22] Ai W, Lo K H, Li X, Kwok C and Pan H 2020 Cavitation erosion damage mechanism of a duplex stainless steel having a ferrite-austenite-sigma-phase triplex microstructure *Journal of Materials Engineering and Performance* **29** 2806

[23] Yin X, Yu S, Wang K, Cheng R and Lv Z 2020 Fluorine-free preparation of self-healing and anti-fouling superhydrophobic Ni_3S_2 coating on 304 stainless steel *Chemical Engineering Journal* **394** 124925

[24] Zhong N F, Yuan R H, He H H, Zhong W H and Wang Y 2021 Electrochemical sensor based on graphene-modified GCE for rapid quantification of benzo(a)pyrene in dark rubber materials *Journal of Physics: Conference Series* **1978** 012020

[25] ISO 9227: 2017. Corrosion tests in artificial atmospheres—Salt spray tests. Geneva: International Organization for Standardization

Simulation system of lighting environment for optical imaging test

Lei Zhang[1], Yu Yang[1], Shuai Zhao[1], Wenbin Xu[3], Bolin Cai[2] and Keyi Wang[1, 4]

[1]Department of Precision Machinery and Precision Instrumentation, University of Science and Technology of China, Hefei 230027, China
[2]School of Internet, Anhui University, Hefei, Anhui, 230039, China
[3]Science and Technology on Optical Radiation Laboratory, Beijing, 100039, China

[4]Corresponding author's e-mail: kywang@ustc.edu.cn

Abstract. In order to overcome the shortcomings of the solar simulator in the aspect of the performance testing for optical imaging system, a lighting environment simulation system is designed. We select dysprosium lamp, xenon lamp and multiple LEDs as light sources of the solar simulator, and the change of the sun zenith angle is simulated by using the symmetrical method of the light source. Also, the light intensity is controlled by adjusting the output current of the power supply. Lighting environment simulation system not only can simulate spectral irradiance and intensity change of real solar, but also can simulate the real sun depending on the season and the angle of illumination period. This system provided a technical support for the performance testing of optical imaging systems.

1. Introduction

Illumination is an important factor affecting the performance of photoelectric imaging system. Therefore, the comprehensive test and evaluation of the performance of photoelectric imaging system requires that the test environment can provide different illumination conditions. In the actual test, the natural lighting environmental conditions of some outfield could not be controlled and repeated, resulting in the difficulty of quantifying the data such as Illumination energy and irradiation angle. Due to the influence of these factors, the reliability of test data is reduced, and the results of multiple tests cannot be compared. Therefore, the way to solve this problem is to use the simulation system of lighting environment indoors as the light source of the irradiation target [1]. Solar simulator is a kind of equipment that can simulate the spectrum and illumination of the sun. At present, it is mostly used in the field of aerospace and solar energy utilization, that is, to test the comprehensive performance of solar devices [2-4]. In recent years, because the development and application of solar energy has become a research hotspot, many researchers have made some progress in theory and technology for solar simulator. For different purposes, many research institutions have developed a variety of advanced solar simulators by using different light sources such as LED and homogenizing rod, as well as optical structure modes such as optical fiber transmission and integrating sphere [5-7]. However, in the test of photoelectric imaging system, the solar simulator is required to not only have large irradiation area and irradiation uniformity in the laboratory, so as to simulate the real solar spectral characteristics and irradiance intensity changes, but also simulate the real solar irradiation angle according to different seasons and time periods. However, the existing solar simulator could not meet all the requirements of these tests [8-10].

Therefore, this paper studies the controllable and repeatable lighting environment simulation technology with dynamic quantitative function. The corresponding system is developed, which enables the outdoor test to carry out simulation test in the laboratory, and provides condition support for verifying the performance of photoelectric imaging system under different lighting conditions.

2. Simulation method of lighting environment

Therefore, this paper studies the controllable and repeatable lighting environment simulation technology with dynamic quantitative function. The corresponding system is developed, which enables the outdoor test to carry out simulation test in the laboratory, and provides condition support for verifying the performance of photoelectric imaging system under different lighting conditions.

2.1. Overall design for simulation system of lighting environment

The lighting environment simulation controls the light source according to the actual sunrise and sunset time of the sun. Thus, different light conditions such as luminous intensity, illumination angle and color temperature are produced, mainly considering spectral characteristics, illumination intensity and zenith angle. The light sources used in the lighting environment simulation system mainly include carbon arc lamp, incandescent lamp, halogen lamp, dysprosium lamp and xenon lamp. Among them, carbon arc lamp will produce highly toxic cyanide gas in ultraviolet, which is not conducive to laboratory testing. The color temperature of incandescent lamp and halogen lamp is low, and the distribution of ultraviolet and visible light is quite different from the actual solar spectrum. The color temperature distribution of xenon lamp is generally consistent with that of the solar spectrum, but there are peaks in the spectrum. Therefore, dysprosium lamp with high luminous efficiency, high color rendering and long-life gas discharge light source is selected as the main light source of the simulator, and xenon lamp and multiple LEDs are used as secondary light sources to adjust the spectral adaptability. The design of the whole simulation system of lighting environment is shown in Figure 1.

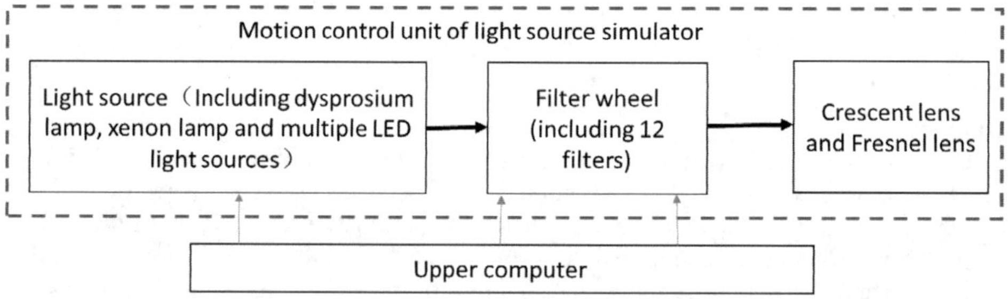

Figure 1. Overall design of solar simulator.

As shown in Figure 1, the whole light source is installed on the control part of the light source motion platform. The upper computer sets the irradiation direction and spectral energy of the simulation system according to the simulation requirements of the experimental user for the solar altitude angle, azimuth angle, spectral characteristics and illuminance value. Under the action of Fresnel lens and crescent lens, the outgoing light of the light source is approximately parallel light. In order to improve the matching degree with the solar spectrum, the combination of multistage filters is adopted. The designed simulation system of lighting environment has simple structure and small volume, and is suitable for illumination angle conversion with servo motor. In order to simulate the large dynamic range illumination in sunny, cloudy days, sunrise and sunset, the power of the light source is adjusted by controlling the current. After calibration with the illuminance meter, the current value required for illuminance simulation in different weather and time periods can be obtained. In order to solve the problem of uneven peripheral light spot, we adopt the dynamic balance control method of light source. The method is as follows: place an illuminance meter in the target area, collect the illuminance of the target area in real time, and transmit the illuminance to the light source controller of the integrated console through the wireless transmission device; The method is as follows: firstly, an illuminance meter is placed in the target area, the illuminance of the target area is collected in real time, and the illuminance is transmitted to the light source controller of the integrated console through the wireless transmission device; Then, the controller converts the actual measured illuminance value into the corresponding current value, calculates the difference between the current value and the real current value, and sends the current difference to the current feedback circuit; Finally, the output current is adjusted by the current feedback device according to the adjustment value, so as to adjust the illumination and complete the dynamic balance control of the light source.

2.2. Simulation of zenith angle change

The position of the sun relative to the target can be characterized by altitude angle and azimuth angle, and changes with time and target position. However, the location of the target is characterized by its longitude and latitude. According to astronomical knowledge, as long as we know the longitude, latitude and time of the target, we can calculate the altitude and azimuth of the sun. In other words, the position of the sun in different regions and times can be accurately simulated.

The arc guide rail of the light source simulating the movement of the sun is designed as a semi open structure, and the angle change in two directions is added to the light source adjustment frame. The azimuth and altitude angles of the sun are simulated by linear motion motor, circular motion motor, the pitch and direction angle of the light source.

The structure diagram of the control part of the light source moving platform is shown in Figure 2, which is mainly composed of the light source transverse guide rail, light source arc guide rail, sliding block, locking mechanism, etc. The light source is fixed on the arc guide rail by the sliding block, and the servo motor drives the steel wire rope along the arc guide rail to realize the movement of the light source, so as to simulate the angular motion track of the sun and ensure that the light source always points to the target. At the same time, the arc guide rail can move laterally driven by the servo motor, so as to ensure that the light emitted by the light source can cover any position and target under various terrain. In order to avoid the motion interference between the moving part of the light source and the multi view detection platform, the arc guide rail of the light source is designed into a semi open structure. In addition, due to the large volume and weight of the light source, the inertial force generated during its movement along the arc track will cause the vibration of the arc guide rail. In order to overcome the influence of inertial chattering force on the light source, the arc track is designed into a double track structure to increase the comprehensive rigidity and transverse overturning moment of the track.

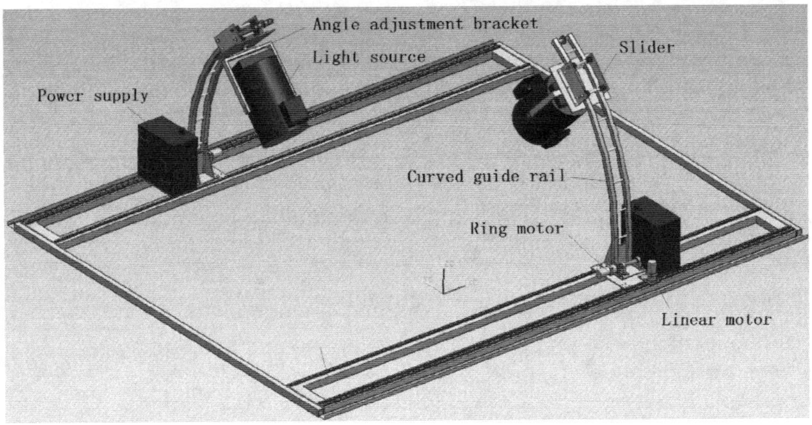

Figure 2. Structure of moving platform control component for light source.

2.3. Software system for simulation system of lighting environment

The illumination environment simulation system controls the illumination change of the simulated light source by adjusting the output current of the light source simulator. In this way, the simulation of illumination change in different regions, seasons and weather conditions can be realized. In addition, we use the symmetrical orbit design of the light source to control the output angle of the light source to simulate the height angle and azimuth of the sun, so as to realize the time control simulation of the lighting environment, as shown in Figure 3.

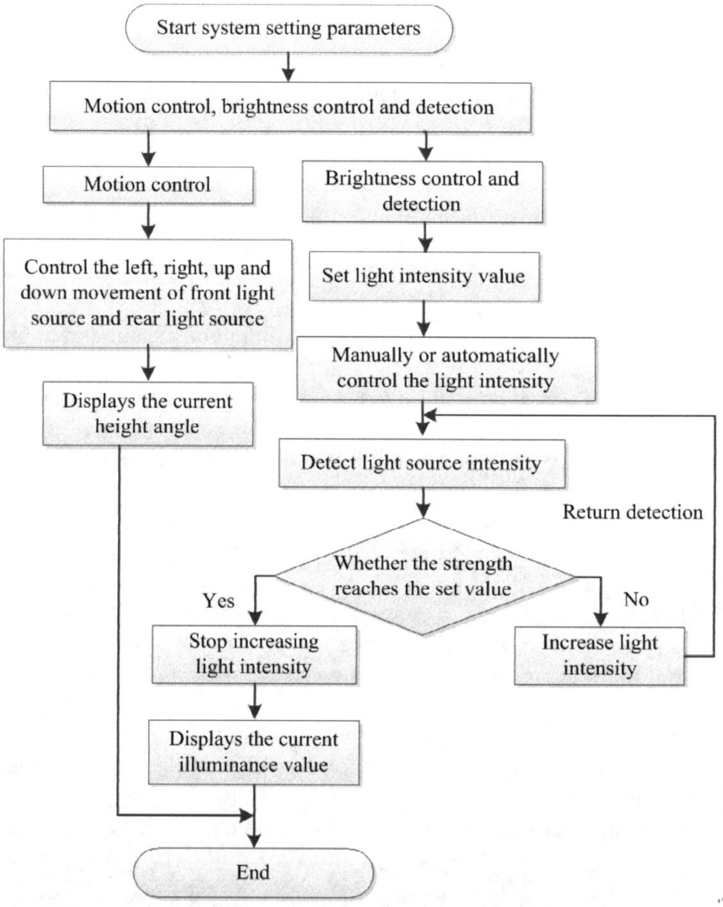

Figure 3. Flow chart of lighting environment control software.

It can be seen from Figure 3 that the change of light source illumination can be controlled automatically or manually to finally reach the set value. Table 1 shows the corresponding relationship between the solar altitude angle, illuminance and lamp current in Hefei at the same time. According to this table, the simulation control of the lighting environment can be realized by setting parameters through the host computer software.

Table 1. Corresponding relationship between illumination and lamp current at different height angles.

Time	Altitude angle	Ground illumination /10^3lx		Lamp current /A	
		Sunny day	Cloudy day	Sunny day	Cloudy day
6：00	9°	8	4	7.3	3.3
8：00	33°	41	10	31.3	7.3
10：00	61°	100	19	81.7	14.5
12：00	80°	122	20	96.1	17.7
14：00	63°	113	21	61	15
16：00	37°	11	12	32	8
18：00	15°	9	4.3	7.8	3.5

2.4. Main performance test of the system

In order to evaluate the spectral mismatch error, the spectral irradiance of the illumination environment simulation system in each wavelength interval is tested by using a special test system for light source, and the test bandwidth is 10 nm. The results show that the mismatch error of spectral distribution can meet the requirements of type A solar simulator with 0.36-0.8μm, and the results are shown in Figure 4. In the test, an integrating sphere with a diameter of 1m is used, and then the light emitted from the simulation system is transmitted to the photodetector through optical fiber at the outlet of the integrating sphere, and converted into electrical signals. Finally, the data analysis is completed by the software.

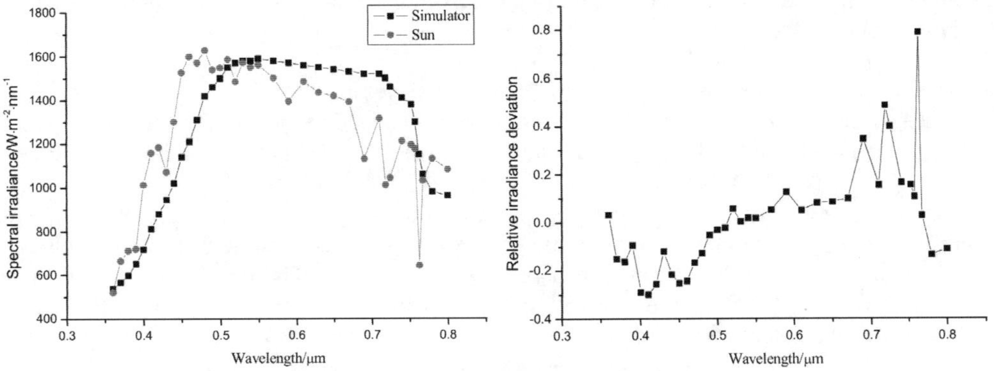

Figure 4. Experimental results of spectral mismatch error test of solar simulator.

In the irradiance non-uniformity measurement, the diameter of the actually measured irradiation surface is 600mm. We divide the irradiation surface into 19 units and test the irradiance of each point respectively. The test schematic diagram and scene are shown in Figure 5, and the test results are shown in Table 2.

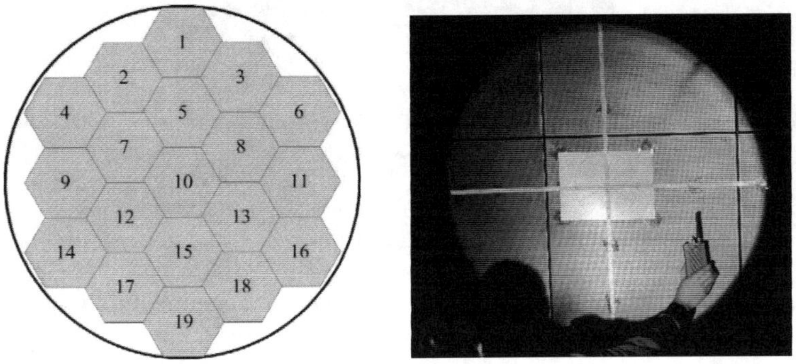

Figure 5. Schematic diagram and experimental scene of irradiance nonuniformity test.

Table 2. Test results of irradiance nonuniformity.

Test point	irradiance / (W·cm^{-2})	Test point	irradiance / (W·cm^{-2})
1	1609	11	1587
2	1608	12	1630
3	1599	13	1628
4	1610	14	1600
5	1621	15	1621
6	1600	16	1597
7	1626	17	1601
8	1629	18	1599
9	1605	19	1609
	1632	—	—

It can be calculated from Table 2 that the irradiance non-uniformity is about 3.7%. In addition, the irradiance of the simulation system decreases gradually from the center to the edge, which is due to the non-uniformity of the light source radiation inside the system. However, under the action of optical integrator and collimator, compared with the Gaussian distribution characteristics of the light source itself, the irradiance uniformity of the whole lighting environment simulation system has been greatly improved. In addition, in the test, the solar zenith angle was changed by rotating the light source angle of the platform, and the output angle of the platform and the output angle of the solar radiometer were recorded at an interval of 1°. By comparing the solar zenith angle data output by the solar radiometer with the platform rotation data, it can be seen that the average deviation between them is 0.039° in the variation range of 0°-60°.

3. Verification for simulation system of lighting environment

In order to verify the simulation effect of the real field sunlight, we use the simulation system to generate the lighting environment under different illumination conditions, and carried out the imaging detection experiments of various targets. The design of the system is shown in Figure 6.

The detection target is a toy car. In the experiment, the illumination value is changed by adjusting the input current of the light source. In the experiment, we set the incident angle of the light source to 55° and the observation angle of the instrument to 40° through software. The lighting conditions are set as cloudy days in mid latitude summer, the illuminance value is 8000-15000lx, and the acquisition integration time of the detector is 500ms.

Figure 6. Lighting environment simulation system.

Under the condition of 9000lx, the original image directly obtained by the camera and the image obtained by the poor illumination enhancement algorithm suitable for the natural environment are shown in Figure 7. In addition, we use the simulation system to generate the test environment under different illumination conditions, and carry out the imaging detection test under poor illumination environment.

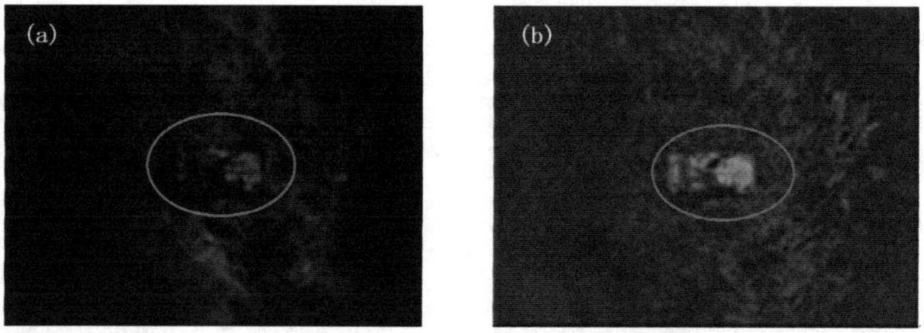

Figure 7. Images of before and after enhancement for simulated vehicle under low illumination. (a) Before enhancement; (b) After enhancement.

It can be seen from the results in Figure 7 that under the simulation system of lighting environment, the effect of the target enhancement algorithm used in the real sunlight background is obvious, and the car image is relatively clear. The result shows that the illumination test environment in the laboratory can meet the requirements of out-field test verification.

4. Conclusion

According to the requirements of photoelectric imaging system performance test, the lighting environment simulation technology in the laboratory is studied, and the corresponding system is developed. We tested the spectral mismatch error, irradiation non-uniformity and angle conversion accuracy of the light source. Experiments show that the low illumination image generated by the simulation system is suitable for the bad illumination enhancement algorithm in the external natural environment. The system can overcome the unpredictability and limitation of solar radiation in the field natural environment. The test conditions such as solar illuminance and illumination angle can be set arbitrarily through the software, and the performance of photoelectric imaging system can be tested repeatedly in the laboratory. As a new type of solar simulator with controllable illumination and irradiation angle, large irradiation range and long-time operation, the simulation system of lighting environment has certain application value in the performance test of photoelectric imaging system. It can make up for the defect that the current solar simulator is mainly aimed at photovoltaic testing, but it is difficult to be applied in the performance testing field of photoelectric imaging system.

Acknowledgments

This work was supported by the National Natural Science Foundation of China (41871229, 52105538).

References

[1] Wang W. Simulate a Sun for Solar Research: A Literature Review of Solar Simulator Technology[J]. KTH Energy and Environmental Technology, 2014: 1-41.

[2] Tawfik M, Tonnellier X, Sansom C. Light source selection for a solar simulator for thermal applications: A review[J]. Renewable & Sustainable Energy Reviews, 2018, 90: 802-813.

[3] Lopez-Fraguas E, Sanchez-Pena J M, Vergaz R. A Low-Cost LED-Based Solar Simulator[J]. IEEE Transactions on Instrumentation and Measurement, 2019: 1-11.

[4] Fauzi F, Tajudin M, Mohamed M F, et al. Assessment of in-house build low cost solar panel simulator[J]. Journal of Physics: Conference Series, 2021, 1878(1): 012038.

[5] Ming S C, Sun G F, Zhang G Y, et al. Continuous Regulation System Design of Wide-Range Irradiance for Solar Simulators [J]. Acta Optica Sinica, 2021, 41(16): 176-184.

[6] Nie F, Yu Y, Bai F, et al. Experimental and numerical investigation on thermal performance of a quartz tube solid particle solar receiver[J]. Solar Energy, 2020, 207: 1055-1069.

[7] Rowe S C, Wallace M A, Lewandowski A, et al. Experimental evidence of an observer effect in high-flux solar simulators[J]. Solar Energy, 2019, 158(dec.): 889-897.

[8] L Martínez-Manuel, Wang W, Pea-Cruz M I. Optimization of the radiative flux uniformity of a modular solar simulator to improve solar technology qualification testing[J]. Sustainable Energy Technologies and Assessments, 2021, 47(19): 101372.

[9] Shevchuk A A, Pastushenko O V, Dvirniy V V, et al. Problems of irradiance characteristics measurement of solar simulators for ground spacecraft tests[J]. Spacecrafts & Technologies, 2020, 4(3): 129-140.

[10] Turek M, Sporleder K, Luka T. Spectral characterization of solar cells and modules using LED-based solar simulators[J]. Solar Energy Materials and Solar Cells, 2019, 194: 142-147.

Journal of Physics: Conference Series **2313** (2022) 012019 doi:10.1088/1742-6596/2313/1/012019

Oscillating charge currents of one-dimensional Hubbard model in electric field

Yong Zheng

School of Physics and Electronics, Qiannan Normal University for Nationalities, Duyun 558000, China

Corresponding author's e-mail: zhengyongsc@sina.com

Abstract. The time evolution properties of charge current for the one-dimensional Hubbard model in an electric field have been studied in a rigorous manner. We find that there is a complete and orthonormal set of time-evolution states for which the charge current can only keep zero or oscillate constantly, differing from the possible picture of damped or over-damped Bloch oscillations due to strong correlations. It is also found that, associated with these states, there is a set of constant phase factors, which are uniquely determined and are very useful on discussing the long-time evolution behaviors of the system.

1. Introduction

The nonequilibrium properties of strongly correlated electron systems are of great interest in condensed matter physics. It is believed that correlated interactions of electrons have a huge effect on the response of a system to external fields, leading to complicated transport behaviors. However, an exact study of these behaviors for a strongly correlated system generally is very hard.

The Hubbard model is perhaps the simplest model to capture physical properties of strongly correlated systems and hence has been extensively studied. Even though, transport properties of this model still have not been well understood. For example, basing on a numerical study, Eckstein and Werner [1] have shown that the on-site correlation of electrons plays an important role in the response of the model to an electric field, and that different magnitudes of the on-site U would lead to different damping behaviors of the charge current, giving damped or over-damped Bloch oscillations. These results have also been supported by the latter study of Mandf [2] via a method of Boltzmann equations. However, there are also many studies (see, e.g., [3-6]) which conclude that, to the extent of linear response, 1D Hubbard model shows ideal conductance; and such behavior is closely related to the integrability, which always results in a nonzero Drude weight in cases away from half filling, though such conclusion is still under debate at half filing [7-11].

In a previous study [12], we have shown that undamped Bloch oscillations, which is an important feature of noninteracting tight-binding systems in an electric field, extensively exist in the Hubbard model. However, the analysis was mainly concentred in the 1D $U \to \infty$ case. Here we want to extend our previous analysis to more general cases. We still focus our discussion mainly on the 1D case, namely, the Hubbard model on a periodic ring, but one will find that the extension to higher dimensions is straightforward.

2. Model and formulation

For our purpose, it is convenient to describe the constant electric field with a time-dependent vector potential along the ring, $\boldsymbol{E} = -1/c\partial_t \boldsymbol{A}(t)$, with $\boldsymbol{A}(t) = -c\boldsymbol{E}t$. The well-known form of the Hubbard Hamiltonian in such a field can be written as $H(t) = -t_0 \sum_{i,\sigma}\{\exp[i\varphi_i(t)]c_{i\sigma}^{\dagger}c_{i+1\sigma} + \mathrm{H.c.}\} + U\sum_i n_{i\uparrow}n_{i\downarrow}$, where $\sigma = \uparrow$ or $\downarrow$, $c_{i\sigma}^{\dagger}$ and $c_{i\sigma}$ the creation and annihilation operators of the spin-σ electron at site i, respectively. $n_{i\sigma} = c_{i\sigma}^{\dagger}c_{i\sigma}$ and t_0 is the hopping integral between nearest-neighbor sites. The additional phase of the hopping integral due to the appearance of the field [13], $\varphi_i(t) = -\frac{e}{\hbar c}\int_{r_i}^{r_{i+1}} \boldsymbol{A}(t)\cdot d\boldsymbol{r} = -e\mathcal{E}at/\hbar$, where a is the lattice constant and $\mathcal{E}$ the magnitude of the electric field. Transforming the Hamiltonian to momentum space, we obtain

$$H(t) = \sum_{k,\sigma} \varepsilon_k(t)c_{k\sigma}^{\dagger}c_{k\sigma} + U\sum_{k_1,k_2,q} c_{k_2-q\downarrow}^{\dagger}c_{k_1+q\uparrow}^{\dagger}c_{k_1\uparrow}c_{k_2\downarrow}, \tag{1}$$

where sums over wave vectors k or q are limited to the Brillouin zone, and $\varepsilon_k(t) = -2t_0\cos(ka + e\mathcal{E}at/\hbar)$. Obviously, $\varepsilon_k((t+\tau) = \varepsilon_k(t)$, where $\tau = h/(e\mathcal{E}a)$; and hence $H(t+\tau) = H(t)$, which is very important for our discussion.

What we are most interested in is the time evolution of the charge current of the system in electric field. The charge current operator in our case can be written as [1,14]

$$\hat{j}(t) = \frac{ea}{\hbar}\sum_{k,\sigma}\frac{\partial\varepsilon_k(t)}{\partial k}c_{k\sigma}^{\dagger}c_{k\sigma}, \tag{2}$$

which is also periodic in time with a period τ.

In principle, to discuss the time-dependent properties of the model, we need a detailed knowledge of the time evolution state $|\Psi(t)\rangle$ of the system in the field, which is determined by the Schrödinger equation

$$i\hbar\frac{\partial|\Psi(t)\rangle}{\partial t} = H(t)|\Psi(t)\rangle. \tag{3}$$

For our case, a direct and exact solution of this equation for $|\Psi(t)\rangle$ seems impossible. However, at any time t, we can expand $|\Psi(t)\rangle$ with the noninteracting particle states in momentum space which can be written as

$$|\psi_n\rangle \equiv \prod_{i=1}^{M_\downarrow} c_{k_{ni}\downarrow}^{\dagger}\prod_{i=1}^{M_\uparrow} c_{k_{ni}\uparrow}^{\dagger}|0\rangle,$$

where $M_\downarrow$ ($M_\uparrow$) is the number of spin $\downarrow$ ($\uparrow$) electrons in the system, and $k_{ni}\downarrow$ or $k_{ni}\uparrow$ the wave vectors of electrons in $|\psi_n\rangle$; that is,

$$|\Psi(t)\rangle = a_1(t)|\psi_1\rangle + a_2(t)|\psi_2\rangle + \cdots + a_N(t)|\psi_N\rangle, \tag{4}$$

with N the total number of the noninteracting particle states involved. Then, the specific form of $|\Psi(t)\rangle$ can be obtained if we can find a way to determine the expansion coefficient $a_n(t)$'s.

Using Eq (4) and the matrix form of the state, we can rewrite Eq. (3) as

$$i\hbar\dot{\boldsymbol{a}}(t) = A(t)\boldsymbol{a}(t), \tag{5}$$

where $\boldsymbol{a}(t) = [a_1(t), a_2(t), \cdots, a_N(t)]^T$, and $A(t)$ is the matrix form of $H(t)$ in the subspace formed by $\{|\psi_n\rangle\}$,

$$A(t) = \begin{bmatrix} H_{11}(t) & H_{12}(t) & \cdots & H_{1N}(t) \\ H_{21}(t) & H_{22}(t) & \cdots & H_{2N}(t) \\ \cdots & \cdots & \cdots & \cdots \\ H_{N1}(t) & H_{N2}(t) & \cdots & H_{NN}(t) \end{bmatrix}, \tag{6}$$

with the element $H_{mn}(t) = \langle\psi_m|H(t)|\psi_n\rangle$. Obviously, $A(t+\tau) = A(t)$ and $A^{\dagger}(t) = A(t)$. Similarly, we can introduce the matrix form of $\hat{j}(t)$ in this subspace, which we denote by $J(t)$, with the matrix element $J_{mn}(t) = \langle\psi_m|\hat{j}(t)|\psi_n\rangle$.

Once Eq. (5) has been obtained, our problem is reduced to a system of linear ordinary differential equations with periodic coefficients, which has already been extensively studied. Although a general discussion of such kind of differential equations can be found in textbooks such as Refs. [15] and [16], or works such as Refs. [17] and [18], for convenience of discussing and reading, especially for pertinence of our problem, we shall proceed with some derivations. Then, it is possible that some of our conclusions below about the system of differential equations have already appeared in textbooks.

From the normal theory of linear ordinary differential equations, starting from a given initial condition $\boldsymbol{a}(0) = [a_{i1}(0), a_{i2}(0), \cdots, a_{iN}(0)]^T$, there exists only one solution of the system. Generally, there are N and only N linearly independent solutions for $\boldsymbol{a}(t)$. Let $\boldsymbol{a}_1(t), \boldsymbol{a}_2(t), \cdots, \boldsymbol{a}_N(t)$ be a system of linearly dependent solutions, where each solution $\boldsymbol{a}_i(t) = [a_{i1}(t), a_{i2}(t), \cdots, a_{iN}(t)]^T$. The matrix

$$\Psi(t) \equiv [\boldsymbol{a}_1(t), \boldsymbol{a}_2(t), \cdots, \boldsymbol{a}_N(t)] = \begin{bmatrix} a_{11}(t) & a_{21}(t) & \cdots & a_{N1}(t) \\ a_{12}(t) & a_{22}(t) & \cdots & a_{N2}(t) \\ \cdots & \cdots & \cdots & \cdots \\ a_{1N}(t) & a_{2N}(t) & \cdots & a_{NN}(t) \end{bmatrix}, \tag{7}$$

as is well known, is called a fundamental matrix of solutions of equation system (5). Then, a general solution of the system (5) can be written as

$$\boldsymbol{a}(t) = c_1\boldsymbol{a}_1(t) + c_2\boldsymbol{a}_2(t) + \cdots + c_N\boldsymbol{a}_N(t), \tag{8}$$

where $c_1, c_2, \cdots, c_N$ are constants.

An important conclusion following from the Hermiticity of $A(t)$ is that if the N linearly independent solutions $\boldsymbol{a}_i(t)$'s are orthonormal at $t = 0$, then they are orthogonal at any time latter. To see this, one just notes that from Eq. (5),

$$-i\hbar\dot{\boldsymbol{a}}^\dagger(t) = \boldsymbol{a}^\dagger(t)A^\dagger(t) = \boldsymbol{a}^\dagger(t)A(t),$$

where $\boldsymbol{a}^\dagger(t) \equiv [\boldsymbol{a}^*(t)]^T$, and then

$$i\hbar[\boldsymbol{a}_i^\dagger(t)\boldsymbol{a}_j(t)]' = i\hbar[\dot{\boldsymbol{a}}_i^\dagger(t)\boldsymbol{a}_j(t) + \boldsymbol{a}_i^\dagger(t)\dot{\boldsymbol{a}}_j(t)] = 0,$$

that is, the scalar product

$$\boldsymbol{a}_i^\dagger(t)\boldsymbol{a}_j(t) = a_{i1}^*(t)a_{j1}(t) + a_{i2}^*(t)a_{j2}(t) + \cdots + a_{iN}^*(t)a_{jN}(t), \tag{9}$$

is time-independent. Hence if $\boldsymbol{a}_i^\dagger(0)\boldsymbol{a}_j(0)$ is 0 for $i \neq j$ and 1 for $i = j$, then so is $\boldsymbol{a}_i^\dagger(t)\boldsymbol{a}_j(t)$. Since $\boldsymbol{a}_i(t)$'s are easily chosen to be orthonormal to each other at $t = 0$ via adjusting their initial conditions, hereafter, the N linearly independent solutions are viewed as orthonormal, and then

$$\Psi^\dagger(t)\Psi(t) = I, \tag{10}$$

where I is the identity matrix. The orthonormality of the fundamental system of solutions is an important base of our discussion.

Another important character of the solutions is related to the periodicity of $A(t)$, from which it follows that if $\boldsymbol{a}(t)$ is a solution of Eq. (5) at time t, so is $\boldsymbol{a}(t + \tau)$. Then both can be expanded with $\boldsymbol{a}_i(t)$'s, since all the solutions, as a function of t, can be expanded with the fundamental system of solutions. As a special case of this conclusion, one can find that $\boldsymbol{a}_i(t + \tau)$'s are also a system of solutions of Eq. (5) at time t. Furthermore, both $\Psi(t)$ and $\Psi(t + \tau)$ can be viewed as the fundamental matrix of Eq. (5) at time t. Since each $\boldsymbol{a}_i(t + \tau)$, as a solution at time t, can be expanded with $\boldsymbol{a}_i(t)$'s in form, we have the well known relation [15,16]

$$\Psi(t + \tau) = \Psi(t)\boldsymbol{C}, \tag{11}$$

where $\boldsymbol{C}$ is a constant matrix.

Noting $\Psi^\dagger(t + \tau)\Psi(t + \tau) = \Psi^\dagger(t)\Psi(t) = I$, one can find that $\boldsymbol{C}^\dagger\boldsymbol{C} = I$. Namely, $\boldsymbol{C}$ is a unitary matrix. Then two points should be noted:

(i) Matrix $\boldsymbol{C}$ can be diagonalized via another unitary constant matrix P,

$$P^\dagger\boldsymbol{C}P = \boldsymbol{C}_d, \tag{12}$$

where $\boldsymbol{C}_d = diag(\lambda_1, \lambda_2, \cdots, \lambda_N)$. To see this, we can introduce two matrices X and Y,

$$X = \frac{1}{2}(\boldsymbol{C} + \boldsymbol{C}^\dagger) \quad \text{and} \quad Y = \frac{1}{2i}(\boldsymbol{C} - \boldsymbol{C}^\dagger),$$

from which, it follows that $\boldsymbol{C} = X + iY, X^\dagger = X, Y^\dagger = Y$ and $[X, Y] = 0$. Hence, the unitary matrix P does exist, and can diagonalize X, Y and $\boldsymbol{C}$ simultaneously.

(ii) $\boldsymbol{C}$, P and hence $\boldsymbol{C}_d$ are unitary. Then, all the diagonal elements λ_n's are complex numbers of modulus unity, that is,

$$\lambda_n = \exp(i\phi_n), n = 1,2,\cdots,N.$$

Hence, $\boldsymbol{C}_d$ can be expressed as

$$\boldsymbol{C}_d = \exp(i\Phi), \tag{13}$$

where $\Phi = diag(\phi_1, \phi_2, \cdots, \phi_N)$, and then

$$P^\dagger \boldsymbol{C} P = \exp(i\Phi). \tag{14}$$

Combining Eqs. (11) and (14), we find that if we define a new matrix

$$\widehat{\Psi}(t) \equiv \Psi(t)P = [\widehat{\boldsymbol{a}}_1(t), \widehat{\boldsymbol{a}}_2(t), \cdots, \widehat{\boldsymbol{a}}_N(t)], \tag{15}$$

then

$$\widehat{\Psi}(t+\tau) = \widehat{\Psi}(t)\exp(i\Phi). \tag{16}$$

Obviously, according to Eq. (15), the matrix $\widehat{\Psi}(t)$ is also a fundamental matrix of Eq. (5); and $\widehat{\boldsymbol{a}}_n(t)$'s, as a new system of fundamental solutions, are also orthonormal to each other, due to that

$$\widehat{\Psi}(t)^\dagger \widehat{\Psi}(t) = [\Psi(t)P]^\dagger \Psi(t)P = I. \tag{17}$$

3. Results and discussion

The most important property of these new solutions is that $\widehat{\boldsymbol{a}}_n(t+\tau)$ differs from $\widehat{\boldsymbol{a}}_n(t)$ only by a constant phase factor, as indicated by Eq. (16),

$$\widehat{\boldsymbol{a}}_n(t+\tau) = \widehat{\boldsymbol{a}}_n(t)\exp(i\phi_n). \tag{18}$$

Namely, $\widehat{\boldsymbol{a}}_n(t)$'s are "periodic-like" fundamental solutions except the additional constant phase factors. This holds the key for our following discussion.

Such result is very interesting when we use $\widehat{\boldsymbol{a}}_n(t)$'s to calculate the instantaneous value of quantum mechanical quantities, such as energy, charge current, to discuss the time dependent properties of the system. For example, the instantaneous energy of the system in the evolution state $\widehat{\boldsymbol{a}}_n(t)$ satisfies

$$\begin{aligned} E_n(t+\tau) &= \widehat{\boldsymbol{a}}_n^\dagger(t+\tau)A(t+\tau)\widehat{\boldsymbol{a}}_n(t+\tau) = \widehat{\boldsymbol{a}}_n^\dagger(t)A(t)\widehat{\boldsymbol{a}}_n(t) \\ &= E_n(t), \end{aligned} \tag{19}$$

namely, the instantaneous energy of such states shows a periodic behavior.

Another time dependent property we interested in is the current response of the system to the electric field. We note that the matrix form of the charge current operator satisfies $J(t+\tau) = J(t)$; then for state $\widehat{\boldsymbol{a}}_n(t)$,

$$j_n(t+\tau) = \widehat{\boldsymbol{a}}_n^\dagger(t+\tau)J(t+\tau)\widehat{\boldsymbol{a}}_n(t+\tau) = \widehat{\boldsymbol{a}}_n^\dagger(t)J(t)\widehat{\boldsymbol{a}}_n(t) = j_n(t), \tag{20}$$

namely, the instantaneous charge current also shows a periodic behavior. Furthermore, using Eqs. (1) and (2), we have

$$\frac{\partial H(t)}{\partial t} = \hat{j}(t)\mathcal{E} \quad \text{or} \quad \frac{\partial A(t)}{\partial t} = J(t)\mathcal{E}.$$

Then,

$$\frac{dE_n(t)}{dt} = \frac{d}{dt}[\widehat{\boldsymbol{a}}_n^\dagger(t)A(t)\widehat{\boldsymbol{a}}_n(t)] = j_n(t)\mathcal{E}, \tag{21}$$

from which, we find that, to guarantee the periodic behavior of $E_n(t)$, $j_n(t)$ can not be a nonzero constant. Therefore, the instantaneous charge current in each state $\widehat{\boldsymbol{a}}_n(t)$ can only keep zero or show a periodic oscillating behavior, which is very similar to the well-known Bloch oscillations in the $U = 0$ case.

The oscillating periodicity of $j_n(t)$ is same to that of the Bloch Oscillations. For a lattice constant $a \sim 10^{-10}$ m and $\mathcal{E} \sim 10^3$ V/m, the oscillating periodicity $\tau = h/(e\mathcal{E}a) \sim 10^{-8}$ s. Hence, the oscillating is very fast. For an insensitive detecting, one may only obtain an average result of the current. Using Eq. (21), we can obtain the average of $j_n(t)$ in a periodicity

$$\overline{j_n} = \frac{1}{\tau}\int_t^{t+\tau} j_n(t)dt = \frac{1}{\tau\mathcal{E}}[E_n(t+\tau) - E_n(t)] = 0, \tag{22}$$

namely, no net nonzero charge current will be detected for long-time averages of these charge-current oscillating states.

While for a general time evolution state, which we can expand with $\widehat{\boldsymbol{a}}_n(t)$'s as

$$\boldsymbol{a}(t) = c_1\widehat{\boldsymbol{a}}_1(t) + c_2\widehat{\boldsymbol{a}}_2(t) + \cdots + c_N\widehat{\boldsymbol{a}}_N(t), \tag{23}$$

the instantaneous energy of the system,

$$\begin{aligned} E(t+\tau) &= \boldsymbol{a}^\dagger(t+\tau)A(t+\tau)\boldsymbol{a}(t+\tau) \\ &= \sum_{m,n} c_m^* c_n \exp[i(\phi_n - \phi_m)]\widehat{\boldsymbol{a}}_m^\dagger(t)A(t)\widehat{\boldsymbol{a}}_n(t) \\ &= E(t) + \Delta E(t), \end{aligned} \tag{24}$$

where $\Delta E(t) = \Sigma_{m\neq n}\{\exp[i(\phi_n - \phi_m)] - 1\}\widehat{\boldsymbol{a}}_m^\dagger(t)A(t)\widehat{\boldsymbol{a}}_n(t)$; and similarly, the instantaneous charge current

$$j(t+\tau) = \boldsymbol{a}^\dagger(t+\tau)J(t+\tau)\boldsymbol{a}(t+\tau) = j(t) + \Delta j(t), \tag{25}$$

where $\Delta j(t) = \Sigma_{m\neq n}\{\exp[i(\phi_n - \phi_m)] - 1\}\widehat{\boldsymbol{a}}_m^\dagger(t)J(t)\widehat{\boldsymbol{a}}_n(t)$.

Hence though the constant phase factors $\exp(i\phi_n)$'s do not lead to obvious effect on the time dependent properties of the system in the "pure" states $\widehat{\boldsymbol{a}}_n(t)$'s, they can give a magnitude adjusting of the instantaneous average of mechanical operators in a general state, which may lead to complicated response behaviors of the system in an electric field. In fact, similar to Eq. (22), the "periodic average" of the charge current now is turned to be

$$\bar{j} = \frac{1}{\tau}\int_t^{t+\tau} j(t)dt = \frac{1}{\tau\mathcal{E}}[E(t+\tau) - E(t)] \propto \Delta E(t), \tag{26}$$

namely, a nonzero average current can be detected now.

We find that the phase factors $\exp(i\phi_n)$'s are important for discussing the long time evolution behavior of the system. Once the state at time t are known, the instantaneous values of energy and current at any integral multiple of τ time latter can still be calculated by using Eqs. (24) and (25) repeatedly.

Our discussion in fact has provided an effective way to solve the time evolution problem of the Hubbard model in an electric field. We need only go ahead as follows:

(i) Solve Eq. (5) in the time interval $[0,\tau]$ with an arbitrary set of N orthonormalized initial state vectors $\boldsymbol{a}_i(0)$'s, to obtain the solutions $\boldsymbol{a}_i(t)$'s and then construct the matrices $\Psi(\tau)$ and $\Psi(0)$.

(ii) Use Eq. (11) to obtain $\boldsymbol{C} = \Psi(0)^\dagger\Psi(\tau)$ and diagonalize $\boldsymbol{C}$ to obtain the phase factor matrix $\boldsymbol{C}_d$ and the matrix P in Eq. (12). Then $\widehat{\boldsymbol{a}}_i(t)$'s for $t \in [0,\tau]$ can be obtained via Eq. (15) or by re-solving Eq. (5) with the initial state vectors given by $\widehat{\Psi}(0) = \Psi(0)P$.

(iii) Any time evolution state for $t \in [0,\tau]$ can be written in the form of Eq. (23), with the coefficients determined at the initial time.

(vi) The states at a time $t > \tau$ can be obtained via Eqs. (18) and (23).

The most remarkable advantage of this procedure is that we need only solve Eq. (5) in a time interval $[0,\tau]$ to obtain a solution in the whole time range. This is very powerful in a numerical solution, since it can restrain the accumulation of errors, especially when the time range for solving is very large.

Obviously, such procedure can also apply to higher dimensions and to other strong correlated models, just requiring that the model Hamiltonian in an electric field is periodic in time. The procedure actually is an effective way to solve temporal-periodicity differential equations, and may find its use even in classical dynamical systems [19].

4. Summary

We have rigorously shown that there exists a complete set of orthonormal time-evolution states for the 1D Hubbard model in an electric field. For these states, once the charge current appears with the electric field at an initial time, it will oscillate constantly with a periodicity $\tau = h/(e\mathcal{E}a)$ and can not decay into a steady one even for very long time as in a normal metal conductor, nor increase constantly with time as in an real ideal conductor. It is very strange that such constantly-oscillated current states can extensively survive with the strong electronic correlation interactions in the system.

The discussion here actually is a direct and detailed extension of our previous study [12], and can be extended to higher dimensions and other more general cases.

Acknowledgments

This research was financially supported by Qiannan Normal University for Nationalities (Grant No. 2020qnsyzd04), and Guizhou Provincial Science and Technology Department.

References

[1] Eckstein M and Werner P 2011 *Phys. Rev. Lett.* **107** 186406
[2] Mandt S 2014 *Phys. Rev. A* **90** 053624
[3] Castella H, Zotos X and Prelovšek P 1995 *Phys. Rev. Lett.* **74** 972
[4] Zotos X, Naef F and Prelovšek P 1997 *Phys. Rev. B* **55**, 11029
[5] Fujimoto S and Kawakami N 1998 *J. Phys. A: Math. Gen.* **31** 465
[6] Kirchner S, Evertz H G and Hanke W 1999 *Phys. Rev. B* **59** 1825
[7] Peres N M R, Dias R G, Sacramento P D and Carmelo J M P 2000 *Phys. Rev. B* **61** 5169
[8] J. M. P. Carmelo, S.-J. Gu, and P. Sacramento 2013 *Annals of Physics* **339** 484
[9] C. Karrasch, Kennes D M and Moore J E 2014 *Phys. Rev. B* **90** 155104
[10] Jin F, Steinigeweg R, Heidrich-Meisner F, Michielsen K and De Raedt H 2015 *Phys. Rev. B* **92** 205103
[11] Ilievski E and Nardis J D 2017 *Phys. Rev. Lett.* **119** 020602
[12] Zheng Y 2021 *Phys. Lett. A* **394** 127112
[13] Peierls R E 1933 *Z. Phys.* **80** 763
[14] Tan L and Callaway J 1992 *Phys. Rev. B* **46** 5499
[15] Pontryagin L S 1962 *Ordinary Differential Equations* (Reading, Massachusetts: Addison-Wesley)
[16] Cronin J 2008 *Ordinary Differential Equations* (Boca Raton: Chapman & Hall/CRC)
[17] Shirley J H 1965 *Phys. Rev.* **138** B979
[18] Barone S R, Narcowich M A and Narcowich F J 1977 *Phys. Rev. A* **15** 1109
[19] El-Nabulsi R A 2014 *Qual. Theory Dyn. Syst.* 13 149; 2018 *Opt. Quant. Electron* 50 170

A study of lightning electromagnetic effect on the metal roof constructions

Biao Zhang[1], Xue Jiao[2, 3], Junchi Zhou[2], Jiayi Mao[1] and Yuzhi Wang[1]

[1]Nanjing Electric Power Design Institute, Nanjing, China
[2]Meteorological Disaster Protection Center of Jiangsu, Nanjing, China

[3]Corresponding author e-mail: 364703295@qq.com

Abstract. Metal roofs are often seen as substitute of traditional materials such as concrete these years. But there have been few researches on the lightning electromagnetic effect on these constructions. This work adopts the Finite Difference Time Domain (FDTD) method to carry out numerical simulation of Lightning Electro-Magnetic Pulses (LEMP) generated by lightning discharge current. A substation was taken as an example, the distribution of LEMP field with metal roof and non-metal roof were calculated separately by 3-Dimentional FDTD method. The results show that the metal roof has a reflection effect on the LEMP field, which makes the LEMP field significantly weakened compared to traditional concrete buildings. But for constructions of windows and doors, the intensity of LEMP field inside the construction is still destructive. This work also discussed effects of doors and windows on the electromagnetic distribution and came to conclusions that the amplitude of LEMP field decreases significantly with distance from the door (or window) increasing. The size of the windows or doors was positively related to the LEMP amplitude in the construction.

1. Introduction

With development of science and manufacturing techniques, a large variety of new materials are used these days. There are many new buildings using metal sheet as their roofs such as gas stations, storage and transportation warehouses, industrial plants and substations [1]. The metal roof has good thermal insulation performance, light weight, good waterproof performance, simple construction and other characteristics [2]. Meanwhile, the whole building can play a certain shielding role to prevent the electrical system in the building from being interfered by other electro-magnetic signals. However, at the same time, the lightning protection of metal-roof buildings has become the focus of architectural attention.

LEMP is a hazardous natural phenomenon with the characteristics of high intensity, high amplitude and high frequency [3]. Its effect can be divided into direct lightning and inductive lightning. Inductive lightning is the electrostatic induction and electromagnetic induction generated on nearby conductors when a lightning discharge happens. The strong electromagnetic field generated by lightning could cause overvoltage on the electric lines, which is directly transmitted to the device. The electromagnetic field also causes the overvoltage that directly damages the devices. Previous studies showed that the weak current equipment (including connections close by) in 2.4 GS magnetic field will be permanently damaged and suffer from false action in 0.07 GS magnetic field [4]. In a word, it is the overvoltage generated by lightning electromagnetic field that damages equipment and circuits.

Journal of Physics: Conference Series **2313** (2022) 012020 doi:10.1088/1742-6596/2313/1/012020

When the building roof and its walls are completely closed metal structure, the whole building constitutes a "Faraday cage". In this case, there is no electromagnetic invasion at all [5]. But this will not actually happen. There have been many researches on the electromagnetic influence of lightning strikes on traditional concrete buildings, but few calculations for metal roof buildings have been seen. Therefore, this study takes a 110kV substation with metal roof as an example to conduct numerical simulation of lightning strike electromagnetic field distribution. Firstly, the lightning channel is modeled through the lightning return strike model, and the distribution of electromagnetic pulse field of ground lightning is calculated. Secondly, based on the structure of the metal roof substation, a frame model of the building structure is established. Then, the distribution and coupling effect of lightning electric field pulse field inside the structure are analyzed by numerical calculation.

2. Numerical computation method and model establishment

2.1. Numerical computation of Lightning Electro-Magnetic Pulses(LEMP) based on the Finite Difference Time Domain (FDTD) method

The FDTD method enjoys a wide applicability on the issues of electromagnetic calculation as well as analysis of lightning electromagnetic pulses [6]. Because FDTD method solves Maxwell equations in the whole computational domain, the calculation of LEMP field is not limited by the distance of lightning discharge channel [7]. It can calculate the field within 10m from the lightning discharge channel. LEMP has greater destructiveness in the nearer regions. So FDTD method is a powerful tool to study the LEMP field distribution near lightning discharge channels. It can not only calculate LEMP conveniently, but also tackle with the coupling problem of LEMP on cables, shielding chambers and other equipment without assistance of other coupling models [8,9]. In this paper, FDTD method is used to study the distribution of LEMP near the return channel of grounding flashes.

The net division of LEMP near lightning discharge channel solved by two-dimensional FDTD method is shown in Figure 1. The lightning discharge channel is simulated by a return stroke model. In this paper, the MTLL model is adopted [10]. We assume that the lightning discharge channel is perpendicular to the ground, the field around the lightning discharge channel is symmetric, and that the magnitude of the field component is independent of the azimuth angle. In other words, it was considered as a two- dimensional problem, which could be solved in cylindrical coordinates.

The difference equations in two-dimensional cylindrical coordinates are as Equation (1)-(3).

$$E_r^{n+1}\left(i+\frac{1}{2},j\right)=\frac{2\varepsilon-\sigma\Delta t}{2\varepsilon+\sigma\Delta t}E_r^{n}\left(i+\frac{1}{2},j\right) -\frac{2\Delta t}{(2\varepsilon+\sigma\Delta t)\Delta z}\cdot\left[H_\varphi^{n+1/2}\left(i+\frac{1}{2},j+\frac{1}{2}\right)-H_\varphi^{n+1/2}\left(i-\frac{1}{2},j+\frac{1}{2}\right)\right] \tag{1}$$

$$E_z^{n+1}\left(i,j+\frac{1}{2}\right)=\frac{2\varepsilon-\sigma\Delta t}{2\varepsilon+\sigma\Delta t}E_z^{n}\left(i,j+\frac{1}{2}\right) +\frac{2\Delta t}{(2\varepsilon+\sigma\Delta t)r_i\Delta r}\cdot\left[r_{i+1/2}H_\varphi^{n+1/2}\left(i+\frac{1}{2},j+\frac{1}{2}\right)-r_{i-1/2}H_\varphi^{n+1/2}\left(i-\frac{1}{2},j+\frac{1}{2}\right)\right] \tag{2}$$

$$H_\varphi^{n+1/2}\left(i+\frac{1}{2},j+\frac{1}{2}\right)=H_\varphi^{n-1/2}\left(i+\frac{1}{2},j+\frac{1}{2}\right)+\frac{\Delta t}{\mu_0\Delta r}\left[E_z^{n}\left(i+1,j+\frac{1}{2}\right)-E_z^{n}\left(i,j+\frac{1}{2}\right)\right] -\frac{\Delta t}{\mu_0\Delta z}\left[E_r^{n}\left(i+\frac{1}{2},j+1\right)-E_r^{n}\left(i+\frac{1}{2},j\right)\right] \tag{3}$$

In the equations above, ε and σ represent the permittivity and conductivity of free space or earth respectively.

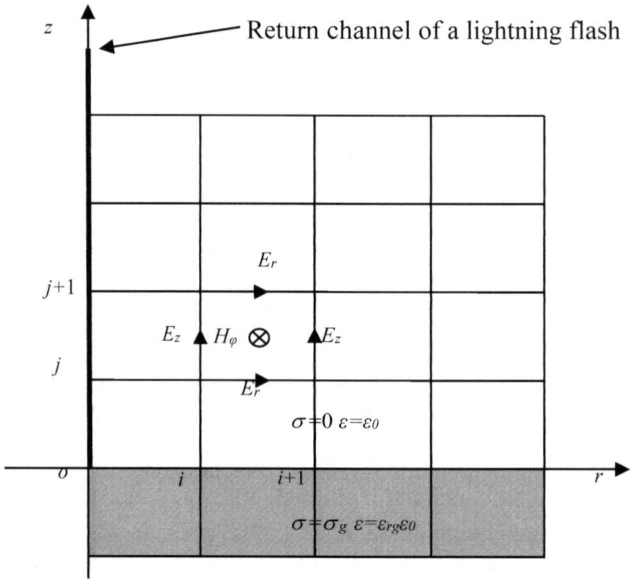

Figure 1. The net division of the two-dimensional FDTD method.

2.2. Model establishment of a lightning stroke on a metal-roof construction based on FDTD method

In this work, we used a simplified three-dimensional model to simulate the effect of a lightning stroke on the metal roof. The distribution of LEMP field in the construction is calculated by 3D FDTD method. The difference equations are shown as Equation (4)-(6).

$$E_x^{n+1}\left(i+\frac{1}{2},j,k\right)=CA\cdot E_x^{n}\left(i+\frac{1}{2},j,k\right)+CB\cdot \left[\frac{H_z^{n+1/2}\left(i+\frac{1}{2},j+\frac{1}{2},k\right)-H_z^{n+1/2}\left(i+\frac{1}{2},j-\frac{1}{2},k\right)}{\Delta y}-\frac{H_y^{n+1/2}\left(i+\frac{1}{2},j,k+\frac{1}{2}\right)-H_y^{n+1/2}\left(i+\frac{1}{2},j,k-\frac{1}{2}\right)}{\Delta z}\right] \quad (4)$$

$$E_y^{n+1}\left(i,j+\frac{1}{2},k\right)=CA\cdot E_y^{n}\left(i,j+\frac{1}{2},k\right)+CB\cdot \left[\frac{H_x^{n+1/2}\left(i,j+\frac{1}{2},k+\frac{1}{2}\right)-H_x^{n+1/2}\left(i,j+\frac{1}{2},k-\frac{1}{2}\right)}{\Delta z}-\frac{H_z^{n+1/2}\left(i+\frac{1}{2},j+\frac{1}{2},k\right)-H_z^{n+1/2}\left(i-\frac{1}{2},j+\frac{1}{2},k\right)}{\Delta x}\right] \quad (5)$$

$$E_z^{n+1}\left(i,j,k+\frac{1}{2}\right)=CA\cdot E_z^{n}\left(i,j,k+\frac{1}{2}\right)+CB\cdot \left[\frac{H_y^{n+1/2}\left(i+\frac{1}{2},j,k+\frac{1}{2}\right)-H_y^{n+1/2}\left(i-\frac{1}{2},j,k+\frac{1}{2}\right)}{\Delta x}-\frac{H_x^{n+1/2}\left(i,j+\frac{1}{2},k+\frac{1}{2}\right)-H_x^{n+1/2}\left(i,j-\frac{1}{2},k+\frac{1}{2}\right)}{\Delta y}\right]$$

$$CA = \frac{1 - \frac{\sigma \Delta t}{2\varepsilon}}{1 + \frac{\sigma \Delta t}{2\varepsilon}}, \quad CB = \frac{\frac{\Delta t}{\varepsilon}}{1 + \frac{\sigma \Delta t}{2\varepsilon}} \tag{6}$$

In the equations above, the values of electrical conductivity σ_g and permittivity ε_g in the earth can be selected according to actual requirements. In this paper, we set σ_g as 0.001S/m and ε_g as 10 [11]. The conductivity σ and permittivity ε_g at the earth-air interface are 1/2 of the sum of the parameters.

The CPML absorption boundary is set on the boundary to avoid reflection at the boundary of the computational domain. The cross-section of 3d computational domain is shown in Figure 2.

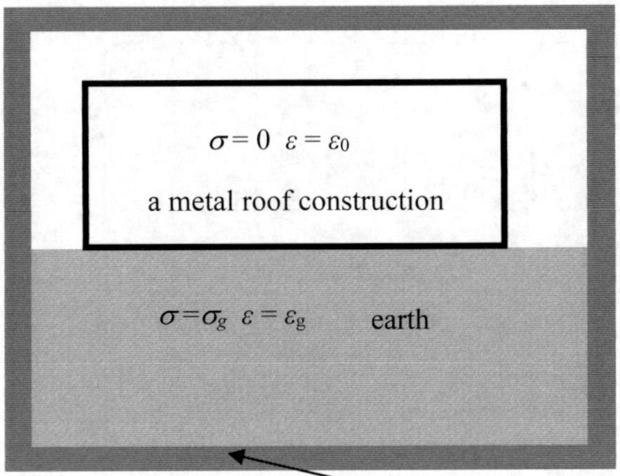

Figure 2. The cross-section of 3d computational domain.

The FDTD difference equations of the E_x on the absorption boundary are listed as Equation (7)-(10) and the difference equations of other components are similar.

$$\begin{aligned} E_x^{n+1}(i+\tfrac{1}{2},j,k) = {} & \frac{2\varepsilon_0\varepsilon_r - \sigma\Delta t}{2\varepsilon_0\varepsilon_r + \sigma\Delta t} E_x^n(i+\tfrac{1}{2},j,k) + \frac{\Delta t / \Delta x}{2\varepsilon_0\varepsilon_r + \sigma\Delta t} \cdot \\ & \Big[(H_z^{n+\frac{1}{2}}(i+\tfrac{1}{2},j+\tfrac{1}{2},k) - H_z^{n+\frac{1}{2}}(i+\tfrac{1}{2},j-\tfrac{1}{2},k)) / \kappa_y \\ & -(H_y^{n+\frac{1}{2}}(i+\tfrac{1}{2},j,k+\tfrac{1}{2}) - H_y^{n+\frac{1}{2}}(i+\tfrac{1}{2},j,k-\tfrac{1}{2}) / \kappa_z \Big] \\ & + \frac{\Delta t}{2\varepsilon_0\varepsilon_r + \sigma\Delta t} \Big[\psi_{e_{xy}}^{n+\frac{1}{2}}(i+\tfrac{1}{2},j,k) - \psi_{e_{xz}}^{n+\frac{1}{2}}(i+\tfrac{1}{2},j,k) \Big] \end{aligned} \tag{7}$$

$$\begin{aligned} \psi_{e_{xy}}^{n+\frac{1}{2}}(i+\tfrac{1}{2},j,k) = {} & b_y \psi_{e_{xy}}^{n-\frac{1}{2}}(i+\tfrac{1}{2},j,k) \\ & + a_y (H_z^{n+\frac{1}{2}}(i+\tfrac{1}{2},j+\tfrac{1}{2},k) - H_z^{n+\frac{1}{2}}(i+\tfrac{1}{2},j-\tfrac{1}{2},k)) / \Delta y \end{aligned} \tag{8}$$

$$\begin{aligned} \psi_{e_{xz}}^{n+\frac{1}{2}}(i+\tfrac{1}{2},j,k) = {} & b_z \psi_{e_{xz}}^{n-\frac{1}{2}}(i+\tfrac{1}{2},j,k) \\ & + a_z (H_y^{n+\frac{1}{2}}(i+\tfrac{1}{2},j,k+\tfrac{1}{2}) - H_y^{n+\frac{1}{2}}(i+\tfrac{1}{2},j,k-\tfrac{1}{2})) / \Delta z \end{aligned} \tag{9}$$

$$b_i = e^{-(\sigma_i/\kappa_i + \alpha_i)(\Delta t/\varepsilon_0)}, \qquad i = x, y, z. \tag{10}$$

In the equations above, κ_i $(i=x, y, z)$ is a real number greater than 1.

In this paper, we set a 110kV substation with metal roof as an example and its simplified model is shown in Figure 3. The height of the structure is 20m and its width and length is 50m and 20m respectively. Theoretically, lightning flashes tend to strike the edges of the building, so we set the simulated lightning discharge channel along the edge.

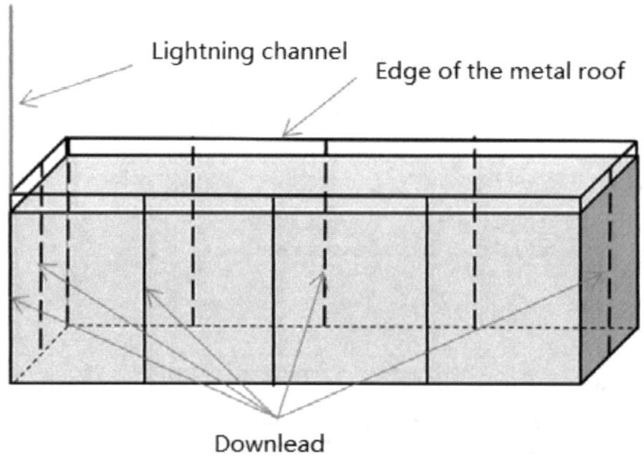

Figure 3. The simplified model of a substation with metal roof.

In the calculation, the base current peak value of lightning discharge channel was set as 30 kA, and the double exponential pulse was used for simulation [12]. The current equation is shown in Equation (11).

$$I = 1.1 \times I_0 \times (\mathbf{e}^{-9.2\times10^4 t} - \mathbf{e}^{-0.5\times10^7 t}) \tag{11}$$

In the equation above, I_0=30 kA. The waveform of the lightning current is shown in Figure 4 and its difference equation is shown in Equation (12). The geoelectric parameter σ_g = 0.001 S/m, ε_{rg}=10[12].

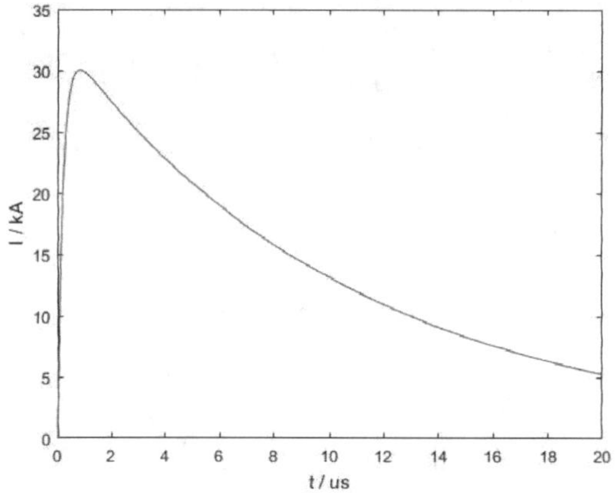

Figure 4. The waveform of the lightning current.

$$E_z^{n+1}\left(i,j,k+\frac{1}{2}\right) = CA \cdot E_z^n\left(i,j,k+\frac{1}{2}\right) + CB \cdot$$

$$\left[\frac{H_y^{n+1/2}\left(i+\frac{1}{2},j,k+\frac{1}{2}\right)-H_y^{n+1/2}\left(i-\frac{1}{2},j,k+\frac{1}{2}\right)}{\Delta x}-\frac{H_x^{n+1/2}\left(i,j+\frac{1}{2},k+\frac{1}{2}\right)-H_x^{n+1/2}\left(i,j-\frac{1}{2},k+\frac{1}{2}\right)}{\Delta y}\right]-\frac{\Delta t}{\varepsilon\Delta x\Delta y}I^{n+1/2} \tag{12}$$

3. Lightning electromagnetic distribution of construction with the metal roof

3.1. Comparision of electromagnetic distribution between metal roof and non-conducting roof

Three-dimensional FDTD is used to calculate the simplified substation model shown in Figure 3. Its size is set as 58m*20m*10m, and the height of the lightning channel is set as 500m. In the calculation, we selected P1 and P2 as evaluation points as shown in Figure 5. P1 is located at the diagonal point of the building horizontally, while P2 is located in the middle. The two points are all located vertically in the middle. The calculation results of the lightning electromagnetic distribution are shown in Figure 6 (The blue line represents electromagnetic distribution of construction with the non-conducting roof and the orange line represents that with metal roof. The pattern on the Y axis direction is similar to that on the X axis direction). It is seen from these pictures that when lightning strikes the building, the metal roof weakens the LEMP field. The metal roof has a reflection effect on the LEMP field at the top of the lightning channel, which makes the LEMP field significantly smaller than that of the non-conducting roof. But we should notice that in occasion of metal roof without complete shielding, the intensity of LEMP field inside the construction is still destructive.

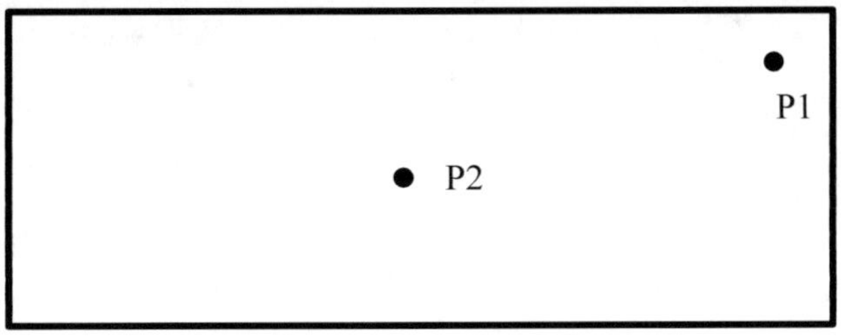

Figure 5. LEMP Evaluation points.

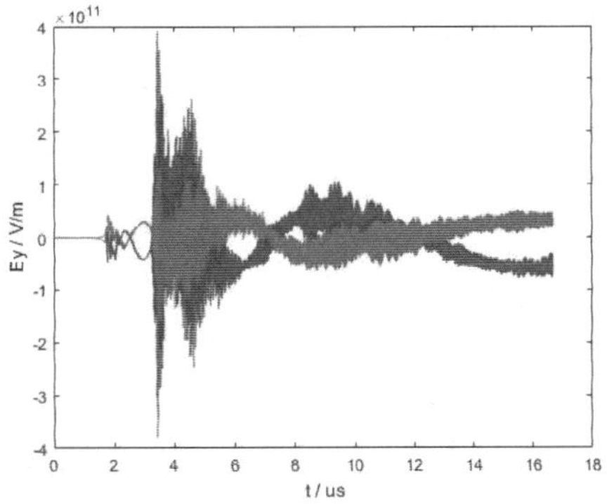

(a) Electric filed on the X axis direction at P1

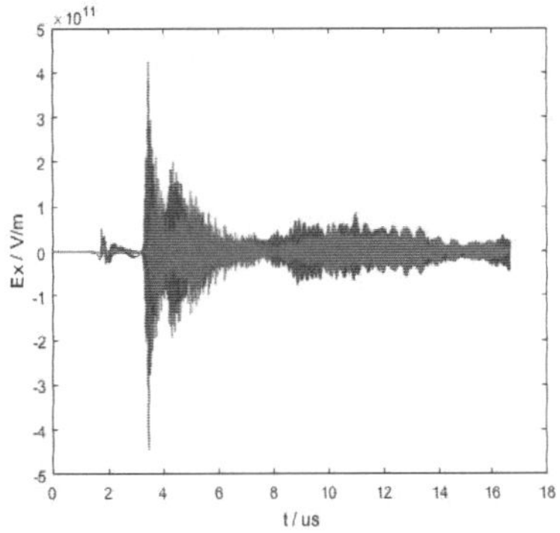

(b) Electric filed on the X axis direction at P2

Journal of Physics: Conference Series **2313** (2022) 012020 doi:10.1088/1742-6596/2313/1/012020

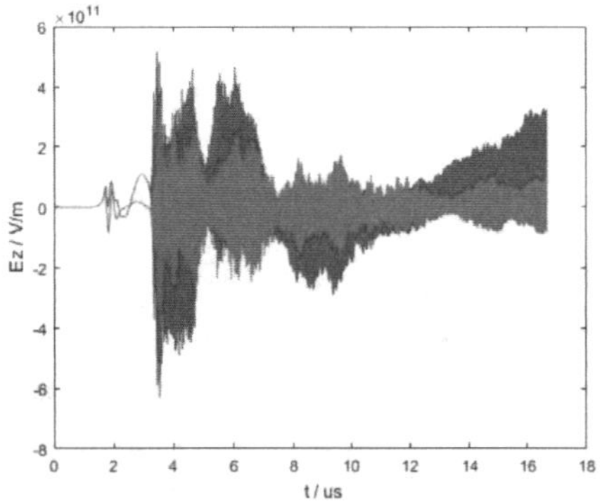

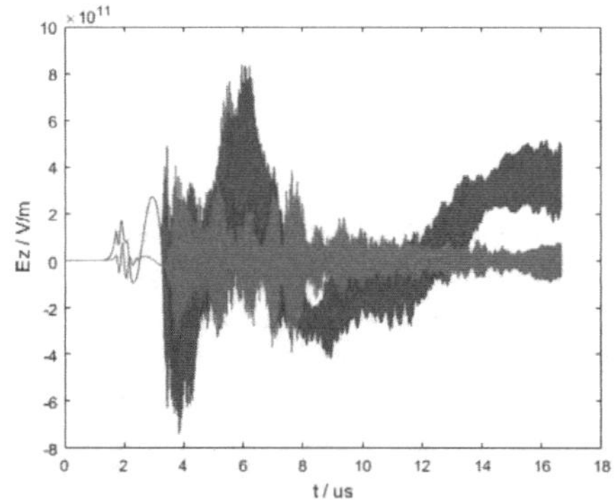

(c) Electric filed on the Z axis direction at P1 (d) Electric filed on the Z axis direction at P2

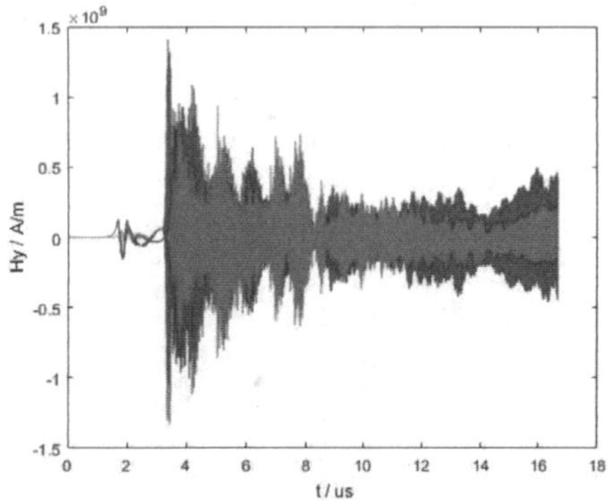

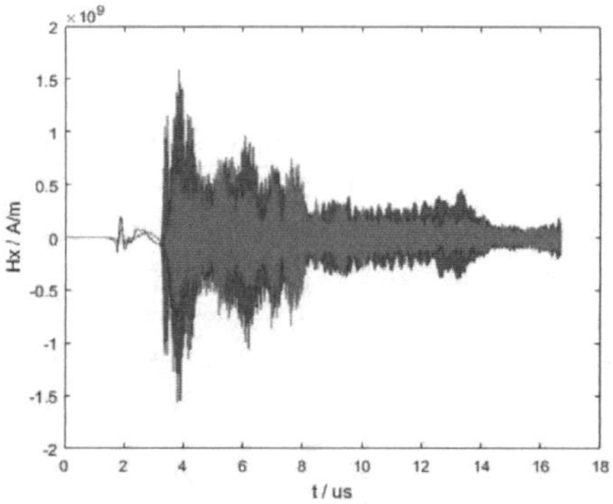

(e) Magnetic filed on the X axis direction at P1 (f) Magnetic filed on the X axis direction at P2

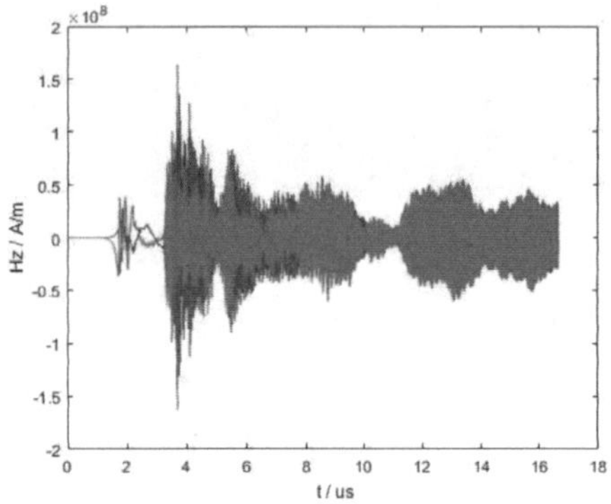

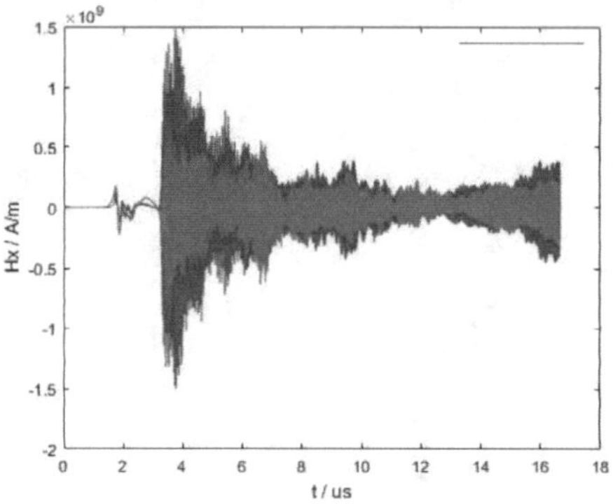

(g) Magnetic filed on the Z axis direction at P1 (h) Magnetic filed on the Z axis direction at P2

Figure 6. The calculation results of the lightning electromagnetic distribution.

3.2. Effect of doors and windows on the electromagnetic distribution

The model shown in Figure 7 is used for calculation. The size of door (or window) is set as 3m*2m. The distance from P1 to the center of the door is 1m. P2 is the central point of the substation. According to the lightning current model mentioned above, the discharge current reaches its peak at 3.3μs. Figure 8 shows the distrubution of LEMP field in the substation at 3.3μs after a lightning strike. As is seen from the figure, the amplitude of LEMP field decreases significantly with distance from the door (or window) increasing. Compared with P1, the coupling field at P2 is greatly attenuated. Then,

Journal of Physics: Conference Series **2313** (2022) 012020 doi:10.1088/1742-6596/2313/1/012020

we further set the door (or window) size as 4m*3m and 6m*5m separately and came to conclusion that the larger the hole size, the larger the coupling field amplitude in the building.

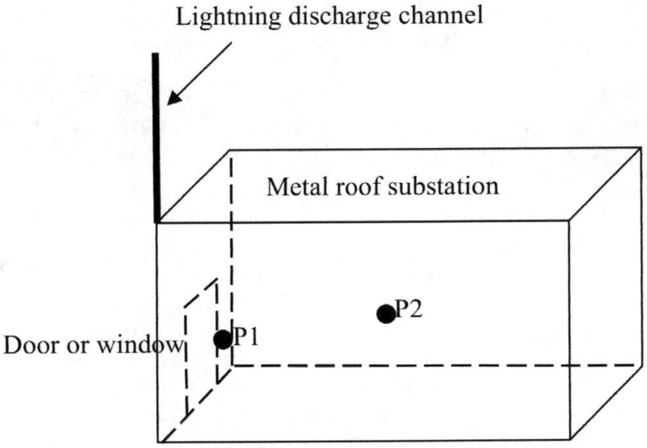

Figure 7. The model of a substaion with a door (or window) struck by lightning.

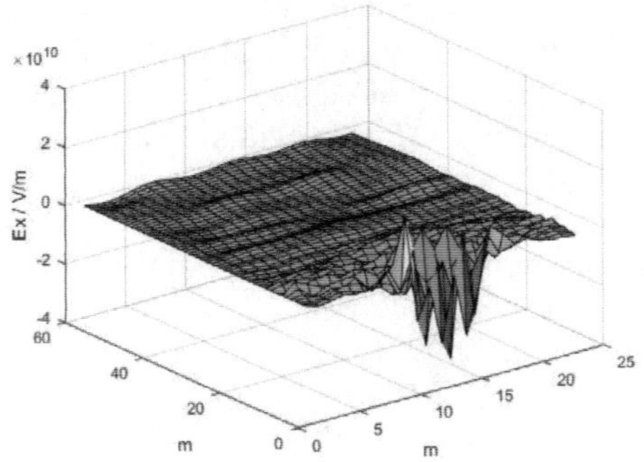

(a) Electric filed on the X axis direction

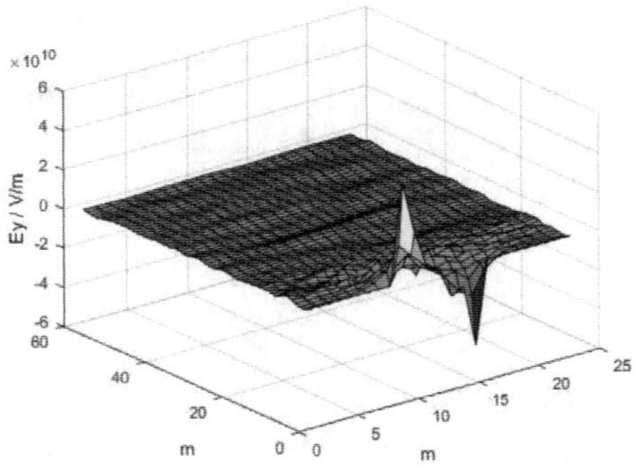

(b) Electric filed on the Y axis direction

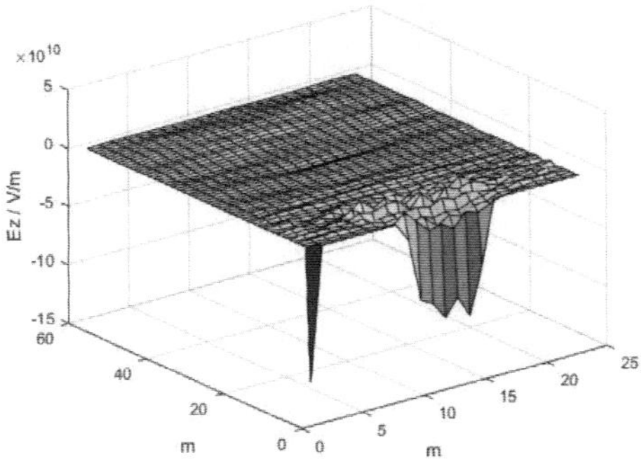

(c) Electric filed on the Z axis direction

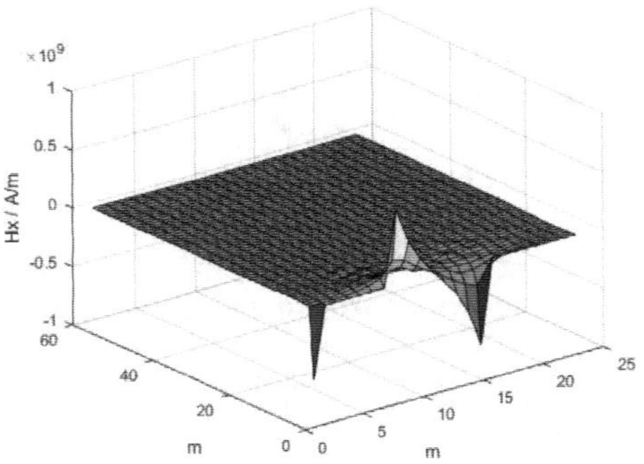

(d) Magnetic filed on the X axis direction

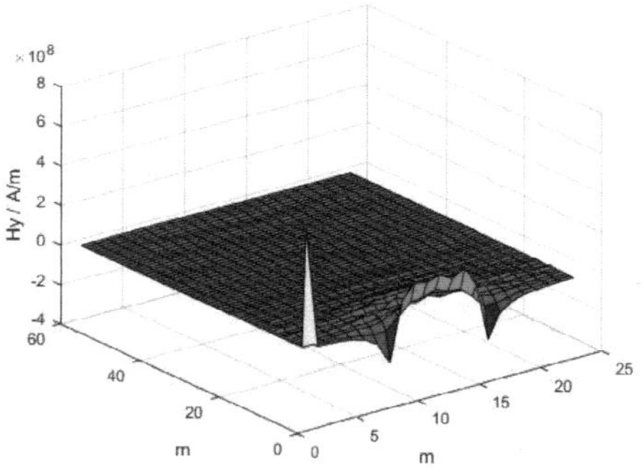

(e) Magnetic filed on the Y axis direction

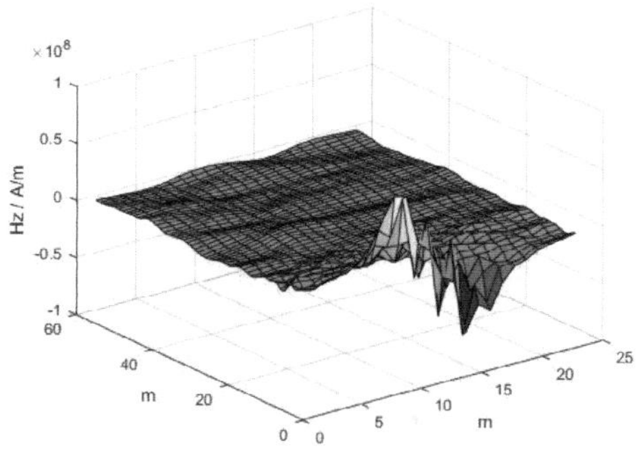

(f) Magnetic filed on the Z axis direction

Figure 8. LEMP distrubution 3.3μs after a lightning strike.

4. Conclusion

In order to study the lightning electromagnetic field distribution of metal-roof constructions, this work adopts FDTD method to carry out numerical simulation calculation of LEMP generated by lightning discharge current. A substation was taken as an example, the distribution of LEMP field with metal roof and non-metal roof were calculated separately for comparison. The results show that when lightning strikes the building, the metal roof weakens the LEMP field. The metal roof has a reflection effect on the LEMP field at the top of the lightning channel, which makes the LEMP field significantly smaller than that of the non-conducting roof. But in occasion of metal roof without complete shielding, the intensity of LEMP field inside the construction is still destructive. Doors or windows are inevitable in most constructions, so this work also analyzed effects of doors and windows on the lightning electromagnetic distribution and came to conclusions that the amplitude of LEMP field decreases significantly with distance from the door (or window) increasing. The size of the windows or doors also counts. It is positively related to the LEMP amplitude in the construction.

Acknowledgments

This paper is co-sponsored by Scientific Research Program of State Grid Nanjing Power Supply Company (SGTYHT/18-JS-202) and Natural Science Foundation of Jiangsu Province (BK20191094).

References

[1] Yan Chen. Research on performance of a kind of integrated metal roof base plate [D]. Harbin Institute of Technology, 2021.

[2] Mao Jie, Gou Jinrui, Peng Qibin. Construction technique of aluminium alloy roofs[J]. Construction Techniques. 2008, 37(7): 69-71

[3] Zhou Junchi, Wang Zhenhui, Feng Minxue, Application of AEF data on lightning prediction around Nanjing[J]. Plateau Meteorology, 2013, 32(5): 1470-1477

[4] V. Cooray. On the concepts used in return stroke models applied in engineering practice. IEEE Trans. Electromag. Compat., 2003, 45(1): 101-108..

[5] S. I. Drabkina. The theory of the development of the spark channel. J.Exper. Theoret. Phys., 1951, 21: 473–483.

[6] B. N. Gorin. Mathematical modeling of the lightning return stroke. Elektrichestvo, 1985, 4:10-16.

[7] V. Cooray. On the concepts used in return stroke models applied in engineering practice. IEEE Trans. Electromag. Compat., 2003, 45(1): 101-108.

[8] V. A. Rakov, A. A. Dulzon. Calculated electromagnetic fields of lightning return stroke. Tekh. Elektrodianm., 1987, 1: 87-89.

[9] C. S. Yang, B. H. Zhou. Calculation methods of electromagnetic fields very close to lightning. IEEE Trans. Electromagn. Compat., 2004, 46(1): 133-141

[10] C. A. Nucci, C. Mazzetti, F. Rachidi, et al. On lightning return stroke models for LEMP calculations. In Proc. 19th Int. Conf. Lightning Protection, Graz, Austria, Apr. 1988.

[11] J. A. Roden, S. D. Gedney. Convolution PML (CPML): An efficient FDTD implementation of the CFS – PML for arbitrary media. Microwave and Optical tech. Letters, 2000, 27(5): 334-339.

[12] T. Noda, S. Yokoyama. Thin wire representation in finite difference time domain surge simulation. IEEE Trans. Power Del., 2002, 17(3): 840–847.

Study on thrust performance of small water rocket launch

Yuqi Yang[1], Longbin Liu[1, 2] and Moyan Chen[1]

[1]College of Aerospace Science and Engineering, National University of Defense Technology, Changsha City, Hunan Province, 410073, China

[2]Corresponding author's e-mail: longbuaa@163.com

Abstract. For water rocket test of low cost, quick assembly and simple, based on the single water rocket, using Bernoulli equation, the ideal gas adiabatic model and mass conservation theorem, this paper derived both the calculation method of thrust rocket the bottle under the different initial water volume ratio of water pressure, and the rule about how the water rocket thrust and pressure performance change. It is found that the pressure, the jet velocity of the water relative to the water rocket body and thrust of water rocket will decrease with the decrease of water amount under different initial charging pressure and different initial water amount. If the initial pressure is higher, the water rocket will gain more energy and launch faster; However, it doesn't mean that the more initial water storage ratio is, the better. The maximum velocity obtained by water rocket under the same initial pressure increases first and then decreases with the initial water storage ratio and the optimal initial water storage ratio increases with the increase of initial pressure. According to data calculation, the numerical value increases with the increase of initial pressure. The optimal performance of water rocket can be obtained by taking the maximum pressure and the corresponding optimal initial injection water when the water rocket body is allowed.

1. Introduction

In order to provide strong theoretical support for water rocket launch, in previous studies, K. Utsav et al. calculated and obtained numerical solutions of water rocket parameters by using partial differential equations which most important finding is that the choked air flow at the exit and occurs after the rocket runs out of water continues to accelerate the rocket considerably [1]. Wang C et al. obtained the time needed to rise to the highest point in the water spraying stage by applying kinematic law in the launching process of water rocket and finally came to the conclusion that the altitude evolution curve with time during the movement of water rocket [2]. Liao L Y et al. using the function principle and ideal gas adiabatic equation, deduced the discharge time and water rocket range and obtained the water rocket firing range and water spray time theory expression which is vital for us to study water rocket better [3].

But theoretical research on water rocket thrust performance is limited, so in this paper the relationship between the volume of water in a water rocket and the initial pressure, and the jet velocity of the water relative to the water rocket body and the thrust of the water rocket is found under the ideal conditions. The relation between different initial pressure of water rocket and corresponding optimal initial water injection is obtained.

2. Notations

As can be seen in the following tables, Table 1 is the constant symbols table and Table 2 is the variable symbols table.

2.1. Constant symbols

Table 1. Constant symbols.

Symbols	Description	Value	Unit
P_a	Ambient pressure	101325	Pa
A_e	Export area	$0.02^2 \times \pi \times 4^{-1}$	m^2
A_p	Cross-sectional area	$0.15^2 \times \pi \times 4^{-1}$	m^2
ρ	The density of water	1000	kg/m^3
V	The bottle volume	0.002	m^3
m_p	The structural mass of the rocket	0.05	kg

2.2. Variable symbols

Table 2. Variable symbols.

Symbols	Description	Unit
P_0	Initial bottle pressure	Pa
P_{in}	Bottle pressure	Pa
P_e	The pressure inside the bottle	Pa
V_e	The water volume	m^3
M	The full mass	kg
m_e	The water mass	kg
u_s	Drop rate of water surface in bottle	m/s
u_e	The jet velocity of the water relative to the water rocket body	m/s
F	Thrust	N
k	The water storage ratio	1

3. Assumptions

It is assumed that the ejection of water from the water rocket is instantaneous. Under this hypothesis, it is unnecessary to consider the decrease of the velocity caused by the gravitational potential energy of the bottle and water in the process of spraying water due to the rising height and the change of spraying speed.

It is assumed that the surface of the water rocket is smooth and the friction in the surface is negligible due to air resistance.

It is assumed that when air and water come into contact, internal compression due to pressure is ignored and water is treated as an incompressible fluid.

4. The simulation condition

4.1. Simulated entity

The entity we want to simulate is 2 L Coca-Cola plastic bottle. Water is injected into the bottle, then a pump is used to inject gas into the plastic bottle to increase atmospheric pressure. In this paper, different initial inflatable pressure $P_0 = (2,3,4)$ bar and different initial water storage ratio $k_0 = (0.20, 0.25, 0.33, 0.50, 0.70)$ were used for data simulation under the range that the water rocket bottle can bear. The shape of water rocket is shown in Figure 1.

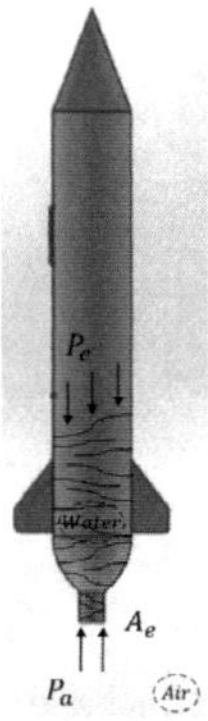

Figure 1. The shape of water rocket.

4.2. Simulation environment

The temperature of simulated environment is 20 °C and the atmosphere is 1 bar.

When the pressure inside the bottle is greater than the external pressure, the water inside the water rocket will be ejected under pressure and cause a force in the bottle so the rocket body will get a thrust and accelerate so that it reaches a height of more than ten meters [4].

5. The simulation design

5.1. The gas pressure in the bottle P_{in}

Assume that a water volume parameter ***k*** represents the ratio of the volume of water in the bottle to the total volume of the bottle so when doing simulations, the initial value of k, k_0 (i.e. the volume of water) can be adjusted according to needs to change the jet speed of the subsequent rocket. Thus, the water injection ratio in the rocket is denoted by k:

$$V_e = kV. \tag{1}$$

In the process of continuous outward spraying of water, the volume of water decreases, the volume of air in the bottle increases and the pressure in the bottle also decreases. Therefore, the volume of air in the bottle V_r can be obtained according to the volume of water in the bottle:

$$V_r = V - V_e = (1 - k)V. \tag{2}$$

According to the adiabatic equation of ideal gas [5], the relationship between the gas pressure in the bottle is as follows:

$$P_0[(1 - k_0)V]^\gamma = P_{in}[(1 - k)V]^\gamma. \tag{3}$$

Where, γ is the specific heat ratio of air, which is generally valued as γ=1.4 [2].
Thus, the gas pressure in the bottle can be expressed as:

$$P_{in} = P_0 \left(\frac{1-k_0}{1-k}\right)^{\gamma}. \tag{4}$$

5.2. *The jet velocity of water relative to the rocket body u_e*

The initial bottle pressure is P_0, the gas pressure in the bottle is P_{in}, the initial volume of the gas is V_0 and the water level is falling at a rate of u_s.

Since the viscosity of the water in the bottle is ignored, assuming incompressibility, it can be obtained according to Bernoulli's equation [6]:

$$P_{in} + \frac{1}{2}\rho u_s^2 = P_a + \frac{1}{2}\rho u_e^2. \tag{5}$$

According to the mass conservation equation, the mass of water flowing out of the lower surface of the bottle is equal to the mass reduction of water on the upper surface, which can be expressed as:

$$A_p u_s \rho dt = A_e u_e \rho dt. \tag{6}$$

That is,

$$u_s = \frac{A_e u_e}{A_p}. \tag{7}$$

By combining equations (5), (6) and (7), the jet velocity of the water relative to the water rocket body at the current moment is:

$$u_e = \left[\frac{2(P_{in}-P_a)}{\rho\left(1-\frac{A_e^2}{A_p^2}\right)}\right]^{1/2}. \tag{8}$$

Substitute the pressure expression into the jet velocity of the water relative to the water rocket body:

$$u_e = \left[\frac{2\left(P_0\left(\frac{1-k_0}{1-k}\right)^{\gamma}-P_a\right)}{\rho\left(1-\frac{A_e^2}{A_p^2}\right)}\right]^{1/2}. \tag{9}$$

5.3. *The thrust F of the rocket*

Let's say the rocket has a uniform external pressure. Assume that fuel inlet momentum is ignored. The fuel inlet velocity is negligible. According to momentum conservation, the formula of thrust received by the rocket is [7]:

$$F = \dot{m}_e u_e + (P_e - P_a)A_e. \tag{10}$$

However, there is a big difference between water rockets and other fuel rockets which is that the medium inside the bottle (water) is incompressible so the water pressure at the spout of a water rocket is consistent with the ambient pressure. Then, the thrust formula of water rocket can be simplified as:

$$F = \dot{m}_e u_e. \tag{11}$$

Instantaneous ejection mass $\dot{m}_e$ is:

$$\dot{m}_e = \rho u_e A_e = A_e \left[\frac{2\rho\left(P_0\left(\frac{1-k_0}{1-k}\right)^\gamma - P_a\right)}{\left(1-\frac{A_e^2}{A_p^2}\right)} \right]^{1/2} . \tag{12}$$

Thus, the total expression of thrust F is:

$$F = 2A_e\left(P_0\left(\frac{1-k_0}{1-k}\right)^\gamma - P_a\right)\left(1-\frac{A_e^2}{A_p^2}\right)^{-1} \tag{13}$$

6. Results and analysis

When the initial pressure and initial water storage ratio in the bottle of 3 *bar* are k_0 = (0.20, 0.25, 0.33, 0.50, 0.70), the gas pressure in the bottle P_{in}, the jet velocity of the water relative to the water rocket body u_e of the water rocket and the thrust F obtained by exhausting water are calculated in dependence with the water storage ratio k through numerical simulation and the curves are shown as follows. Figure 2 presents the gas pressure in the bottle, Figure 3 presents the variation of rocket velocity and Figure 4 presents the thrust variation with water storage ratio.

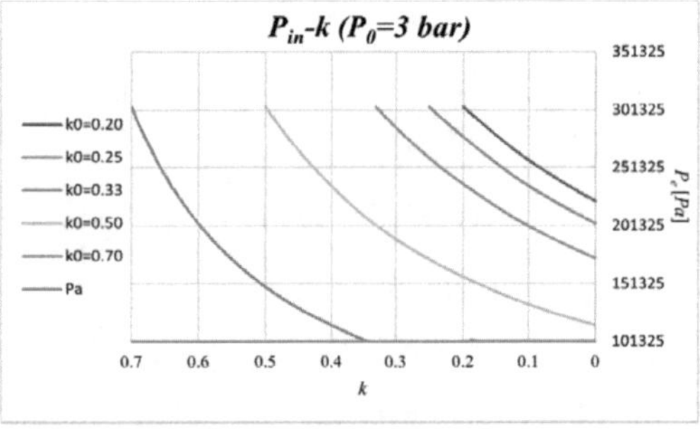

Figure 2. Gas pressure in the bottle with water storage ratio (3 *bar*).

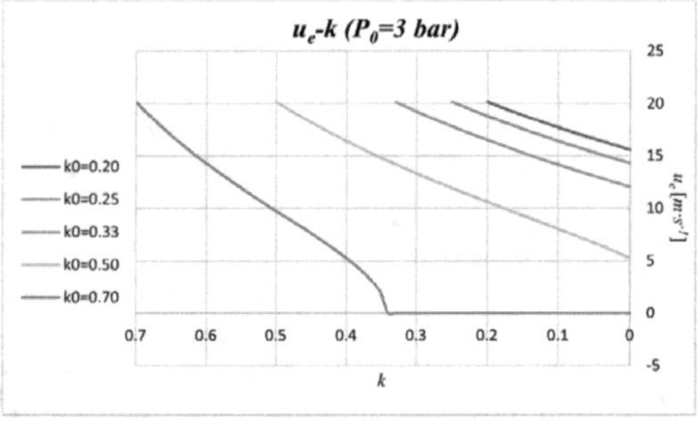

Figure 3. Variation of rocket velocity with water storage ratio (3 *bar*).

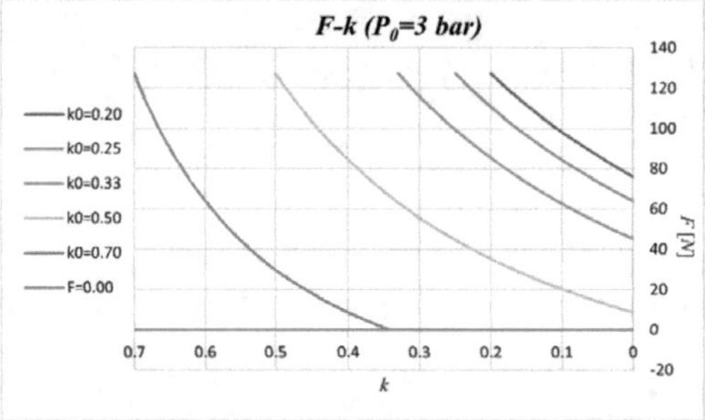

Figure 4. Thrust variation with water storage ratio (3 bar).

As can be seen from the figure, the pressure in the bottle, the jet velocity of the water relative to the water rocket body and the obtained thrust all decrease with the spraying process, which is consistent with the actual situation of water rocket launch. From the perspective of initial water storage ratio k_0, when the volume of remaining water in the bottle is the same, the larger the initial water storage ratio k_0 is, the lower the calculated pressure in the bottle will be and the spray speed and thrust will be smaller.

By observation, when $k_0 = 0.7$, $P_{in} = P_a$ occurs around $k = 0.35$ in the process of spraying water. In this instant, the water rocket will not be able to eject gas because the pressure in the bottle is equal to the ambient pressure, resulting in the situation of $u_e = 0$ and $F = 0$. The water will never be able to eject and will no longer produce thrust. In this case, after the initial storage of water in the bottle as a propellant, according to the momentum conservation law, the ejection of water will give the rocket thrust and energy. Some of the water remains inside the water rocket, adding weight to the water rocket and becoming the resistance of the subsequent stage of the water rocket's ascent.

Therefore, the optimal water volume of the water rocket should be the critical value of water energy ejection. The following is the result curve of outlet pressure P_{in} simulated by different initial internal pressure $P_0 = (2, 3, 4)$ *bar* as is shown in the Figure 5.

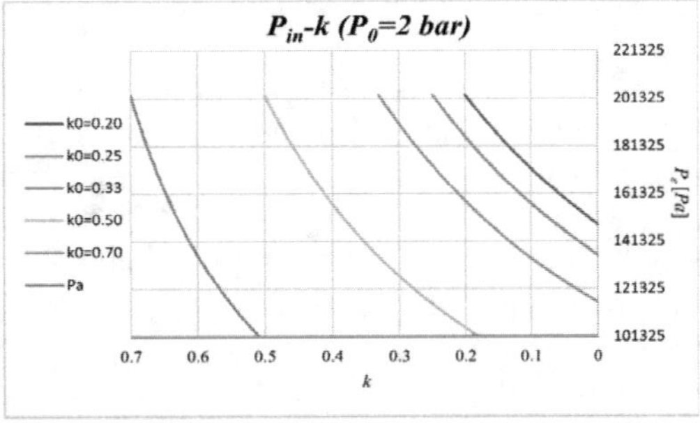

Figure 5. Gas pressure in the bottle with water storage ratio (2 bar).

As shown in the figure, the pressure in the bottle will decrease from the initial $P_0 = 2$ *bar* no matter how much water is injected initially. However, as the proportion of initial injection increases, the pressure in the bottle eventually decreases until it is equal to the ambient pressure. In both cases of initial water storage ratio $k_0 = 0.50$ and $k_0 = 0.70$, there will be residual water in the bottle when the pressure inside the bottle is equal to the ambient pressure. And in the case that the initial pressure inside the bottle is 2 *bar*, with the increase of k_0, the volume of remaining water becomes larger and larger, so the resistance caused by the remaining water will become more and more. In the three cases of initial water storage ratio $k_0 = (0.20, 0.25, 0.33)$, there is no residual water, but the pressure in the bottle is still greater than the ambient pressure after all water injection, which is equivalent to waste part of the pressure. By reading the graph, it can be easily found that the optimal initial water storage ratio is $k_0 \in (0.33, 0.5)$ when $P_0 = 2$ *bar*.

Compared with Figure 2, when the initial pressure increases to 3 *bar*, there is no residual water at $k_0 = 0.50$ and the volume of residual water at $k_0 = 0.70$ also decreases. In general, the pressure after water injection at $P_0 = 3$ *bar* is generally greater than the pressure at $P_0 = 2$ *bar*. As the initial pressure increases, the optimal initial injection volume also increases [8]. It can be reflected from the side that under the condition of pressure increase, we can choose to increase the initial injection water to make full use of the kinetic energy generated by pressure injection water. In the case of $P_0 = 3$ *bar*, there is an optimal initial storage amount $k_0 \in (0.50, 0.70)$. It is shown in the Figure 6.

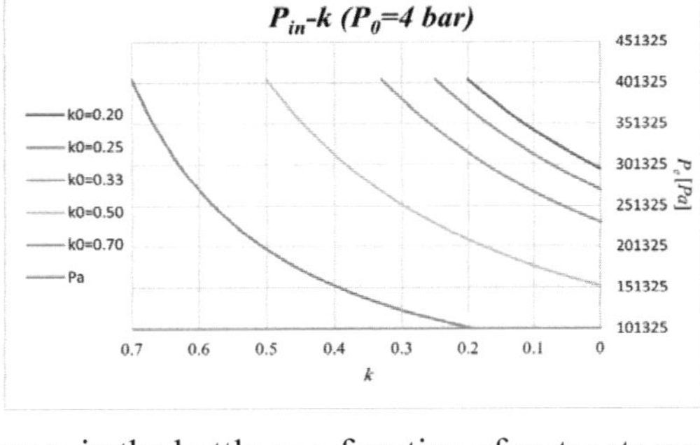

Figure 6. Pressure in the bottle as a function of water storage ratio (4 *bar*).

The diagram of $P_0 = 4$ *bar* can confirm the correctness of the above analysis law. Similarly, in the case of $P_0 = 4$ *bar*, the optimal initial storage amount is $k_0 \in (0.50,0.70)$.

It should be noted that the values solved in this paper are all simulated under ideal conditions and without considering the loading weight of water rocket. Practical factors need to be added in order to obtain the real optimal initial injection amount.

7. Conclusion

In this paper, different initial charging pressure $P_0 = (2, 3, 4)$ *bar* and different initial water storage ratio $k_0 = (0.20, 0.25, 0.33, 0.50, 0.70)$ were used for data simulation under the range that the water rocket bottle can bear.

A higher initial pressure in the bottle is better, but a larger initial injection volume is not for higher attitude. If the initial water injection is not optimized and too much water is added, the water injected into the bottle will not completely provide thrust for the launch of the water rocket, but the gravity of water will produce resistance for the water rocket. Too little water is added to take full advantage of the momentum and thrust generated by the pressure so an optimal initial injection volume should be found. The optimal initial injection volume is affected by other factors, and the optimal value will increase with the increase of the initial pressure in the bottle.

However, in order to obtain the optimal initial water storage ratio k_0, it is necessary to conduct experiments under actual conditions and consider more complete factors to obtain the optimal water storage corresponding to different initial pressures.

Acknowledgements

This work was supported by the Young Talents Project of the School of Aerospace Science and Engineering in National University of Defense Technology.

References

[1] K Utsav, R Alnouti and M. Muflehi 2020 "Prediction of Water Rocket Performance," 2020 Advances in Science and Engineering Technology International Conferences (ASET), pp. 1-5.
[2] Wang C, Jin Y N, Xie M and Chen Y C. 2017(06) Theoretical study of water Rocket motion. *Chinese Physics Bulletin*, pp. 24-27. (in Chinese)
[3] Liao L Y and Yang X D 2016 Physics and engineering, **26**(03), pp. 82-86. (in Chinese)
[4] Information on http://www.aircommandrockets.com/water.htm
[5] Information on https://baike.baidu.com/item/ / 5803033 adiabatic equation? Fr = Aladdin(in Chinese)
[6] Yi W B, Meng Q C, Deng H, Wang C, Zhang Z H and Gu J N. 2021 Some problems in teaching design of Bernoulli Equation. *Mechanics in Engineering*, **43**(06), pp. 967-972. (in Chinese)
[7] Information on https://baike.baidu.com/item/ Law of momentum conservation /552651?fr=aladdin (in Chinese)
[8] Information on http://www.aircommandrockets.com/flying_higher.htm

Simulation of gradient floating field limiting rings for 4H-SiC power devices

Xiuxiu Gao[1] and Xiaoping Dai

CORESING SEMICONDUCTOR TECHNOLOGY CO. LTD, Zhuzhou, China

[1]Corresponding author's e-mail: 1507978182@qq.com

Abstract. In order to have higher breakdown voltage (BV) for 4H-SiC power devices, a novel edge termination structure with gradient floating field limiting rings (FFLR) is proposed. The device blocking performance between conventional FFLR and proposed structures are compared by 2-D numerical device simulation. The BV can be increased by 10% and 20% of parallel plane breakdown voltage with the same P-well and epitaxial parameters when step is 0.2μm and 0.4μm.

1. Introduction

The most basic requirement of all semiconductor power devices is the ability to withstand a larger terminal voltage with minimum leakage current in the off state. The maximum terminal voltage that the device can withstand is blocking voltage (BV). BV is usually determined by the material characteristics and device design. For power devices made of chips by cutting wafers, BV decreases due to the 3-D effects of field crowding at the junction periphery [1], which is usually relieved by special edge termination. Therefore,the edge termination in high-voltage power devices include 4H-SiC is a key design area [2]. The simple and popular way is to implement multiple FFLRs [3-4] and junction termination extension (JTE), its modified forms [5-15] have been studied. FFLR includes a series of concentric p+ rings around the main junction and can manipulate the depletion contour to be smooth. FFLRs requires a small ring spacing between the main junction and the first ring [16]. However, there is still a lot of room to further improve the device blocking capability with the existing process.

In this work, a novel gradient FFLRs(G_FFLRs) is proposed. The device blocking performances between conventional FFLRs(C_FFLRs) and proposed structures are systematically compared by 2-D device simulation to prove the effectiveness of the new edge termination design.

2. Conventional and Proposed FFLRs

Schematic diagram of the conventional and gradient FFLRs as termination structures of 4H-SiC power device are shown in Figure 1a and 1b. The structure consists of 0.02 Ω·cm substrate, a 0.5μm thick/ 1×10^{18} cm^{-3} doped n-type buffer layer, a 11μm thick, $8\times10^{15}cm^{-3}$ doped n-type epitaxial layer. The junction depth of P_0 is 1.2μm. The ring space of the conventional and gradient FFLRs are the same, and the initial spacing of the first ring S_1 is 1.2μm. The expansion ratio of ring spacing XFFLR=Sn+1/Sn=1.05, and the period of adjacent field rings Pn=6μm. The proposed concept consists of gradually deepened rings. In Figure 1(b), the junction depth and doping distribution of p_1~p_n are the same as that of the main junction p_0, the junction depth from p_{n+1} is gradually deepened in a manner that $Xj(p_{n+i})=Xj(p_n)+i\times step$, where $Xj(p_n)$ and $Xj(p_{n+i})$ is the junction depth of FFLRs p_n and p_{n+I}, step is the incremental junction depth, and the minimum step is equal to "0", which corresponds to a conventional FFLRs.

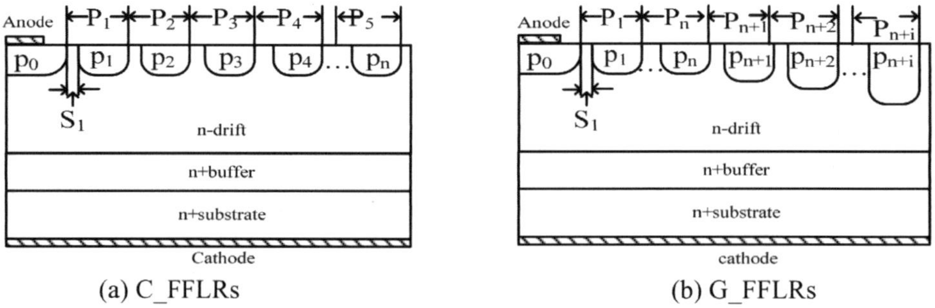

Figure 1. The schematic cross-section of FFLRs.

3. Results and discussion

3.1. Blocking characteristic comparison of conventional and proposed FFLRs

The simulation results of electric field contours, electric field cut along depth minus 0.1μm and BV for C_FFLRs and G_FFLRs with 0.1μm, 0.2μm, 0.3μm, 0.4μm step are shown in Figure 2(a)-2(h). All simulation keep the same FFLRs spacing and doping profile, but changes the $Xj(p_{n+1})$~$Xj(p_{n+i})$. Gaussian doping profile with peak position y=0μm and peak concentration $5\times10^{19}cm^{-3}$ was used.

Figure 2(a) shows that the space charge region of the drift region tilts from the vicinity of the seventh FFLRs, that means the n of G_FFLRs is 6. It can be seen that the BV is improved by introducing G_FFLRs, as shown in Figure 2(g). As a result, the BV can be increased by 10% and 20% of parallel plane breakdown voltage with the same P-Well and epitaxial parameters when step is 0.2μm and 0.4μm. Adding G_FFLRs redistributes the electric field to makes full use of the drift region,as shown in Figure 2(b-f).Smoother distribution of electric field can be seen with the proposed G_FFLRs from Figure 2(g), indicating this structure to be more sustainable to high electric field, which results in enhanced BV. The improvement is pronounced with 0.4μm step on the premise that the FFLRs total width keep the same.

3.2. Effects of FFLRs width on blocking characteristic of conventional FFLRs

Figure 3 shows the electric field contours and BV for C_FFLRs with 126.3μm, 134.2μm, 142.2μm FFLRs width which corresponding to 0μm, 0.2μm, 0.4μm step of the G_FFLRs, which is used for comparison with the same width of G_FFLRs with step 0.2um and 0.4um. It keeps the spacing of FFLRs which is described in section 1.

From Figure 3, it is demonstrated that C_FFLRs with 126.3μm, 134.2μm, 142.2μm width show very similar behaviors and BV. Compared with 126.3μm wide C_FFLRs, BV can be improved 9V and 24V higher by 134.2μm, 142.2μm wide C_FFLRs, which means that the influence of the change of G_FFLRs width caused by lateral diffusion of different junction depth on BV can be ignored.

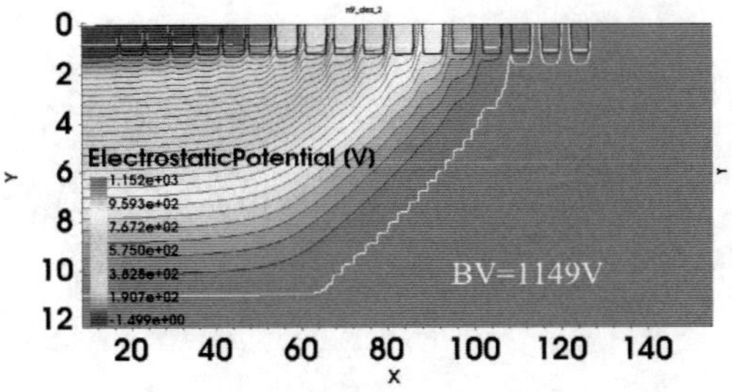

(a) C_FFLRs (potential)

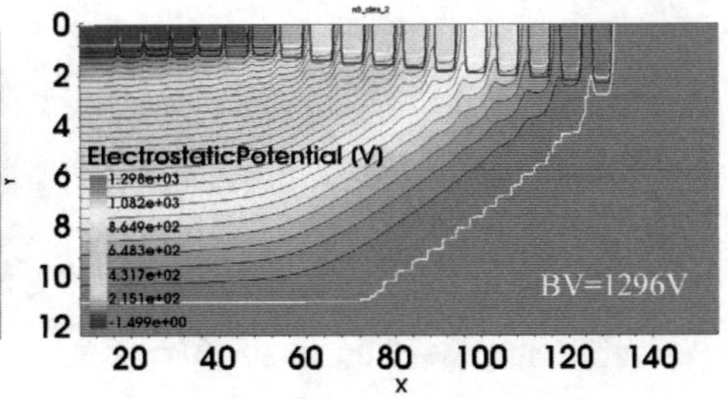

(b) G_FFLRs, step=0.1μm (potential)

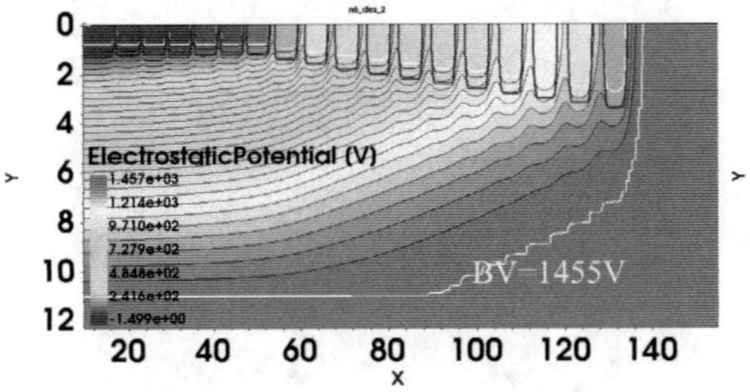

(c) G_FFLRs, step=0.2μm (potential)

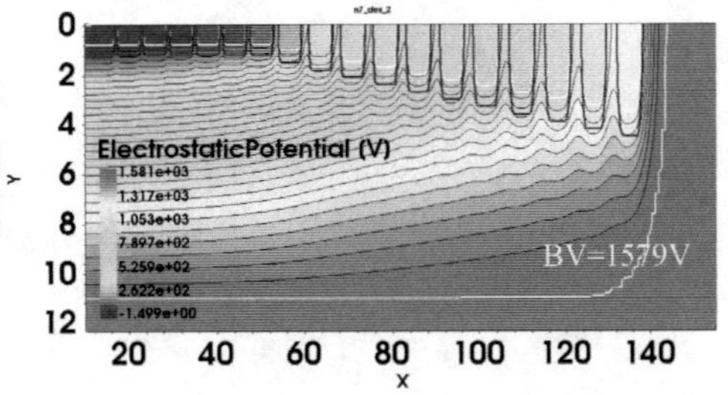

(d) G_FFLRs, step=0.3μm (potential)

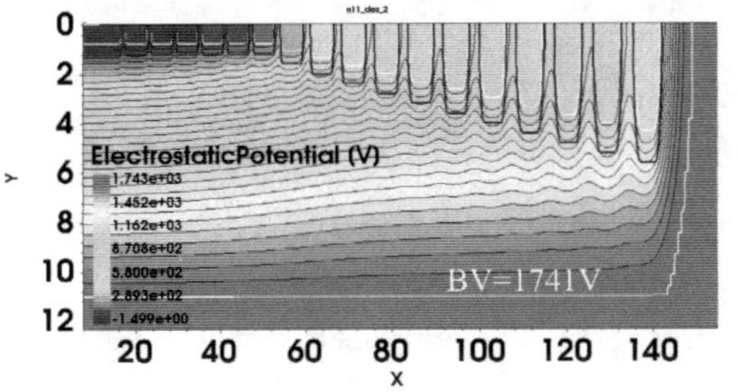

(e) G_FFLRs, step=0.4μm (potential)

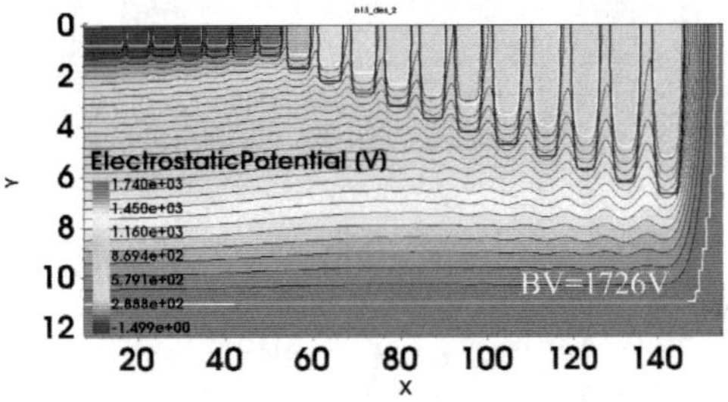

(f) G_FFLRs, step=0.5μm (potential)

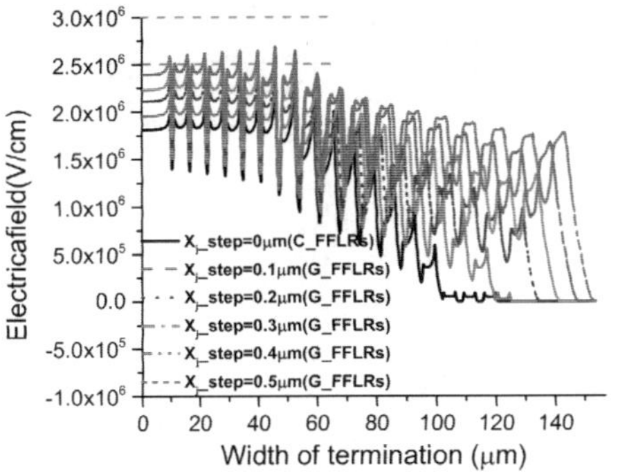

(g) Electric Field distribution~step relationship

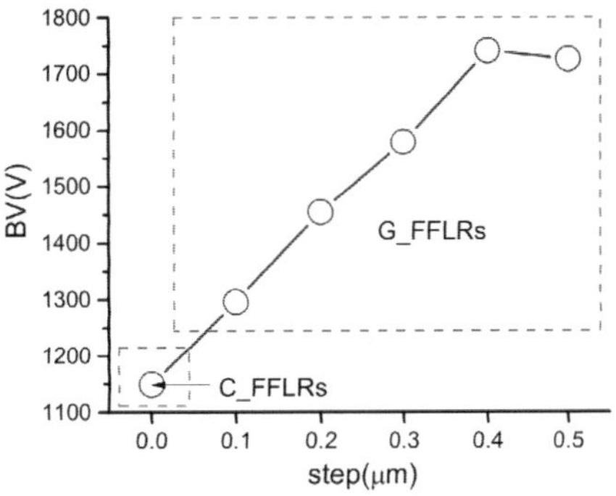

(h) BV~step relationship

Figure 2. Simulation results of conventional and gradient FFLRs.

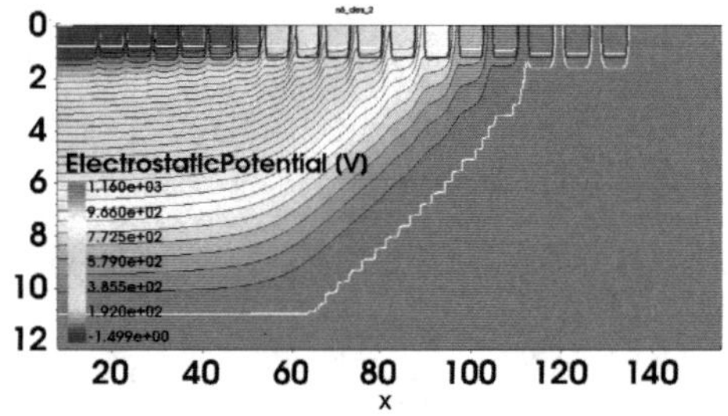

(a) C_FFLRs width 134.2μm (potential)

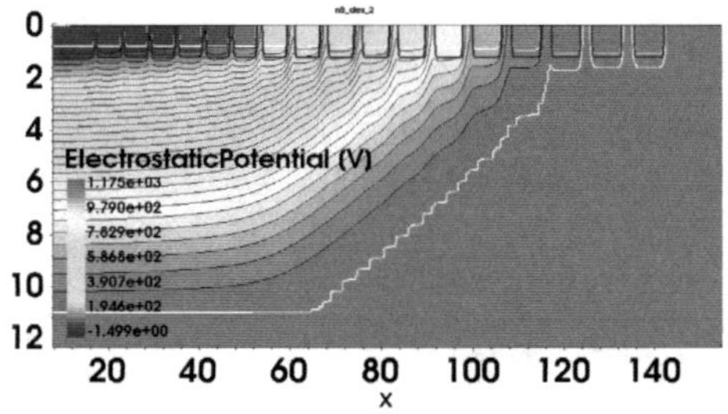

(b) C_FFLRs width 142.2μm (potential)

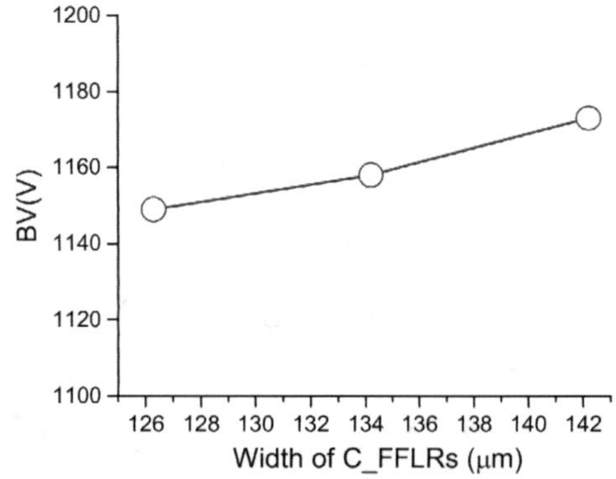

(c) BV~FFLRs width relationship

Figure 3. Simulation results of electric static potential distribution and BV in C_FFLRs with different ring width.

3.3. Effects of doping profile on blocking characteristic

Simulated doping profile along cut lines X=100μm, Y=0~4μm and BV with various junction depth and doping profile are depicted in Figure 4. The doping profile is Gaussian distribution, the peak position is at half of the junction depth.

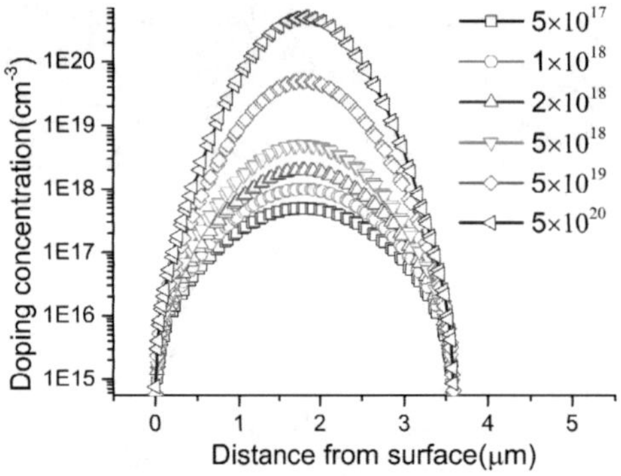

(a) Doping Profile

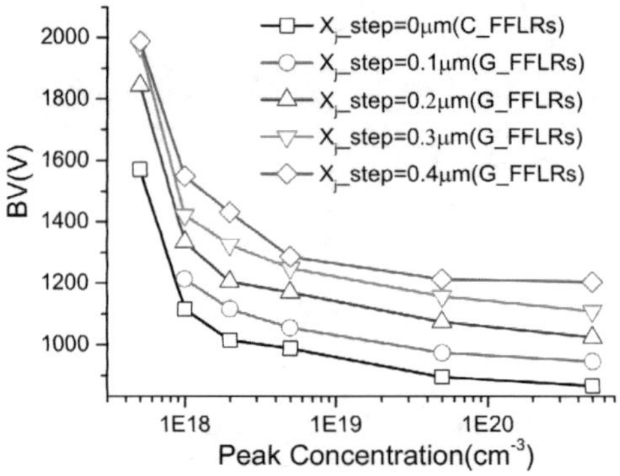

(b) BV~Peak Concentration relationship

Figure 4. Simulation results of BV in conventional and gradient FFLRs with different doping profile and step of junction depth.

As can be seen from Figure 4, BV increase about 100V with the junction depth step increases by 0.1μm at the same peak concentration. BV increase from about 500V with peak concentration $5\times10^{17}cm^{-3}$ to about 10V with peak concentration $5\times10^{20}cm^{-3}$ at the same junction depth. The peak concentration and junction depth changed in G_FFLRs at the same time, showing a comprehensive effect on BV.

3.4. Blocking characteristic of trench gradient FFLRs

Normally, ion implantation process is used to form the heavily doped P-well rings in the FFLRs structure for high-voltage SiC power device. Nowadays, the achievable depth of p-type implantation is normally around 1~1.2μm max, and the depth of p-type implantation can reach 1.7um if the implantation energy is 1.2MeV. In order to make G_FFLRs structure realistic,a trench gradient FFLRs(TG_FFLRs) structure are proposed.

Schematic diagram of the TG_FFLRs as termination structures of 4H-SiC power device is shown in Figure 5a. The doping profile and dimension of G_FFLRs and TG_FFLRs with 0.4μm step are the same except for the trench, the depth from trench surface $X_{jT(n+i)}$ are 1.6μm for P_{14}-P_{17} and 1.2μm for P_7-P_{13}, respectively. The trench width is equal to ion implantation window.

The simulation results of electric field contours, electric field cut along junction depth minus 0.1μm and BV for G_FFLRs and TG_FFLRs are shown in Figure 5b-5d.

As can be seen from Figure 5b-5d, TG_FFLRs show the same edge termination effect as G_FFLRs, which will be more achievable in terms of process implementation.

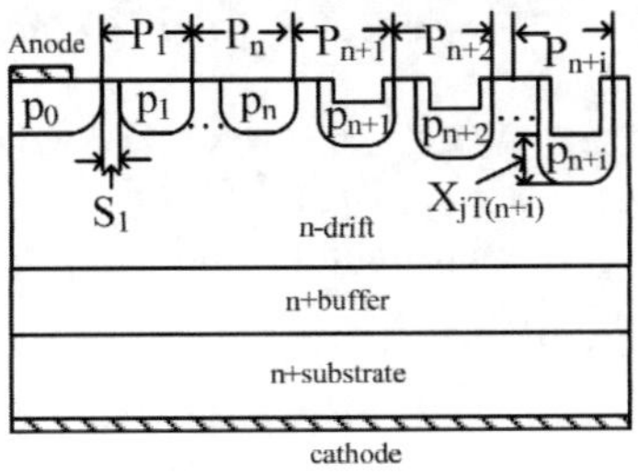

(a) TG_FFLRs

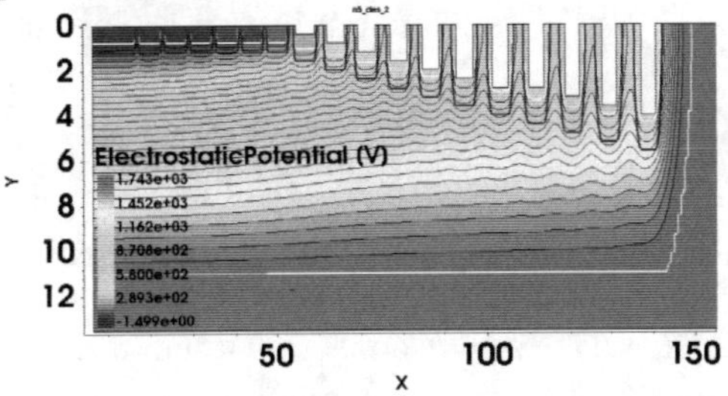

(b) Electric potential distribution of TG_FFLRs, step=0.4μm

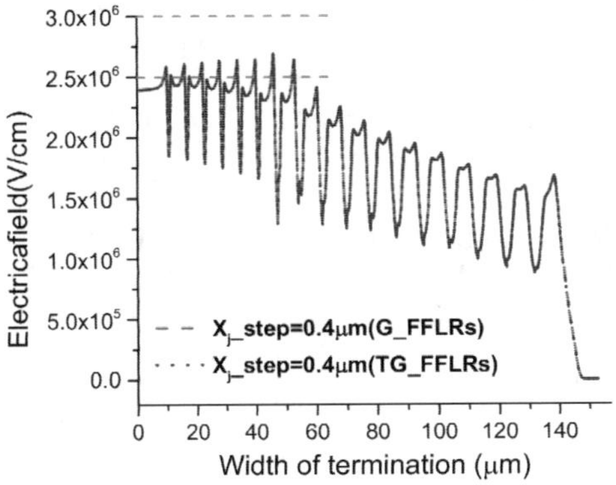

(c) Electric field of G_FFLRs and TG_FFLRs

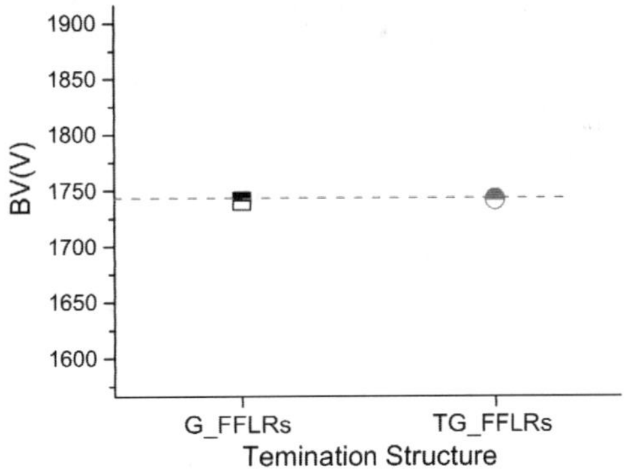

(d) BV of G_FFLRs and TG_FFLRs

Figure 5. The schematic cross-section of TG_FFLRs and blocking simulation results of G_FFLRs and TG_FFLRs.

4. Conclusion

In summary, a novel gradient floating field limiting rings termination structure is proposed and simulated to exhibit BV improved significantly. The ion implantation processes include implantation tilt and trench dimension need to be optimized next.

Acknowledgments

Thanks for supported by the National Science and Technology to Promote the Economy 2020 Key Projects (No. 2020YFF0426582).

References

[1] B. J. Baliga 2008 Fundamentals of Power Semiconductor Devices[M]. New York, NY: Springer, Chap. 3, pp. 91–155.

[2] David C. Sheridan, Guofu Niu, J. Neil Merrett 2001 Comparison and optimization of edge termination techniques for SiC power devices[C]//Proceedings of the 13th International Symposium on Power Semiconductor Devices & ICs, Osaka, Japan. DOI: 10.1109/ISPSD.2001.934587

[3] Hao Yuan, Yancong Liu, Yanjing He 2007 Characteristic and Robustness of Trench Floating Limiting Rings for 4H-SiC Junction Barrier Schottky Rectifiers[J]. IEEE Electron Device Letters, 41(7): 1056-1059. DOI: 10.1109/LED.2020.2993590

[4] A. Mihaila, V.K. Sundaramoorthy, R. Minamisawa 2016 A novel edge termination for high voltage SiC devices[C]//28th International Symposium on Power Semiconductor Devices and ICs (ISPSD), Prague, Czech Republic. DOI: 10.1109/IS PSD.2016.7520818

[5] Reza Ghandi, Benedetto Buono, Martin Domeij 2009 High-voltage 4H-SiC PiN diodes with etched junction termination extension[J]. IEEE Electron Device Letters, vol. 30, no. 11, pp. 1170–1172.

[6] J. H. Zhao, P. Alexandrov, and X. Li 2003 Demonstration of the first 10-kV 4H-SiC Schottky barrier diodes[J]. IEEE Electron Device Letters, vol. 24, no. 6, pp. 402–404 .

[7] Toru Hiyoshi, Tsutomu Hori, Jun Suda 2008 Simulation and experimental study on the junction termination structure for high-voltage 4H-SiC PiN diodes[J]. IEEE Trans. Electron Devices, vol. 55, no. 8, pp. 1841–1846.

[8] X. Wang, J. A. Cooper 2004 Optimization of JTE edge termination for 10 kV power devices in 4H-SiC[J]. Mater. Sci. Forum, vol. 457–460, pp. 1257–1262.

[9] Woongje Sung, Edward Van Brunt, B. J. Baliga 2011 A New Edge Termination Technique for High-Voltage Devices in 4H-SiC–Multiple-Floating-Zone Junction Termination Extension[J]. IEEE Trans. Electron Devices, vol. 32, no. 7, pp. 880-882. DOI: 10.1109/LED.2011.2144561

[10] V. Soler, M. Berthou, A. Mihaila 2015 Experimental analysis of planar edge terminations for high voltage 4H-SiC devices[C]//European Solid State Device Research Conference (ESSDERC), Graz, Austria. DOI: 10.1109/ESSDERC.2015.7324715

[11] Gan Feng, Jun Suda, Tsunenobu Kimoto 2012 Space-Modulated Junction Termination Extension for Ultrahigh-Voltage p-i-n Diodes in 4H-SiC[J]. IEEE Trans. Electron Devices, vol. 59, no. 2, pp. 414-418. DOI: 10.1109/TED.2011.2175486

[12] V. A. K. Temple and W. Tantraporn 1986 Junction termination extension for near-ideal breakdown voltage in pn junctions[J]. IEEE Trans. Electron Devices, vol. ED–33, no. 10, pp. 1601–1608.

[13] W. Sung and B. J. Baliga 2016 A near ideal edge termination technique for 4500V 4H-SiC devices: The hybrid junction termination extension[J]. IEEE Electron Device Lett., vol. 37, no. 12, pp. 1609–1612, Dec. 2016, doi: 10.1109/LED.2016.2623423.

[14] Hao Yuan, Qingwen Song, Xiaoyan Tang 2016 Trench Multiple Floating Limiting Rings Termination for 4H-SiC High-Voltage Devices[J]. IEEE Trans. Electron Devices, vol. 37, no. 8, pp. 1037–1040. DOI: 10.1109/led.2016.2581183

[15] Xiao Zou, Ruifeng Yue, Yan Wang 2016 Etched Junction Termination Extension with Floating Guard Rings and Middle Rings for Ultrahigh-Voltage 4H-SiC PiN Diodes[C]//IEEE International Conference on Electron Devices and Solid-State Circuits (EDSSC). Hong Kong, China. DOI: 10.1109/EDSSC.2016.7785297

[16] Woongje Sung, B. J. Baliga 2016 A Near Ideal Edge Termination Technique for 4500V 4H-SiC Devices: The Hybrid Junction Termination Extension[J]. IEEE Electron Device Letters, 37(12): 1609 -1612. DOI: 10.1109/LED.2016.2623423

Analysis of discharge characteristics of pulsed plasma power source in the microsecond scale

Qian Zhang[1, 2], Zhiheng Liu[1], Deling Lin[1], Yuzhe Hu[1], Borui Zheng[1] and Taifei Zhao[1]

[1]School of Automation and Information Engineering, Xi'an University of Technology, Xi'an 710048, People's Republic of China

[2]Corresponding author's e-mail: 2200321172@stu.xaut.edu.cn

Abstract. The characteristics test of plasma discharge is an essential physical basis for studying plasma flow control. The power consumption of pulsed plasma aerodynamic actuation is 30% lower than when the actuation voltage is a continuous sine wave. This paper uses a self-developed pulsed plasma power source to superimpose DC. The power source conducts the discharge measurement on the sliding discharge plasma. The experimental results show that the power source effectively outputs 0-20 kV pulsed high voltage. The superimposed DC power source generates a negative DC high voltage, accelerating the drift speed of positive ions in the plasma and forming a stable and uniform surface discharge. But the essence of pulse discharge is still filamentary discharge. In a cycle, the plasma discharge is mainly concentrated on the rising edge of the pulse. In contrast, the pulse current is bipolar, and the negative pulse current has a more obvious growth trend, which is more effective for improving the flow control ability.

1. Introduction

As a new type of active flow control technology, plasma flow control technology has the advantages of simple structure, short response time, and wide actuation frequency. According to the actuation voltage waveform of the plasma power source, it is mainly divided into millisecond plasma discharge, nanosecond plasma discharge, and microsecond plasma discharge [1]. The research on millisecond plasma discharge is the most extensive in domestic research, but when the flow velocity is high, the control effect will be lost [2]. Debien et al. tested the characteristics of millisecond plasma discharge, showing that plasma discharge under millisecond actuation has two stages, corresponding to two current peaks. The first stage is in the rising period of voltage, and the discharge is accompanied by intense flow transport produces a larger current; in the second stage, during the voltage drop period, the current is smaller, but accompanied by a stronger glow [3-5]. Liang H et al. compared the energy conversion process of pulse-actuated plasma at three different time scales of the nanosecond, microsecond, and millisecond. They found that the energy utilization rate of nanosecond pulse is the highest, followed by microsecond pulse, and millisecond pulse is the worst [6]. Although nanosecond pulsed plasma has high instantaneous energy, it will produce substantial electromagnetic interference (EMI), which makes the measurement and control instruments fail, which is difficult in practical engineering applications [7].

Comprehensively considered, the time scale of microsecond plasma power source is between milliseconds and nanoseconds, which can reduce electromagnetic interference while producing good control effects, and has broad application prospects [8]. In this paper, a microsecond high-voltage pulsed plasma power source will be developed. The discharge uniformity of the sliding discharge plasma

Journal of Physics: Conference Series **2313** (2022) 012023 doi:10.1088/1742-6596/2313/1/012023

actuator can be effectively maintained. Compared with the millisecond power source, the microsecond pulsed plasma power source has a larger current in the second discharge than the first time. Compared with nanosecond power, microsecond pulsed plasma supply can release all the energy in a short time with weaker electromagnetic interference.

2. The basic principle of the main circuit

The main circuit structure of the power source is shown in Figure 1. The input of the power source is connected to the 220 V power frequency AC mains through the AC pressure regulating valve. The current and voltage of the AC power change with time according to the law of the sine function, and the AC power changes smoothly and is not easy to produce high voltage. The input AC mains is full-wave rectified by the classical bridge rectification method. When the current direction of the input terminal is up positive and down negative, the diodes D1 and D4 are turned on. When the current direction of the input terminal is up and down, the diode D2 and D3 are turned on. and the direction of the current flowing through the capacitor C1 is up positive and down negative. The input power frequency 220 V AC signal is rectified and filtered to convert the waveform into a DC signal. The commonly used filter circuits are divided into capacitor filtering and inductor filtering. The former is suitable for high load voltage and little load change. At the same time, the latter is ideal for low voltage, high current, and enormous current transformation, the initial voltage of sliding discharge plasma discharge is about 7 kV. The output current of the power source is less than 10 A, therefore, the capacitive filtering method is selected. To make the output DC as smooth as possible, it is necessary to control the ripple coefficient as small as possible.

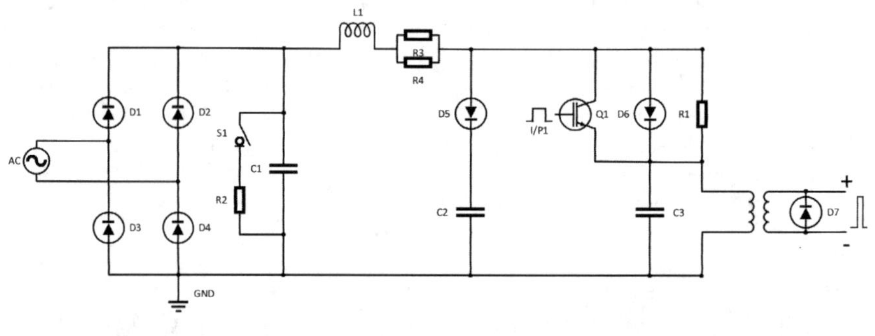

Figure 1. Plasma power main circuit structure.

The capacitor C2 is charged by the unidirectional conduction circuit formed by the inductor L1 and the diode D5. The waveform of the capacitor charge and discharge is pulsed by using the fast recovery characteristics of the IGBT transistor (Insulated Gate Bipolar Transistor) turning on and off in a short time scale. The capacitor can be quickly charged and discharged to form a pulse waveform. Figure 2 shows the equivalent circuit diagram of the IGBT transistor. The value of the gate voltage determines the working state of the IGBT transistor. The IGBT driver board mainly provides the gate voltage for the IGBT transistor.

The IGBT driver board is powered by 12 V DC power source, mainly to amplify the power output by the signal generator pulse through KA102, which is the driver of IGBT transistor and the control core of IGBT driver board. The IGBT driver can control the IGBT transistor by outputting the driving signal, the maximum output power of each channel is 4.5 W, and the current of the output is 30 A, and the output charge is 40 μc. As shown in Figure 2, the structure diagram of the IGBT driver board is shown in Table 1. The driving parameters are shown in Table 1, where V_S is the voltage of the input pulse, I_S is the current of input pulse, Q is the actual driving charge required by the IGBT transistor. τ is the duty cycle, T_r is the rising delay, T_f is the falling delay, P_W is the output pulse width, f is the working frequency.

Table 1. IGBT drive parameters.

V_S	I_S	Q	τ	T_r	T_f	P_W	f
20 V	10 mA	40 μc	0.1%	0.5 μs	0.5 μs	2 μs	1 kHz

$$P_o = Q \times f \times V_s \tag{1}$$

In Equation (1), Q is the actual driving charge required by the IGBT transistor, f is the working frequency, V_S is the input pulse voltage amplitude, According to Equation (1), The output power of the IGBT driver board is about 0.8 W.

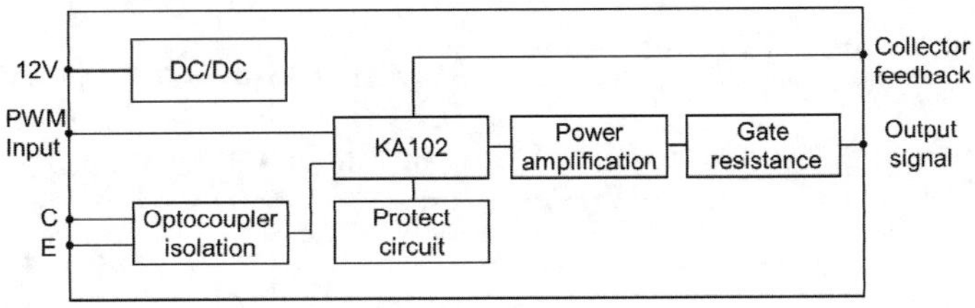

Figure 2. Structure block diagram of IGBT driver board.

Considering that the IGBT will have a high reverse peak current during the experiment, the IGBT protection circuit is composed mainly of a diode D6, a resistor R1, and a capacitor C3. Resistor R1 and diode D6 are connected in parallel with the IGBT. The fast recovery diode D6 can limit the voltage on the resistor R1. The fast recovery diode selection model is MZK300TS120S, the current is 160 A, the voltage is 1600 V, and the resistance value is 300 Ω , leaving as much margin as possible. The capacitor C3 absorbs the energy of the signal control part. When the IGBT transistor is not working, the capacitor C3 is charged by the current from the collector, and at the same time, it suppresses the current abrupt increase when the IGBT is working, and effectively alleviates the device waste during the working process of the IGBT. Capacitor C3 can be estimated by equation (2):

$$C_3 = \frac{I_c(t_r + t_f)}{V_{ce}} \tag{2}$$

In equation (2), I_c is the threshold current of the collecting electrode, t_r is the current rise threshold time, t_f is the current fall threshold time, and V_{ce} is the voltage between V_c and V_e.

The high-voltage output circuit consists of a pulse transformer and a high-voltage diode. The transformer converts the input low-voltage into a high-voltage output, and finally outputs a high-voltage pulse with a peak value of 0-20 kV. Figure 3(a) shows the physical image of the power source, and (b) is the output voltage-current waveform diagram. Through the electrical measurement system in Figure 4, the voltage and current of the output terminal of the power source are tested at the same time. It can be observed that the pulse voltage width is about 1.5 μs, and the output power of the power source can be calculated according to the equation (3).

$$P_t = \frac{1}{T}\int_0^T u_t i_t dt \tag{3}$$

In equation (3), P_t is the average power of the actuator, T is the discharge period, u_t and i_t are the real-time voltage and current respectively, and t is the time. The calculated output power of the power source is about 1.5 W.

Journal of Physics: Conference Series **2313** (2022) 012023 doi:10.1088/1742-6596/2313/1/012023

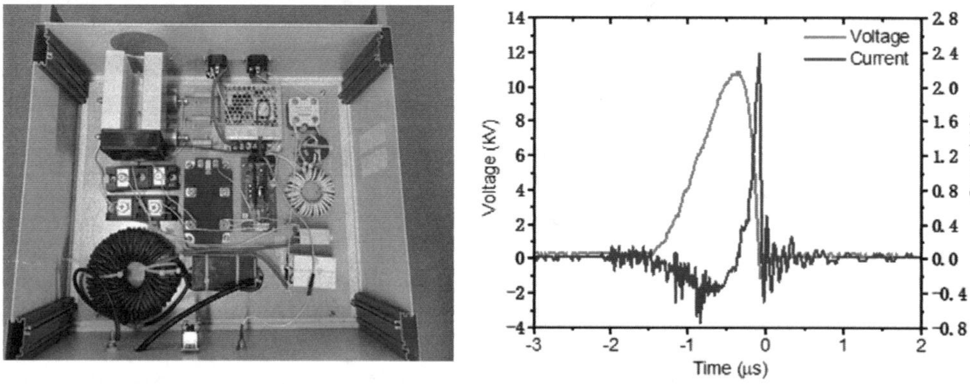

(a) Physical diagram of microsecond pulse plasma power source.

(b) Typical V-I Waveform.

Figure 3. Physical diagram and typical V-I waveform of a microsecond pulsed plasma power source.

3. Results and discussion

The conventional dielectric barrier discharge plasma actuator (DBD-PA) is composed of an exposed electrode, a covered electrode and an insulating dielectric layer in the middle, as shown in Figure 4 is the schematic diagram of DBD-PA, the yellow arrow is the direction of the induced flow, and also the direction of the plasma generation area. the working principle of the plasma actuator is mainly by applying high voltage between the exposed electrode and the covered electrode. When the voltage increases to a certain value, the air will be ionized to generate plasma ($V_{\mu s}$=7 kV in this paper is the initial voltage of the plasma actuator), the plasma actuator will be broken down if the voltage is continuously increased. while the sliding discharge plasma actuator (SD-PA) adds an exposed electrode on the basis of the conventional configuration, as shown in Figure 5 is the schematic diagram of SD-PA.

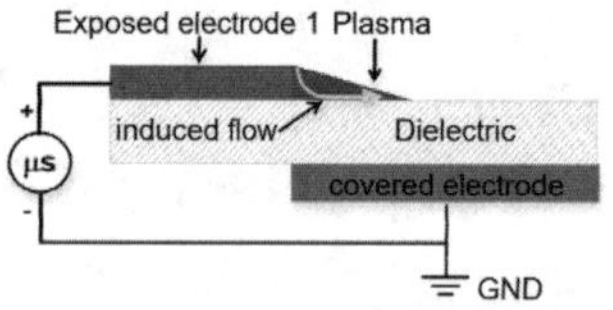

Figure 4. Schematic diagram of DBD-PA.

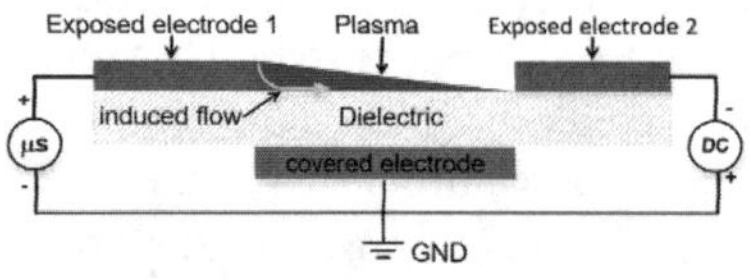

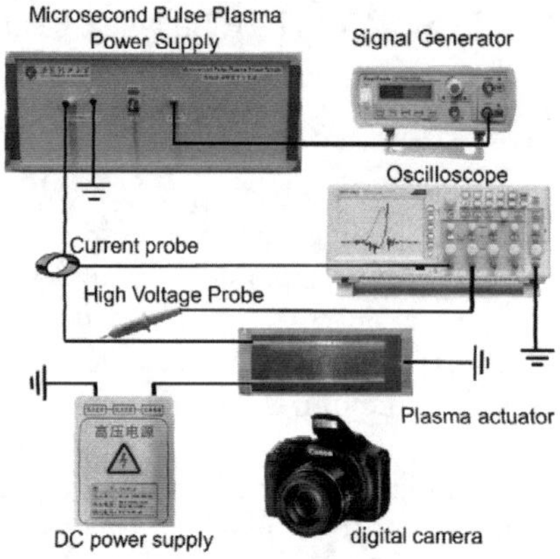

Figure 5. Schematic diagram of SD-PA. **Figure 6.** Electrical measurement system.

In this experiment, the self-developed pulse plasma power source and DC power source are used to conduct the sliding discharge plasma test experiment. The experimental parameters have been repeatedly tested for at least 3 times. Since the experimental process is manual pressurization, the error range is within 0.1%, which belongs to the normal range. The measurement system is shown in Figure 6, including a plasma power source, DC power source, signal generator, current probe, high voltage probe,

Journal of Physics: Conference Series **2313** (2022) 012023 doi:10.1088/1742-6596/2313/1/012023

oscilloscope, digital camera, etc. Figures 4 and 5 are the schematic diagram of the plasma actuator, which adds an exposed electrode compared to the traditional DBD, which consists of two exposed electrodes and a covered electrode. The electrodes are made of copper, the insulating medium layer is made of polyimide (PI) film material, and the distance between the two exposed electrodes is 20 mm. As shown in Figure 6, the experimental measurement system is shown. The signal generator provides the driving signal for the plasma power source. The output of the plasma power source is connected to the exposed electrode 1 of the actuator. At the same time, the exposed electrode 2 is connected to the GND of the DC power source. The output is connected, and the voltage probe and the test port of the current loop are respectively connected to the high-voltage output end of the plasma power source and the output end of the DC power source, and the other end is connected to the oscilloscope, and the voltage-current waveform can be obtained through the oscilloscope.

Using a digital camera to capture the plasma discharge image, and obtain the discharge effect as shown in Figure 7 and Figure 8. Figure 7 shows the conventional DBD-PA discharge effect without DC power source, and Figure 8 shows the SD-PA discharge effect with the addition of DC power source. Compared with DBD-PA, SD-PA adds an exposed electrode driven by a DC power source. Under the actuate of DC power source, the number of positive particles in the plasma is increased, and its movement speed is also accelerated, thereby a large-area and uniform stable plasma discharge is formed on the surface of the insulating dielectric layer [9-10]. It can be found that SD-PA can effectively increase the plasma discharge area, and there is a purple filamentous discharge channel near the electrode, indicating that the discharge form of SD-PA is filamentary discharge.

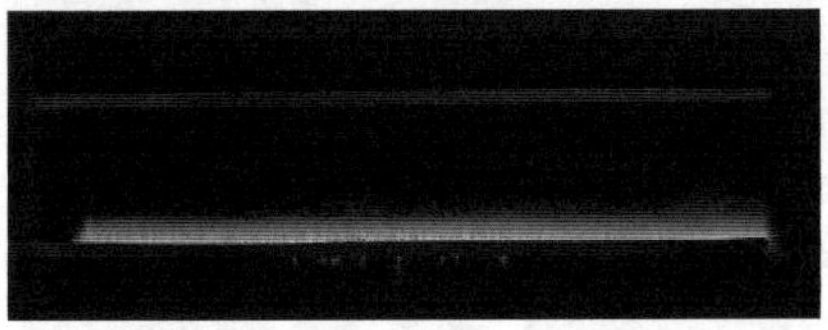

Figure 7. Discharge diagram of DBD-PA.

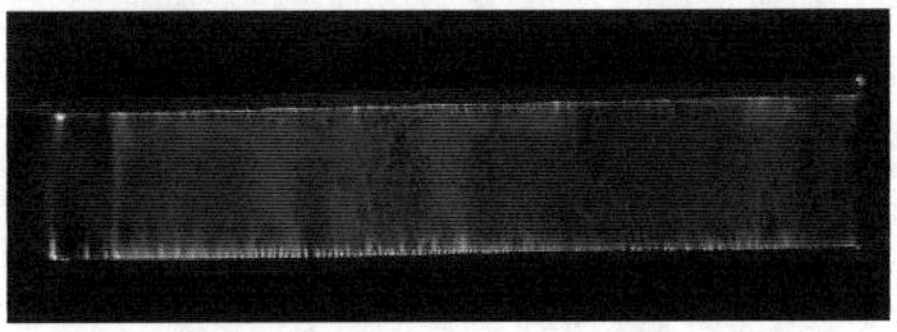

Figure 8. Discharge diagram of SD-PA.

Table 2. Experimental conditions.

Case	$V_{\mu s}$, kV	$I_{\mu s}$, A	V_{DC}, kV
1	6.92	1.52	-11.9
2	8.12	2.24	-12.1
3	8.95	3.12	-12
4	9.97	4.32	-12.1

Keeping the actuation frequency f =1000 Hz, the experimental conditions are shown in Table 2, and the waveforms of different discharge voltages and currents within a single pulse scale are shown in Figures 9 and 10. Since the entire discharge cycle is quite different from the pulse width of the microsecond pulse (1.5 μs), To display the waveform shape, only the waveform change within 4 μs is shown in the figure, and the voltage and current at other times in the cycle are approximately 0. It can be seen from the figure that the voltage waveform generated by the discharge is a high-voltage pulse similar to a triangular wave. The peak-to-peak value of the discharge voltage is V1=6.92 kV, V2=8.12 kV, V3=8.95 kV, V4=9.97 kV, and the pulse width is about 1.5 μs. The peak-to-peak values of the discharge current are I1=1.52 A, I2=2.24 A, I3=3.12 A, and I4=4.32 A, respectively. There is a current peak value on the rising edge and falling edge of the pulse voltage respectively, which corresponds to the two discharges of the sliding discharge plasma actuator. At the same time, the discharge currents are all bipolar, the positive current peak value can reach 1.5 A, and the peak value of the negative current can reach 3 A, indicating that under one voltage pulse, the sliding discharge plasma actuator has two discharges, corresponding to the rising and falling edges of the voltage pulse, respectively.

This is because at the beginning of the plasma discharge, the gas is broken down under the barrier of the medium, forming a micro-discharge channel, and the space charge is transported in the discharge channel and accumulated on the medium. This process corresponds to the discharge process when the voltage rises; the space charge is in the accumulation on the medium forms a reverse electric field. When the voltage begins to drop rapidly, the reverse electric field increases rapidly, causing the gas to reverse breakdown and discharge, thereby forming a negative current peak at the falling edge of the voltage. This discharge regularity is similar to millisecond plasma power [11-12], but unlike millisecond plasma power, which has a stronger first discharge, Microsecond pulsed plasma power source has a larger current in the second discharge than the first.

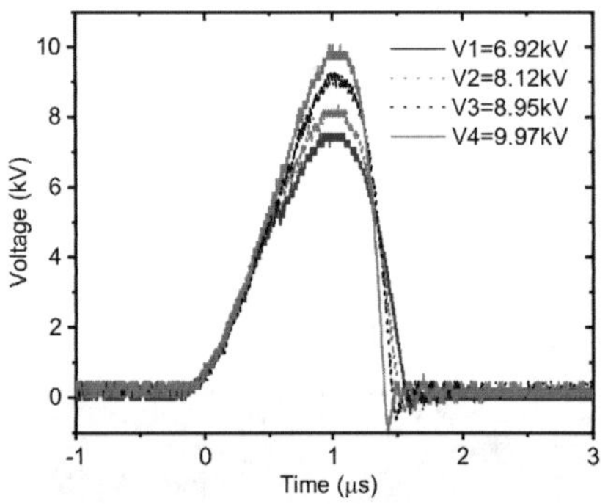

Figure 9. Microsecond sliding discharge output voltage waveform.

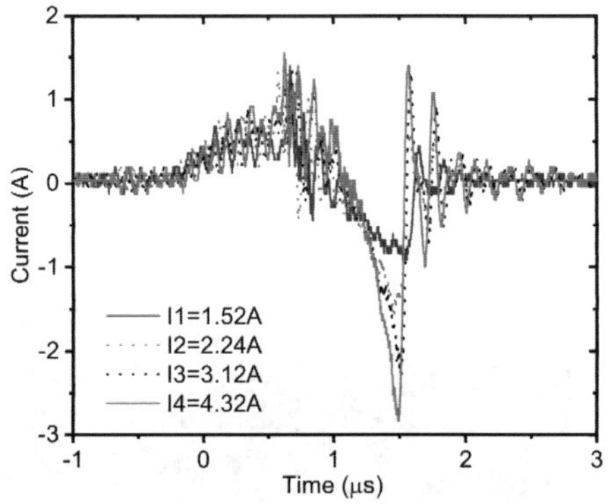

Figure 10. Microsecond sliding discharge output current waveform.

4. Conclusion

In this paper, a high-voltage plasma power source with a pulse width of microseconds is independently developed by using the principle of capacitor charging and discharging, combined with the fast recovery characteristics of IGBT transistor that are turned on and off in a short time. By studying the plasma discharge characteristics under the actuation of pulsed high voltage power source, it is found that the pulsed power source can effectively output pulsed high voltage actuation plasma actuator. At the same time, it is found that the discharge characteristics of the pulsed power source and the millisecond power source are significantly different. The first discharge energy of the millisecond plasma power source (at the rising edge of the pulse) is larger, while the second discharge of the pulsed plasma power source (at the falling edge of the pulse) is more intense. And the microsecond plasma power source can concentrate a small amount of energy and release it in a short period of time, thereby forming a large instantaneous power and generating a strong instantaneous disturbance to the flow field. In our future research, the microsecond plasma power source is mainly applied to the fields of suppression of flow separation and turbulent drag reduction experiment.

References

[1] Shi Y W, Zhou R Y, Cui X L 2018 *J. Transactions of China Electrotechnical Society.*, 33(22): 5371-5380.
[2] Kelley C L, Bowles P O, Cooney J 2014 *J. AIAA journal.*, 52(9): 1871-1884.
[3] Debien A, Benard N, Moreau E 2012 *J. Journal of Physics D: Applied Physics.*, 45: 215201
[4] Debien A, Benard N, Moreau E 2011 R. In Proceeding of 2nd ISNPEDADM new electrical technologies for environment.
[5] Debien A, Benard N, David L, Moreau E 2012 *J. Applied Physics Letter.*, 100: 013901

[6] Liang H, Li Y H, Jia M, Zhang S Z and Bai J X 2010 *J. High Voltage Engineering.*, 36:3054-3058.
[7] Roupassov D V 2009 *J. AIAA Journal.,* 47(1): 168-185.
[8] Su Z，Li J，Liang H 2018 *J. Chinese Physics B.*, 27(10).
[9] Zheng B R, 2018 *J. AIAA Journal.*, 57: 467-471.
[10] Zheng B R, 2020 *J. AIAA Journal.*, 58: 733-746.
[11] Liang H. 2008 D. Air Force Engineering University.
[12] Moreau E, Louste C, Touchard G. 2008 *J. Journal of Electrostatics.*, 66: 107-114.

Numerical analysis and optimization of radio-frequency ion trap potential

Yuhan Liu[1], Lijun Du[1, 2, 3], Yuling He[1], Shiyi Huang[1], Qiang Zhang[1], Keliang He[1], Yulong Tang[1], Yansong Meng[1, 2] and Shenghua Zhai[1]

[1] China Academy of Space Technology (Xi'an), Xi'an 710100, China
[2] National Key Laboratory of Science and Technology on Space Microwave, China Academy of Space Technology (Xi'an), Xi'an 710100, China

[3] Corresponding author's e-mail: dulj@cast504.com

Abstract. The trap of multi-ion frequency standard is an ideal candidate for a new generation of ground and space-borne timekeeping systems, which has high signal-to-noise ratio, low frequency drift rate and excellent long-term frequency stability. The non-ideal characteristic of radio-frequency potential is one of the dominant factors that restrict the performance of ion frequency standard. In this study, the finite element numerical analysis method is used to quantitatively analyse and optimize the electrical potential field of ion trap, and an analytical model of ideal trapping potential is established. The typical transient numerical potential surface of trapping potential is used for decoupling analysis of various electrical potential effects. Based on the optimal structure parameters, the numerical model and analytical model of non-standard ion trap are established to verify the validity and accuracy of the numerical analysis method. This method can provide a quantitative solution for the high precision reconstruction of the trapping potential model of heterotype ion traps.

1. Introduction

Radio-frequency (RF) ion trap system can effectively isolate the coupling effect between ions and the environment, which has extensive and important application value in ion frequency standard [1-3], ion trap quantum computation [4], mass spectrometry analysis [5-6] and precision spectroscopy research [7]. Compared with the neutral atom system, the trap of multi-ion frequency standard can achieve very low frequency drift rate and excellent long-term frequency stability and can obtain high signal-to-noise ratio (SNR) transition lines. It is an ideal candidate for a new generation of ground and space-borne timekeeping atomic clocks [8]. Multi-ions have a three-dimensional structure distribution in the RF confinement field and are very sensitive to the non-ideal properties of the trapping potential [9-10]. RF anharmonic potential can aggravate the dynamic effect of ion random walk, shorten the quantum coherence time, and affect the accuracy of clock frequency measurement and the performance of frequency standard [11].

Based on Finite Element Method (FEM), this study uses COMSOL Multiphysics (COMSOL for short) software to simulate and study the radial transient trapping potential effect of ion traps, in order to quantitatively analyze the distribution characteristics of RF electrical potential in ion traps. The radial potential model and numerical analysis boundary conditions of the RF ion traps were established according to the actual pole material, spatial distribution and charging properties of the ion trap. The steady-state electric potential numerical solution of the radial plane model of the radiofrequency ion trap was obtained by calculating the ion activity region step by step according to Maxwell's equations. According to the analytical physical model of the trapping potential, the decoupling and reliability of the potential effects are evaluated. On the basis of the optimal well structure parameters, the numerical model and analytical model of the non-standard well were compared, and the validity of the finite element numerical analysis method in the high precision reconstruction of the heterotype ion trap model clock was verified.

2. Finite element modeling of RF ion trap potential

The radial plane structure of ion trap is shown in Figure 1. The ion trap consists of four circular cross-section electrodes. In Figure 1, r_e is the radius of the electrode pole, r_0 is the shortest distance from the ion trap center to the electrode surface, and D is the radius of the ion trap. The inner region of the black dotted circle, whose radius is the distance between the center of the pole and the center of the trap ($r_e + r_0$), is the active region of the ions.

The natural boundary condition and voltage source excitation condition of transient electric field are set as followed. The initial potential of all region is 0 V. Consider the two opposite electrodes as one group. The voltage of $U_0 \cdot \cos(\Omega t)$ with opposite phase is applied to the boundary of the two groups of electrodes. RF voltage amplitude U_0 is set as 1000 V, Ω as RF frequency, t as 0 s. The outer boundary of the ion trap is set as 0 V. Set the electrode material as molybdenum, and the rest of the area as vacuum. As shown in Figure 1, the center of the ion trap is taken as the origin of the rectangular coordinate system, the direction of a group of electrodes loaded with forward voltage is X-axis, the direction of the other group of electrodes is Y-axis, then the radial plane rectangular coordinate system is established. Taking the center of ion trap as the origin of polar coordinate system, the polar axis is recombined with the X axis of rectangular coordinate system, and the positive direction of Angle θ is taken as counterclockwise direction, then the radial plane polar coordinate system is established. In the division and optimization of spatial grid, the smaller the grid is, the higher the calculation accuracy is. Therefore, two levels of grid accuracy are set in this paper: the mesh fineness of the central potential field region which has a great influence on ion activity is 10 μm, and which of the other regions is 50 μm, so as to optimize the numerical weight and the calculation amount of the core region. The radius of the electrode (r_e) is 0.005 m, $D = 3.54\ r_0$. Parametric scanning is used to set multiple groups of equally spaced structural parameters k, which is equal to r_e / r_0, and the spatial grid is divided respectively. Based on Maxwell's equations, the steady-state approximate solutions of all sub-regions in the ion trap, that is, the potential field data of each point in the trap, are calculated to form the transient state energy surface of the trapped region, and the factors of the radial potential field are studied. In this study, the potential field region with the origin as the center of the circle and the radius of 1 mm is called the central potential field region, which basically covers the active region of the trapped ions.

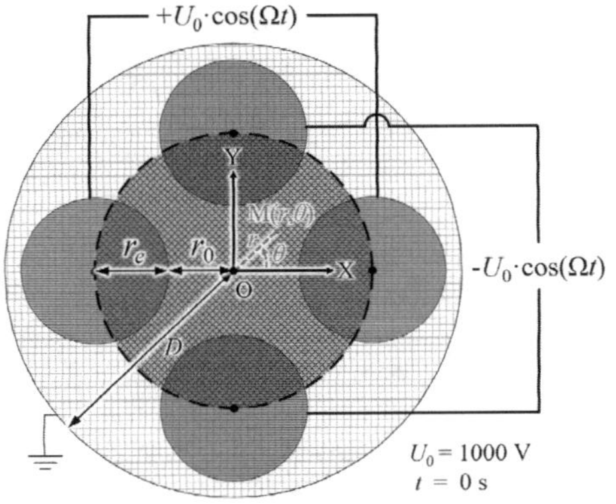

Figure 1. The radial plane structure of ion trap.

3. Optimization of core parameters of RF ion trap potential field

In this section, the potential energy surface is decoupled, evaluated and optimized according to the radial transient state energy surface of the ion trap region. The finite element numerical analysis results are compared with the analytical model results to verify the credibility and universality of the finite element numerical analysis method applied to the analysis of the ion radio frequency incarceration potential.

3.1. Matching numerical solution to analytic series form

The infinite series expansion method is a general method to solve the ion trap potential function. The potential effect of each order can be analyzed numerically. Based on the infinite series expansion method, the potential function of any point in the quadrupole is:

$$\begin{aligned}\phi(r_i,\theta_i) &= U_0\sum_{n=0}^{\infty}C_{2(2n+1)}\left(r_i/r_0\right)^{2(2n+1)}\cos\left[2(2n+1)\theta_i\right] \qquad i=1,2,3,\ldots,N \\ &= C_2U_0\left(r_i/r_0\right)^2\cos\left(2\theta_i\right)+C_6U_0\left(r_i/r_0\right)^6\cos\left(6\theta_i\right)+C_{10}U_0\left(r_i/r_0\right)^{10}\cos\left(10\theta_i\right)+\ldots \\ &= C_2U_0\left(r_i/r_0\right)^2\cos\left(2\theta_i\right)+\phi'(r_i,\theta_i)\end{aligned} \tag{1}$$

Here, (r_i,θ_i) is the polar coordinates of that point, r_0 is the minimum distance between ion trap center and electrode surface, U_0 is the amplitude of the RF voltage applied to the electrode pole. $C_{2(2n+1)}$ is the $2(2n+1)$ order coefficient, the magnitude of which is linearly related to the strength of the $2(2n+1)$ order potential, and satisfies $C_2 + C_6 + C_{10} + C_{2(2n+1)} + \ldots = 1$. $\phi'(r_i,\theta_i)$ is the high order anharmonic potential of the point. N is the number of finite elements in the transient state energy surface. According to Equation (1), we match the potential energy surface of all orders, and obtain the potential function of the second-order potential, the sixth-order potential and the tenth order potential:

$$\phi_2(r_i,\theta_i)=C_2U_0\left(r_i/r_0\right)^2\cos\left(2\theta_i\right) \tag{2}$$

$$\phi_6(r_i,\theta_i)=C_6U_0\left(r_i/r_0\right)^6\cos\left(6\theta_i\right) \tag{3}$$

$$\phi_{10}(r_i,\theta_i)=C_{10}U_0\left(r_i/r_0\right)^{10}\cos\left(10\theta_i\right) \tag{4}$$

On the basis of Equation (1), the anharmonic potential of the quadrupole ion trap is evaluated and optimized by using the root mean square minimum of the second-order potential residual as the standard. The residual effective potential expression of the ion active region is as follows:

$$e=\left\{\sum_{i=1}^{N}\left[\frac{\phi(r_i,\theta_i)-C_2U_0\left(r_i/r_0\right)^2\cos\left(2\theta_i\right)}{U_0}\right]^2\bigg/N\right\}^{1/2}=\left\{\sum_{i=1}^{N}\left[\frac{\phi'(r_i,\theta_i)}{U_0}\right]^2\bigg/N\right\}^{1/2} \tag{5}$$

Equation (5) is used to describe the intensity of the normalized anharmonic potential in the ion active region.

Finite element analysis is used to calculate the potential field data of each point in the ion trap, and the transient state energy surface of the trapped region is established. Then we post-process the potential field data of the central region. According to Equations (1) and (5), the relationship between the sixth order potential coefficient and residual effective potential and the structural parameters is shown in Figure 2. When the structural parameter (k) is 1.1451479, the sixth-order potential is suppressed to the greatest extent, while the strength of the tenth and fourteenth order anharmonic potential is not the lowest. In order to concentrate the spatial RF energy on the second-order potential, we need to further reduce the random micro-motion effect of all the anharmonic potential on the ion, so that the overall anharmonic potential can be suppressed to the maximum extent. At this point, the structural parameter (k) is 1.1451447.

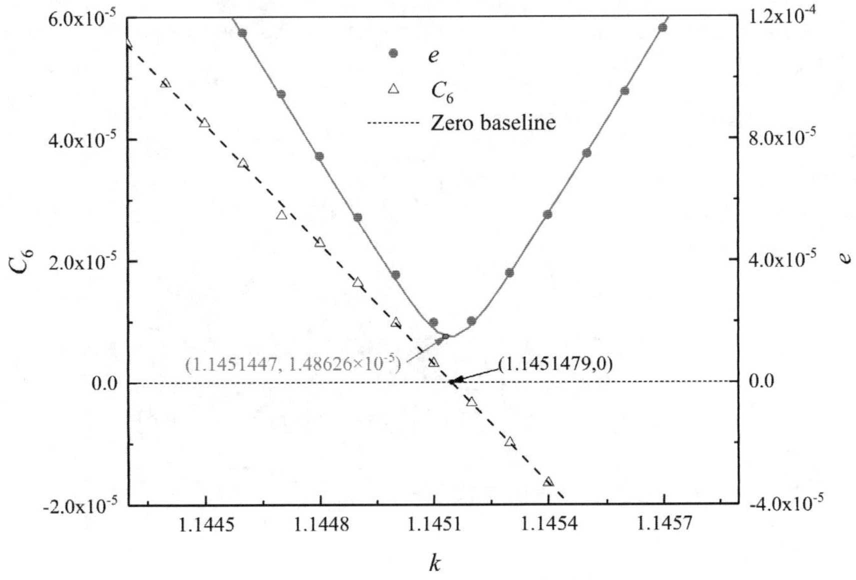

Figure 2. The relationship between the sixth order potential coefficient and residual effective potential and the structural parameters.

3.2. Decoupling and credibility assessment of order effects of trapping potential

It is assumed that the boundary of the radial active region of ions is a circular region with a radius of d mm from the center of ion trap, and the position of the point (X = d mm, Y = 0 mm) can represent the worst trap characteristics of the ion. Figure 3 shows the numerical curve of the total potential along the X-axis and the fitting distribution of each order of potential.

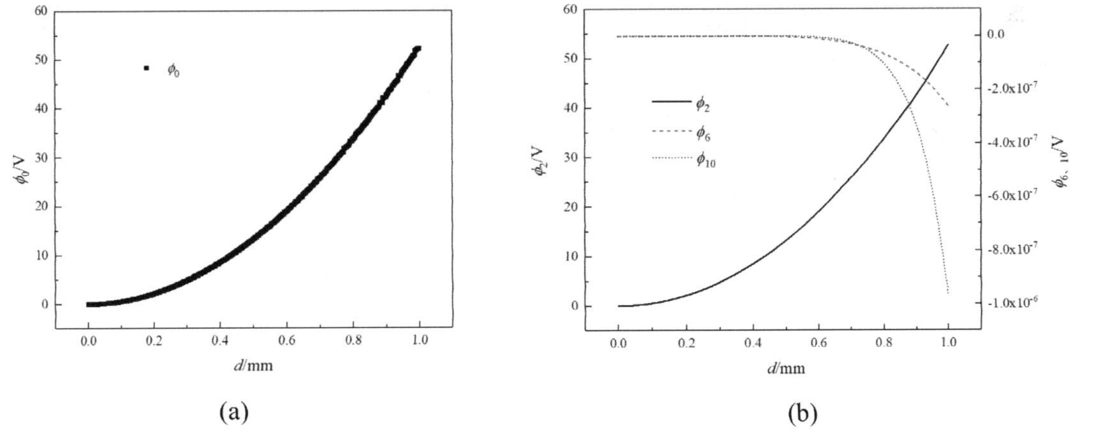

Figure 3. Actual data of total potential and fitting curves of second, sixth and tenth order potential.

The residual of the transient potential except below the tenth order in polar coordinates is defined as ϕ_e:

$$\phi_e(r_i,\theta_i) = \phi(r_i,\theta_i) - [\phi_2(r_i,\theta_i) + \phi_6(r_i,\theta_i) + \phi_{10}(r_i,\theta_i)] \qquad i = 1,2,3,...,N \tag{6}$$

The lower ϕ_e is, the closer it is to white noise, which indicates that the decoupling fitting of potential field is more accurate. The relationship between ϕ_e and the horizontal coordinate d of the captive position is shown in Figure 4.

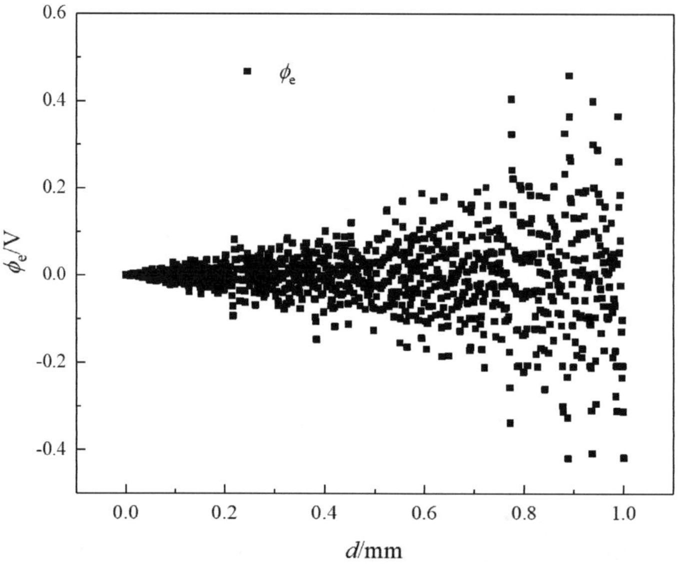

Figure 4. Residual potential distribution diagram (except below the tenth order).

On the one hand, as shown in Figure 4, the residual potential of numerical calculation can be controlled within the range of ± 0.5 V, which is far less than the total boundary potential of 52 V shown in Figure 3(a). On the other hand, the residual potential always meets the approximate white noise characteristics with an average of 0 in different confinement regions, indicating that the typical results obtained by finite element analysis are highly matched with the ideal analytical model, achieving the evaluation of the trapping potential is reliable.

3.3. Application of finite element method in the analysis of nonstandard ion trap potential

Solid Angle through light is an important parameter to evaluate the light transmission of ion trap and the efficiency of light manipulation. The larger light transmission angle is, the more beneficial it is to efficient light pumping and light detection. In the process of increasing light transmission angle of cylindrical ion trap, it will inevitably reduce the trapping efficiency and intensify the heating effect of anharmonic potential. In this study, on the basis of the structural model shown in Figure 1, without changing the structural parameters of the ion trap, the cylindrical ion trap electrode radians are cut to increase the angle and suppress the anharmonic potential. As shown in Figure 5, the shaded part is the simplified electrode after cutting, α is the cutting angle of the pole in the cylindrical ion trap, and β is the light transmission angle in the non-standard ion trap, which is increased compared with the cylindrical pole.

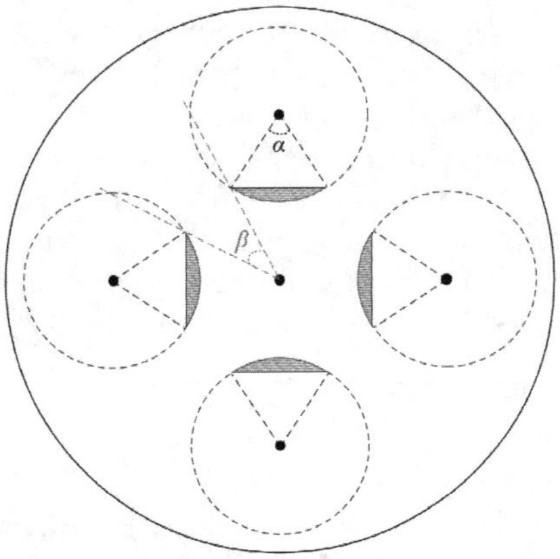

Figure 5. Shape improvement model of ion trap.

The following is a study of the relationship between the magnitude of each potential and the cutting angle based on the numerical analysis method. Finite element simulation parameters and data processing are the same as described above. Taking sixth-order potential analysis as an example, the typical values and fitting curves of total potential, sixth-order potential and its proportion in the central potential region are obtained, as shown in Figure 6.

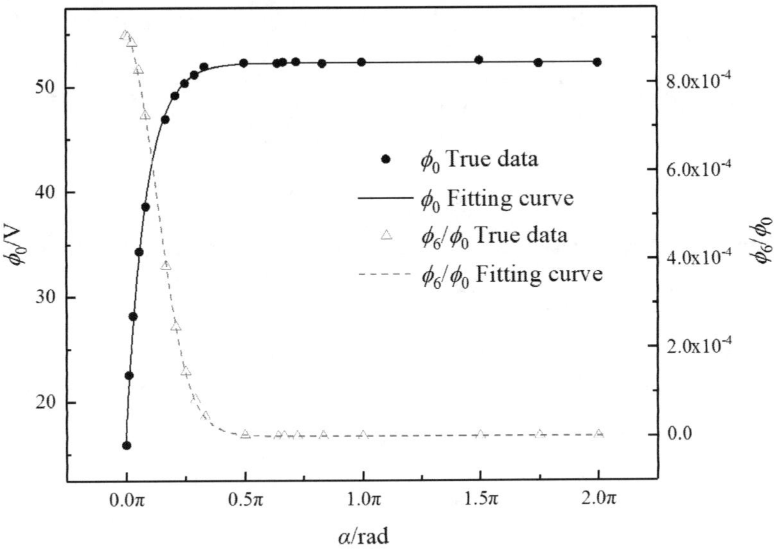

Figure 6. True data and fitting curve of total potential and the proportion of sixth-order potential.

Compared with the central potential field of the complete cylindrical pole, the influence factor of the cutting pole on the total potential has an exponential relationship with the cutting angle, that is, the loss coefficient of the total potential on the cutting angle is about $0.66944 \times 0.0235^{\alpha}$. The sixth order potential analytical model is the product of the total potential loss term and the sixth order potential proportion term. Figure 7 shows the relationship between the typical value of the sixth order potential and the cutting angle. It can be seen that when the cutting angle is above π/2, it has little influence on the electric potential. When the cutting angle is below π/4, the influence of sixth-order potential

Journal of Physics: Conference Series **2313** (2022) 012024 doi:10.1088/1742-6596/2313/1/012024

increases rapidly, and a maximum value appears near 0.074π. This is because with the decrease of the cutting angle, the overall value is on the order of 10^{-4}, although the proportion of the sixth order potential increases sharply. Meanwhile, the total potential drops more rapidly and dominates. The combined effect of the two mechanisms shows that the sixth order potential reaches an extreme value at 0.074π. In application, the minimum cutting angle should be above π/4. In a word, the numerical results of the sixth-order potential finite element method basically match the analytical model. The finite element numerical analysis method is adaptable and analytically accurate in the evaluation of non-standard ion trap structures, which is helpful for the high-precision reconstruction of various kinds of non-standard trapping potential models in engineering applications.

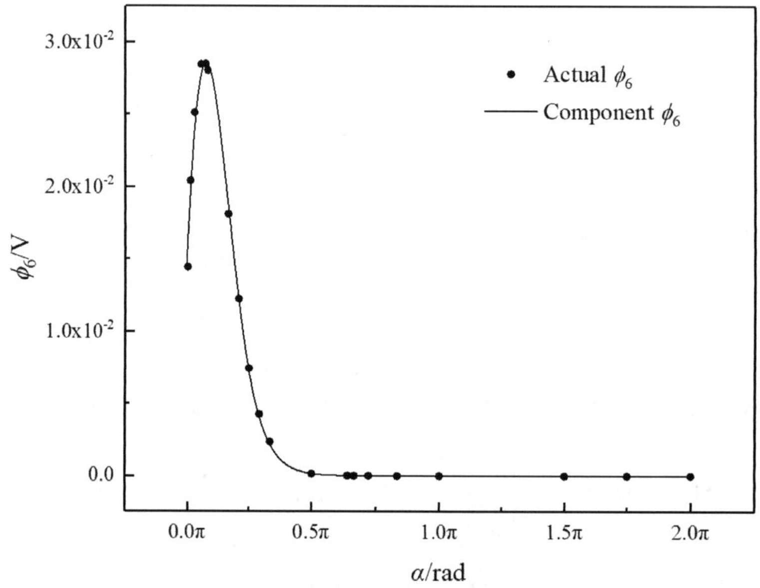

Figure 7. Actual data and component curve of the sixth-order potential.

4. Conclusion

In this study, the finite element numerical analysis method is used to analyze the electrical potential field of ion trap. Based on the infinite series expansion method, an analytical physical model of central potential is established. By decoupling and extracting each order of potential, this study proposes a standard to evaluate the anharmonic potential strength of ion trap, and the optimum structural parameters (k = 1.1451447) are obtained. By means of calculating the residual potential except below the tenth order, it is found that the residual potential is low and always meets the approximate white noise characteristics with an average of 0. This result indicates that the calculation results of the finite element numerical analysis method are highly matched with the physical analytical model, and the reliability of the analytical method is verified. Considering the need of large light transmission angle in practical application, potential energy surface decoupling and evaluation of non-standard ion trap structure are also carried out in this study. Taking sixth-order potential suppression as an example, it is verified that the finite element numerical analysis method also has strong adaptability and high analysis accuracy in the evaluation of non-cylindrical ion trap structure. It is helpful to reconstruct and evaluate the trapping potential models of different types of ion traps in engineering. However, this method is limited by the mesh fineness. In theory, the smaller the finite element is, the higher the accuracy of this method is. The quantified boundary value of the mesh fineness will be studied and improved in the future, which will have a more accurate guiding significance for the engineering application, and can avoid the shortage or waste of the amount of calculation.

Acknowledgments
This work was supported by the the Equipment Pre-research Foundation (Grant No. 6142411196406), Foundation for Distinguished Young Scholars of China Academy of Space Technology (Grant No. Y21-RCWY1-01) and Key R&D Program of Shaanxi Province (Grant No. 2017ZDXM-GY-113).

References
[1] Rosenband T *et al* 2008 Frequency ratio of Al^+ and Hg^+ single-ion optical clocks; Metrology at the 17th decimal place *Science* **319** 1808-12
[2] Schmidt P O, Rosenband T, Langer C, Itano W M, Bergquist J C and Wineland D J 2005 Spectroscopy using quantum logic *Science* **309** 749-752
[3] Chou C W, Hume D B, Koelemeij J C J, Wineland D J and Rosenband T 2010 Frequency Comparison of Two High-Accuracy Al^+ Optical Clocks *Physical Review Letters* **104** 4
[4] Monz T *et al* 2011 14-Qubit Entanglement: Creation and Coherence *Physical Review Letters* **106** 130506
[5] Germann M, Tong X and Willitsch S 2014 Observation of electric-dipole-forbidden infrared transitions in cold molecular ions *Nature Physics* **10** 820-824
[6] Deb N, Pollum L L, Smith A D, Keller M, Rennick C J, Heazlewood B R and Softley T P 2015 Coulomb crystal mass spectrometry in a digital ion trap *Physical Review A* **91** 6
[7] Chou C-w, Kurz C, Hume D B, Plessow P N, Leibrandt D R and Leibfried D 2017 Preparation and coherent manipulation of pure quantum states of a single molecular ion *Nature* **545** 203-207
[8] Burt E A *et al* 2021 Demonstration of a trapped-ion atomic clock in space *Nature* **595** 43
[9] Lin G D and Duan L M 2016 Sympathetic cooling in a large ion crystal *Quantum Information Processing* **15** 5299-5313
[10] Du L J, Meng Y S, He Y L and Xie J 2021 Quantitative evaluation and regulation of cold-ion system intrinsic micromotion by numerical simulation *International Journal of Mass Spectrometry* **468** 116626
[11] Drewsen M and Brøner A 2000 Harmonic linear Paul trap: Stability diagram and effective potentials *Physical Review A* **62** 045401

The dominant mechanism of gold film damage induced by near-threshold femtosecond laser: non-thermal deformation or thermal ablation

Ke Zhang[1] **and Jingya Sun**[1, 2, 3]

[1]Laser Micro/Nano Fabrication Laboratory, School of Mechanical Engineering, Beijing Institute of Technology, Beijing, 100081, P. R. China
[2]Yangtze Delta Region Academy of Beijing Institute of Technology, Jiaxing 314019, P. R. China

[3]Corresponding author's e-mail: sjy@bit.edu.cn

Abstract. Near-threshold femtosecond laser produces nanostructures on the gold film, which has received much attention in applications. However, the theoretical researches on the early stage of gold film excitation induced by the femtosecond laser are scarce, because of the ultrafast and non-equilibrium characteristics. In this study, stress and temperature distributions in different thicknesses of gold film induced by the near-threshold fluences were calculated. At the fluence below the melting threshold, we found that only the 30 nm gold films might be damaged by the non-thermal deformation. At the fluence of the melting threshold, the dominant damage mechanism shifted from non-thermal deformation to thermal melting as the gold film thickness increased. Our study provided a way to distinguish the dominant damage mechanism of gold film and it was valuable for the theoretical study of near-threshold processing.

1. Introduction

Nanostructures on metal thin film can induce the surface or localized plasmon excitation [1], which has a wide range of applications in sensing [2], detection [3], and other fields. Near-threshold femtosecond laser excitation of metal thin film has been proved to produce various nanostructures, such as hollow bumps [4], nano-jets [5], etc. Compared with traditional processing methods, the advantages of the modulation of energy deposition and the processing without a photolithographic mask make near-threshold femtosecond laser processing an efficient and controllable method for precision fabrication of nanostructures on metal thin film. However, the mechanisms of thin film deformation induced by femtosecond laser are complex.

Various models have been proposed to explain the deformation, such as the Marangoni convection [6] and the thermoplastic deformation with the basis of continuum simulations [7], which focus on the later stage of film deformation. Because the ultrafast and nonequilibrium characteristics of femtosecond laser at the early stage impede the application of the continuum simulations [8], the exploration of dominant mechanisms at the early stage is important. In addition, at near-threshold fluence, it is difficult to determine whether the film damage is caused by the thermal ablation or ultrafast deformation.

In this work, the thermal and mechanical processes within gold films of different thicknesses below or equal to the threshold were simulated by the ultrafast thermoelasticity model with dynamic optical parameters. We proposed a method to determine whether thermal ablation or mechanical deformation caused the film damage at the early stage, and explain the cause of damage induced by fluence below the melting threshold. These results reveal the dominant mechanism of gold film damage and are valuable for near-threshold femtosecond laser processing of metal thin film.

2. Model and method

The deformation of metal film induced by femtosecond laser heating is an ultrafast non-equilibrium process, so we used the ultrafast thermoelasticity model to describe the ultrafast deformation and the heat transfer process. The one-dimensional ultrafast thermoelasticity model [9] was shown as follows:

$$C_e\frac{\partial T_e}{\partial t}=\nabla(k_e\nabla T_e)-G(T_e-T_l)+S(z,t) \tag{1}$$

$$C_l\frac{\partial T_l}{\partial t}=G(T_e-T_l)-(3\lambda+2\mu)\alpha T_l\varepsilon_{kk} \tag{2}$$

$$\rho\frac{\partial^2 u}{\partial t^2}=(\lambda+2\mu)\frac{\partial^2 u}{\partial z^2}-(3\lambda+2\mu)\alpha_0\frac{\partial T_l}{\partial z}+2\Lambda_{zz}\left(T_e\frac{\partial T_e}{\partial z}\right) \tag{3}$$

The model equations (1) and (2) describe the heat transfer process. T represents the temperature, C is the heat capacity, and k is the thermal conductivity. The subscripts e and l represent "electron" and "lattice". G is the electron-phonon coupling factor. S represents the laser source term. Equation (3) describes the ultrafast deformation of the lattice. u is the lattice displacement. Femtosecond laser irradiation caused the extremely high temperatures of electrons, which affected the optical properties, heat capacity, and thermal conductivity of the metal. Therefore, the calculation of parameters that could be covering a wide range of the electron temperature was essential.

For the electron thermal conductivity, the calculation related to the electron and lattice temperatures was shown as follows:

$$K_e=\chi\frac{(\vartheta_e^2+0.16)^{\frac{5}{4}}(\vartheta_e^2+0.44)\vartheta_e}{(\vartheta_e^2+0.092)^{\frac{1}{2}}(\vartheta_e^2+\eta\vartheta_l)} \tag{4}$$

where ϑ_e and ϑ_l are the normalization of T_e and T_l with the Fermi temperature. χ and η are the material constant.

The calculation of electron heat capacity was shown as follows:

$$C_e(T_e)=C_{eo}T_e \tag{5}$$

C_{eo} is the electron heat capacity constant.

The calculation of the electron-phonon coupling factor was shown as follows:

$$G=G_{rt}\left[\frac{A_e}{B_l}(T_e+T_l)+1\right] \tag{6}$$

A_e and B_l are the material constant for electron relaxation time, G_{rt} is the factor G at room temperature.

The laser source term S in equation (1) is calculated by the following equation:

$$S(t,z)=\frac{0.94F}{t_p\,\delta(t,z)}\times\left(1-R(t)\right)\times exp\left[-2.77\left(\frac{t}{t_p}\right)^2-\frac{z}{\delta(t,z)}\right] \tag{7}$$

where F represents the laser fluence; t_p is the pulse duration.

To accurately describe the surface optical parameters after femtosecond laser irradiation, we used the Drude model and Fresnel expression to calculate the reflectivity R and the optical penetration depth δ, which was used by Jiang et al in the improved two-temperature model [10]. According to the Drude model, the dielectric function of metal was shown below.

$$\epsilon(t,z)=\epsilon_1(t,z)+i\epsilon_2(t,z)=1+(\frac{n_e e^2}{m_e\epsilon_0})(\frac{-\tau_e^2(t,z)+(i\tau_e(t,z))/\omega}{1+\omega^2\tau_e^2(t,z)}) \\ =1+\omega_p^2(\frac{-\tau_e^2(t,z)+i\tau_e(t,z)/\omega}{1+\omega^2\tau_e^2(t,z)}) \tag{8}$$

The complex refractive index can be defined as

$$\left(\frac{\mathbf{c}}{\mathbf{v}}\right)=\boldsymbol{f}=(f_1+if_2)=\sqrt{\epsilon}=\sqrt{\epsilon_1+i\epsilon_2} \tag{9}$$

where **c** represents the speed of light in a vacuum and **v** represents the speed of light in a material. f_1 and f_2 are the real and imaginary parts. The penetration depth is the reciprocal of the absorption coefficient. The reflectivity R and the absorption coefficient α were obtained by the Fresnel expression:

$$R(t)=\frac{[f_1(t,0)-1]^2+f_2^2(t,0)}{[f_1(t,0)+1]^2+f_2^2(t,0)},\alpha(t,z)=\frac{2\omega f_2(t,z)}{c}=\frac{4\pi f_2(t,z)}{\lambda} \tag{10}$$

The electron relaxation time used in the Drude model was shown as follows:

$$\tau_e=\frac{1}{A_e T_e^2+B_l T_l} \tag{11}$$

The meaning and the value of the parameters in equations (1)-(8) was listed in Table 1.

Table 1. The physical and mechanical parameters of gold.

Name	Symbol	unit	value
lattice thermal conductivity	k_l	$Wm^{-1}K^{-1}$	315
lattice heat capacity	C_l	$J\ m^{-3}\ K^{-1}$	2.5×10^6
electron heat capacity constant	C_{e0}	$J\ m^{-3}\ K^{-2}$	70
material constant for k_e	χ	$Wm^{-1}K^{-1}$	353
material constant	η	/	0.16
G at room temperature	G_{rt}	$Wm^{-3}K^{-1}$	2.2×10^{16}
material constant A for τ_e	A_e	$K^{-2}\ s^{-1}$	1.2×10^{-7}
material constant B for τ_e	B_l	$K^{-1}\ s^{-1}$	1.23×10^{11}
Lame constant	λ	MPa	1.38×10^{11}
shear modulus	μ	MPa	2.63×10^{10}
thermal expansion coefficient	α_0	m/m	14.2×10^{-6}
hot-electron blast force constant	Λ_{zz}	$J\ m^{-3}\ K^{-2}$	70
density	ρ	kg/m^3	1.93×10^4

Figure 1(a) showed the schematic of the Spatial calculation zone. The front and back surfaces were assumed without heat loss during the ultrashort duration of the femtosecond laser, so the boundary and initial conditions are as follows:

$$T_e(z,0)=T_l(z,0)=T_{lo} \tag{12}$$

$$q(0,t)=q(L,t)=0 \tag{13}$$

L is the thickness of the film. q means the heat transfer at the boundary. T_{lo} sets the room temperature (300 K). For the mechanical boundary, static initial conditions and unconstrained boundary conditions were set.

$$\sigma_{zz}(z,0)=0 \tag{14}$$

$$u_z(z,0)=\frac{\partial u_z(z,0)}{\partial t}=0 \tag{15}$$

$$\sigma_{zz}(0,t)=\sigma_{zz}(L,t)=0 \tag{16}$$

σ_{zz} is the stress at the depth direction. The numerical solution of the model is solved by the forward differential format. Simulations were performed using three thicknesses of the gold film (30 nm, 60 nm, 100 nm) irradiated by a femtosecond laser of 100 fs pulse width. The time and space grids were shown in Figure 1(b). the time increment was 0.2 fs and the space increment was 1 nm. To solve the V-shaped oscillation problem in the calculation, the artificial viscosity was set as the follows:

$$\Pi=\omega_l \rho V_s \Delta z \left|\frac{\partial u_z}{\partial z}\right| - \rho\left(\omega_Q \Delta z\right)^2 \left|\frac{\partial u_z}{\partial z}\right| \frac{\partial u_z}{\partial z} \tag{17}$$

u_z is the lattice velocity. V_s is the sound speed. The constants ω_Q=2 and ω_l= 0.1-0.3.

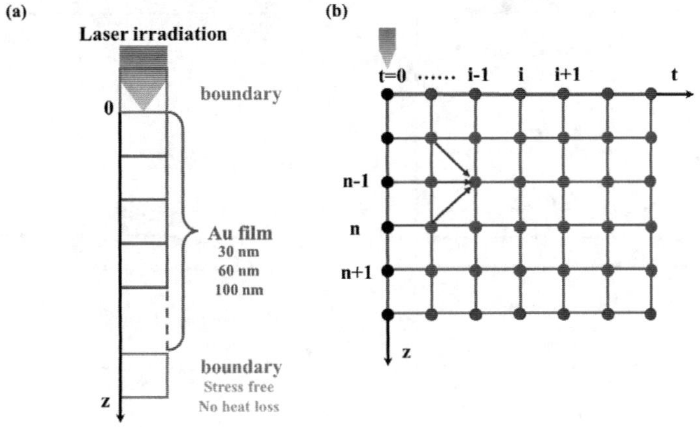

Figure 1. (a) Schematic diagram of the calculation zone. (b) The differential grid is used in the calculations. The initial and boundary conditions are indicated by the black and blue dots.

3. Results and discussion

In this work, the variations of temperature, reflectivity, and stress were induced by fluences below and equal to the melting threshold. The melting threshold fluence is 0.45 J/cm^2 [10], which referred to the fluence caused by the surface exactly to the melting point. The fluence below the threshold used for the simulation was 0.36 J/cm^2.

3.1. Surface temperature and optical properties

The dynamic optical parameters were chosen in the calculation as shown in **Figure 2**. After laser excitation, the surface reflectivity of the gold film decreased, which lead to more absorption of energy. As shown in **Figure 2**(a), the reflectivity decreased to a lower degree at a higher fluence. **Figure 2**(b) shows the variation of the optical penetration depth of the surface. Similar to the consequences of reflectivity, laser excitation led to an increase in penetration depth, which resulted in the absorption of energy. The reflectivity and penetration depth calculated by the Drude model was more accurate compared to the constant values. Especially for the simulations with near-threshold fluence, the constant optical parameters might make the temperature calculation result smaller than the reality.

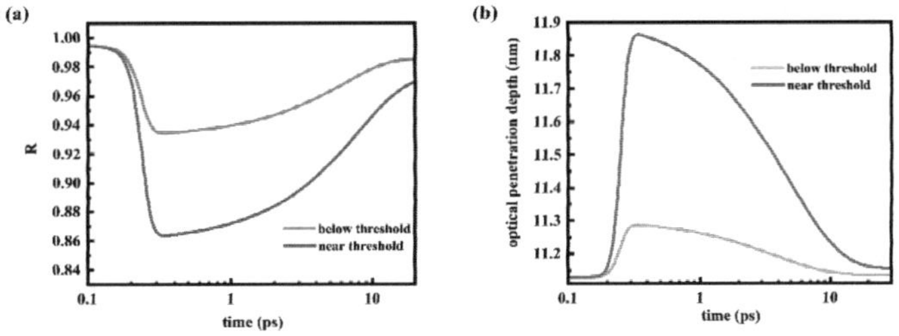

Figure 2. Variation of optical parameters after laser irradiation, 0 ps was defined as the moment when the laser pulse reaches the material. (a) Reflectivity variation as a function of time. (b) Surface optical penetration depths as a function of time.

The variations of the electron and lattice temperature on the surface induced by different fluences were shown in Figure 3(a) and (b). Because of the significant difference between the thermal capacity of the electron and the lattice, the electron temperature was excited to an extremely high level in a very short time. In contrast, the lattice temperature increased at a slow rate, which eventually equaled the electron temperature as the electron-lattice thermal coupling.

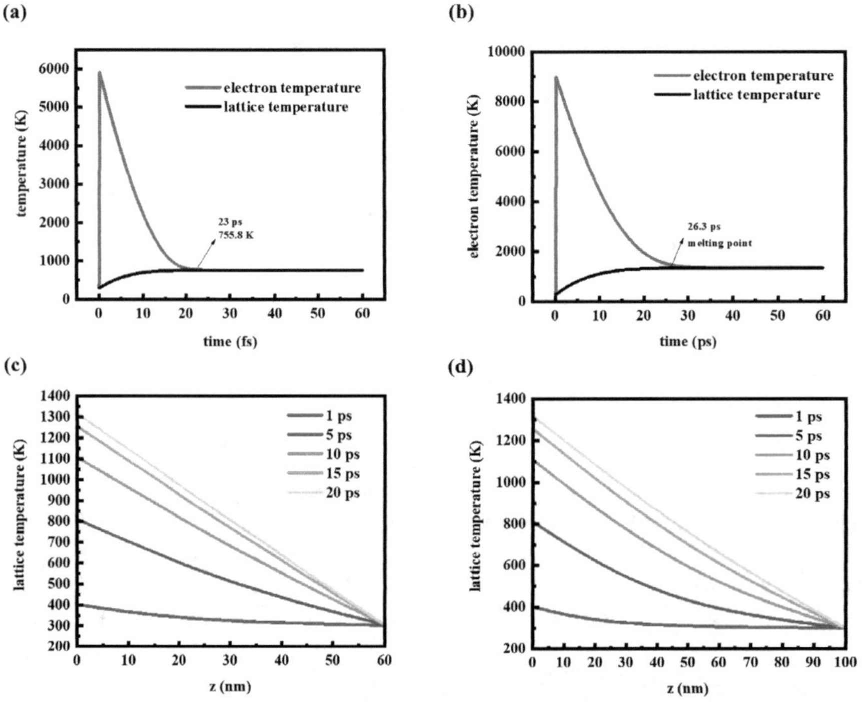

Figure 3. (a)(b) Variation of surface electron and lattice temperature as the function of time induced by the fluence below the melting threshold and the fluence at the melting threshold. (c) (d) Lattice temperature distribution in the depth direction induced by the fluence at the melting threshold of 60 nm gold film and 100 nm film.

At low energy excitation, the electron temperature equilibrated with the lattice temperature at 23 ps as shown in Figure 3(a). The equilibrium lattice temperature is 755.8 K, which did not reach the melting point of gold at 1337.33 K. The variation of the surface electron/lattice temperature at the threshold fluence was shown in Figure 3(b). The equilibrium lattice temperature reached 1337.3 K at the surface. Compared to low energy, the electron and lattice temperature reaches equilibrium at 26 ps, and the final equilibrium temperature exactly reached the melting point. With the increase of fluence, the equilibrium time was delayed and the equilibrium temperature rose. However, the surface temperature did not change with the thickness of the gold film as shown in Figure 3(c) and (d). Compared with the 100 nm film, the temperature gradient in the depth direction was larger in the 60 nm film, which might cause more drastic lattice deformation.

3.2. Stress distribution of gold film below the melting threshold

The temperature simulations indicated that the electrons could reach a very high temperature within the optical penetration depth because the heat capacity of the lattice is about two orders of magnitude of the electron. The rapid increase of electron temperature causes the generation of hot electron blast forces, which may destroy the film below the melting point. Thus, in this section, the stress distribution in different thicknesses of gold films below the melting threshold was calculated, which can describe the effect of stress.

The film damage of the fluence below the melting threshold has been reported widely. In general, this damage is considered to be the deformation caused by the stress, but when and where the damage in the film starts is still uncertain. Figure 4 showed the stress distribution of different thickness gold films induced by the below threshold fluence. The stress variation can be divided into two stages. In the first stage: after laser excitation, the compressive stress was transferred to the internal (1 ps -10 ps), as shown in Figure 4. The second stage: stress rebounded. After reaching the bottom boundary, the stress wave bounced back and tensile stress is formed, as shown by the yellow solid line in Figure 4 (b)(c). If the stress exceeded the tensile strength of gold (1.24 GPa) at the second stage, it may produce peeling damage to the substrate, as is shown in Figure 4(b) and (c). If the stress didn't exceed the limit, the tensile stress propagated forward, then produced the stress damage in the middle or surface layer, as shown in Figure 4(a) at 22 ps.

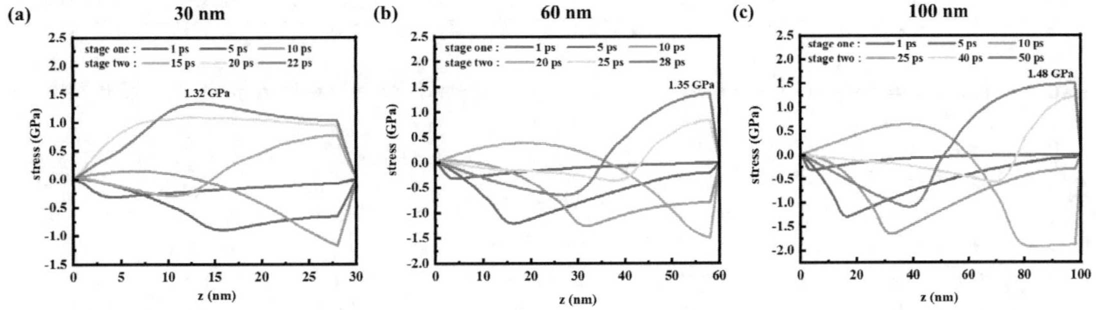

Figure 4. The stress distribution in different thicknesses of gold films at different times induced by the fluence below the melting threshold.

However, the model didn't consider the energy loss during the stress wave propagation. As shown in Figure 4(c), the stress didn't reach above 1.48 GPa until 50 ps. Therefore, we assumed that if the stress could not destroy the film before the lattice-electron equilibrium, the non-thermal damage didn't occur. As shown in Figure 3(a), the lattice-electron equilibrium was built at 23 ps and the lattice temperature didn't reach the melting point. In the thickness of 100 nm/60 nm gold film, the stress didn't reach the tensile strength before 23 ps, so the damage didn't occur. By analogy, the 30nm gold film would be destroyed in the middle of the film because the stress exceeded the limit and occurred before 23 ps as is shown in Figure 4(a).

By calculating the stress distribution of gold films with different thicknesses induced by the femtosecond laser below the melting threshold, it was concluded that although the material has not reached the melting point, non-thermal damage was probable to occur. As the time of reaching the tensile strength became shorter as the thickness decreases, the possibility of the non-thermal damage increases.

3.3. The dominant damage mechanisms near the melting threshold

The stress distribution of different thicknesses at a fluence of melting threshold was shown in **Figure 5**. The trend of stresses was identical to the below the threshold situation. As shown in the yellow lines in **Figure 5**, the stress reached tensile strength at the bottom of the material in the second stage. At the melting threshold, whether the film produced damage of non-thermal deformation or thermal ablation depends on the duration of time to reach the melting point or stress tensile strength.

As shown in Figure 3(b), the surface temperature reached the melting point at 26.3 ps. At this time, the 30 nm film has already reached the tensile strength, so the dominant damage mechanism is non-thermal deformation. The 60 nm film reached the tensile strength exactly at 26 ps, so the damage mechanism might be a synergistic effect of thermal ablation and non-thermal deformation. The 100 nm film reached the tensile strength at 40 ps, so the dominant damage mechanism was thermal melting.

By calculating the stress distribution of different thicknesses induced by the femtosecond laser at the melting threshold, it is concluded that the dominant mechanism of damage depends on the speed of achieving the stress or temperature limit. If the stress reached the tensile strength first, the film was damaged by non-thermal deformation. Thinner films were more susceptible to non-thermal damage than the thick ones.

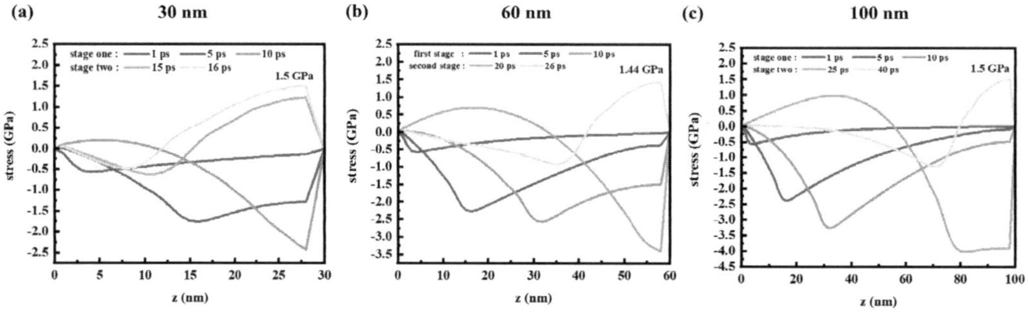

Figure 5. The stress distribution in different thicknesses of gold films at different times induced by the fluence at the melting threshold.

4. Conclusion

In summary, we calculated the temperature and stress distribution in the early stage of gold films induced by femtosecond laser by using the ultrafast thermoelasticity model. We obtained the changes in the material surface reflectivity and optical penetration depth after femtosecond laser excitation, which made the calculation results more accurate. There is no difference in the surface temperature of different films excited by the same laser fluence, but the internal lattice temperature gradient is slightly different. Only 30 nm of gold films could be damaged by non-thermal deformation at the fluence below the melting threshold. At the melting threshold fluence, the damage of non-thermal deformation occurred in the 30 nm gold film, because the stress peak reached the tensile strength before the lattice temperature was above the melting point. The 60 nm gold film exceeded the tensile strength when the melting point was achieved, so the damage mechanism might be a combination of thermal and non-thermal effects. The 100 nm gold film exceeded the tensile strength after the melting point was achieved, so thermal ablation occurred. Our results illustrated the early mechanism of material deformation. which was valuable for the near-threshold femtosecond laser processing of gold films.

References

[1] Schwind M, Kasemo B and Zorić I 2013 *Nano Lett.* **13** 1743–50
[2] Mcfarland A D and van Duyne R P 2003 Single Silver Nanoparticles as Real-Time Optical Sensors with Zeptomole Sensitivity *Nano Lett.* **3(8)** 1057–1062
[3] Devaux E, Ebbesen T W and Weeber J-C 2003 *Appl. Phys. Lett.* **83** 4936
[4] Nakata Y, Okada T and Maeda M 2003 *Jpn. J. Appl. Phys., Part 2* **42** L1452
[5] Kuznetsov A I, Koch J and Chichkov B N 2009 *Appl. Phys. A: Mater. Sci. Process.* **94** 221–30
[6] Meshcheryakov Y P and Bulgakova N M 2006 *Appl. Phys. A* **82** 363–8
[7] Korte F, Koch J and Chichkov B N 2004 *Appl. Phys. A: Mater. Sci. Process.* **A79(4/6)** 879-881
[8] Ivanov D S, Rethfeld B, O'connor G M, Glynn T J, Volkov A N and Zhigilei L v 2008 *Appl. Phys. A* **92** 791–6
[9] Chen J K, Beraun J E, Grimes L E and Tzou D Y 2002 *Int. J. Solids Struct.* **39(12)** 3199-3216
[10] Jiang L and Tsai H L 2005 *J. Heat Transfer* **127** 1167–73

Journal of Physics: Conference Series **2313** (2022) 012026 doi:10.1088/1742-6596/2313/1/012026

Theoretical simulation on Fraunhofer diffraction of arbitrarily shaped aperture

Yuanshuo Liu[1], Luyan Wang[1], Jia Dong[1], Jihao Xia[1], Lei Yang[1] and Yi Jin[1, 2]

[1]Department of Physics, University of Jinan, 250022 Jinan, P. R. China

[2]Corresponding author's e-mail: ss_jiny@ujn.edu.cn

Abstract. The diffraction refers to the phenomenon that when light encounters an obstacle during propagation, it will deviate from the original propagation path and propagate behind the obstacle. In the study of diffraction and the relevant, we often encounter the questions about the prediction or estimation on the diffraction patterns of apertures with various shapes. Yet the specially designed optical experiment is very time consuming and laborious for this simple purpose. In this paper, the laser far-field diffraction is studied based on Kirchhoff diffraction theory. The intensity distribution is described by using a two-dimensional matrix and then the mathematical representations of the diffraction pattern of arbitrary aperture are provided. Visualization of calculated data is achieved by computer programming, and high-precision diffraction patterns are obtained. The above theoretical analysis method presented in this paper has higher universality. When the calculation infinitesimals are small enough and the computer graphics display function is sufficient, the diffraction pattern with high resolution can be obtained efficiently compared with the experiment.

1. Introduction

Diffraction of light refers to the phenomenon that when light encounters an obstacle in the process of propagation, it will deviate from the original propagation path and propagate behind the obstacle. In modern optics and even modern physics, the diffraction has been more and more widely used. In diffraction research, optical experiments are often required. However, optical experiments have high requirements on the environment, besides the production of diffraction screens is cumbersome and diffraction image has poor sharpness in many cases. The above difficulties hinder the intuitive learning and further exploration of diffraction phenomena.

In optics, the basis of diffraction theory is the Huygens principle. This principle assumes that every spot in the light wave surface can be regarded as a secondary light source which emits secondary waves, and these secondary waves are superimposed to obtain a light wave propagating forward [1]. Using this principle, diffraction images and light intensity distributions of apertures of various shapes can be calculated theoretically.

At present, many literatures [2-5] have used different methods, such as Fourier transform, irregular aperture segmentation synthesis, etc., to carry out Fraunhofer diffraction modelling of the apertures like single-slit, multi-slit, cross-shaped, #-shaped, arbitrary regular polygons and so on. Referring to previous researches, in this paper we propose another theoretical method for apertures of arbitrary shapes, in which the complex amplitude contribution of spherical waves of point light sources at each spot of the aperture screen can be deduced to mathematical representations on the basis of Kirchhoff diffraction theory. Besides the fast Fourier transform and the inverse fast Fourier transform are used to

Journal of Physics: Conference Series **2313** (2022) 012026 doi:10.1088/1742-6596/2313/1/012026

obtain the complex amplitude distribution of the optical screen, and finally take the square of the magnitude of the complex amplitude to visualize the data. Finally with computer software, the data can be visualized. The above method raised in this paper can be applied to apertures of any shape, which provides a very convenient and efficient tool for the relevant research and teaching.

2. Fundamental

The wave surface of the light emitted by a monochromatic light source P_0 is a spherical wave, Q is any point on the aperture plane, P is the observation point behind the aperture light field, r_0 and r are the distance from P_0 and Q to P, respectively. When the centre of the rectangular coordinate system coincides with the centre of the spherical wave, the complex amplitude of monochromatic spherical wave at any point P in the light field is:

$$\psi(Q) = A\exp(ik_0 r_0)/r_0 \tag{1}$$

Where A is the amplitude value at P_0, $k_0 = 2\pi/\lambda$. According to Kirchhoff diffraction formula, the complex amplitude of point P can be obtained:

$$\psi(P) = \frac{1}{i\lambda}\int_{S_0}\psi(Q)K(\theta)\frac{\exp(ik_0 r)}{r}dS' = \frac{A}{i\lambda}\int_{S_0}\frac{\exp[ik_0(r+r_0)]}{rr_0}dS' \tag{2}$$

In the above formula, $K(\theta) = [\cos(\vec{n},\vec{r}) - \cos(\vec{n},\vec{r}_0)]/2 \approx 1$.

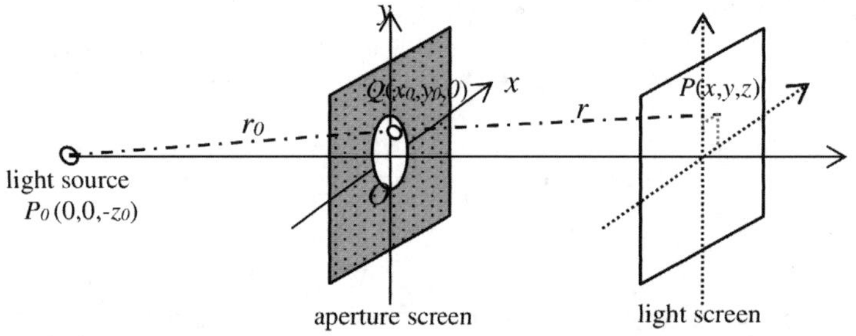

Figure 1. Schematic diagram of coordinate system and physical quantities.

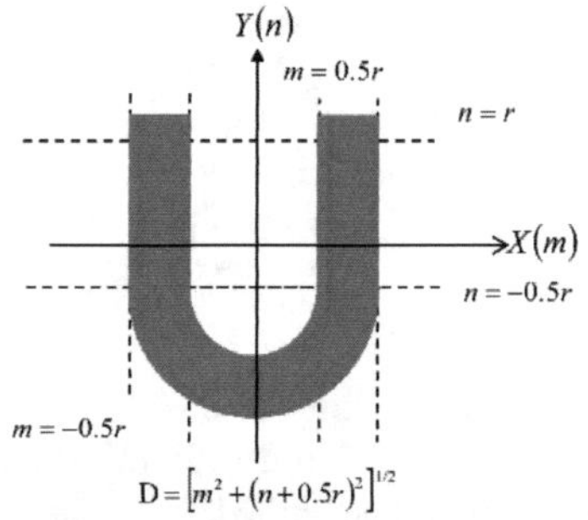

Figure 2. Representation of "U" shaped aperture in coordinate axis.

It can be seen in Figure 1, $r_0 = (z_0^2 + x_0^2 + y_0^2)^{1/2}$, $r = [z^2 + (x-x_0)^2 + (y-y_0)^2]^{1/2}$. The r in the denominator of formula (2) can be directly replaced by z approximately, but since λ in the index is too small, a small change in r may cause a phase change far greater than 2π. Therefore, it cannot be directly replaced by z approximately [4].

We expand r by binomial to obtain the following formula:

$$r = z\left[1 + \frac{(x-x_0)^2 + (y-y_0)^2}{2z^2} + \frac{[(x-x_0)^2 + (y-y_0)^2]^2}{8z^4} + \cdots\right] \tag{3}$$

By paraxial approximation [2], only the first two items can be preserved:

$$r = z\left[1 + \frac{(x-x_0)^2 + (y-y_0)^2}{2z^2}\right] \tag{4}$$

So, the Q-point complex amplitude is:

$$\psi(x, y) = \frac{\exp(ikz)}{i\lambda z} \iint_{-\infty}^{+\infty} \psi_0(x_0 + y_0) \exp\left[ik \frac{(x - x_0)^2 + (y - y_0)^2}{2z} \right] dx_0 dy_0 \tag{5}$$

We take $h(x, y) = \frac{\exp(ikz)}{i\lambda z} \exp\left[\frac{ik(x^2 + y^2)}{2z} \right]$, then formula (5) can be transformed into:

$$\psi(x, y) = \iint_{-\infty}^{+\infty} \psi_0(x_0, y_0) h(x - x_0, y - y_0) dx_0 dy_0 = \psi_0(x, y) * h(x, y) \tag{6}$$

Next, we make a fast Fourier transform on it [6]:

$$\psi(\xi, \ \eta) = F[\psi_0(x, y)] \cdot F[h(x, y)] \tag{7}$$

The complex amplitude distribution of the light screen is obtained by inverse fast Fourier transform:

$$\psi(x, y) = F^{-1}[\psi(\xi, \eta)] \tag{8}$$

3. Calculation program

3.1. Mathematical description of apertures

In this program, we use Δx to represent the side length of the square micro-element taken, and a micro-element represents a pixel. The aperture screen is represented by a two-dimensional matrix, and the value of Δx depends on the order of the matrix. Here $(x, y) = (m, n)$ represents an element in the matrix, representing the position of the square element. For example, the curve $m^2 + (n + 0.5r)^2 = (0.35r)^2$ (r is a constant) represents the set of all points in the two-dimensional matrix where the element (m, n) satisfies the functional relationship $x^2 + (y + 0.5r)^2 = (0.35r)^2$, and r is a parameter of the function, representing the width of r pixels.

The letter-shaped hole 'U' in Figure 2 is used as an example for description below. The blue shaded part in the figure represents the light-transmitting area of the aperture screen, and the rest are opaque areas. In the programming process, only the light-transmitting part contributes to the complex amplitude of the dot on the light screen, which is represented by '1'; the remaining opaque areas do not contribute to the complex amplitude of the dot on the light screen, which is represented by '0'. Except for the small hole, the rest of the aperture screen is opaque to light, so after presetting the points on the entire aperture screen to be '0', we can use the inequality about m and n to represent the 'U' -shaped hole:

$$\begin{cases} D = \left[m^2 + (n + 0.5r)^2\right]^{1/2} \\ (0.3r \le m \le 0.6r \cap -0.5r \le n \le r) = 1 \\ (0.3r \le D \le 0.6r \cap n \le -0.5r) = 1 \\ (-0.6r \le m \le -0.3r \cap -0.5r \le n \le r) = 1 \end{cases} \tag{9}$$

In this way, we complete the mathematical representation of the 'U' shaped hole. This representation method can play a role in any shape, and readers can design functions according to their own needs to get the hole shape they want. In the programming calculation, just add the two-dimensional matrix representing the aperture screen to the algorithm, and then the ψ_0 of each point on the aperture screen can be represented.

3.2. Diffraction simulation

Firstly, we set the following basic parameters: the parameter related to the overall hole width is r, the number of split pixels of aperture screen and light screen is N, the amplitude value of the point light at the unit circle is a_0, the side length of the aperture screen is d_0, the side length of light screen is d, the distance from the light source to the aperture screen is z_0, the distance from aperture screen to light screen is z and the laser wavelength is λ. In addition, the aperture screen is divided into several micro-elements, the shape of each hole s is described, and the complex amplitude distribution ψ' of the spherical wave emitted by the point light source on the left side of the aperture screen is calculated. The complex amplitude distribution ψ_0 on the right side of the aperture screen can be obtained by multiplying the two. Divide the light screen into several micro elements, and take the fast Fourier transform of ψ_0 and h respectively and multiply. Then reverse fast Fourier transform the product result $\psi(x, y)$ to obtain the complex amplitude distribution $\psi(x, y)$ on the light screen. Finally, the square of the absolute value of the complex amplitude is taken to obtain the diffraction image function $I(x.y)$.

```
[m,n]=meshgrid(linspace(-N/2,N/2,N)); n=-n;
x0=linspace(-d0/2,d0/2,N);
y0=linspace(-d0/2,d0/2,N);
r0=zeros(N,N);s=zeros(N,N);
      D1=(m.^2+n.^2).^0.5;
      D2=0.5*n.^2+0.5*m.^2;
      s(find(D1<=r&D2>=0.354025*r.^2))=1;
      s(find(D1>=r))=0;
      s(find(D2<=0.354025*r.^2))=0;
%Divide the aperture screen into several micro elements and draw the aperture shape of the aperture screen

for i=1:N
  for j=1:N
    r0(i,j)=sqrt(((x0(i)).^2+(y0(j)).^2+z0^2));
    Psi_l(i,j)=a0/r0(i,j)*exp(1j*k*r0(i,j));
  end
end
Psi_r=Psi_l.*s;
%Complex amplitude distribution of each point on the left and right of aperture screen
x1=linspace(-d/2,d/2,N);
y1=linspace(-d/2,d/2,N);
[x,y]=meshgrid(x1,y1);
%Divide the light screen into several micro elements

U=fftshift(fft2(Psi_r));
h=exp(1j*k*z)*exp((1j*k*(x.^2+y.^2))/(2*z))/(1j*lamda*z);
H=fftshift(fft2(h));
A=fftshift(ifft2(H.*U));
axis image;
colormap(hot)
I=abs(A).^2;
%The diffraction image function I is calculated by fast Fourier algorithm
```

In this code, the wavelength is $\lambda = 650\text{nm}$, the distance from the aperture screen to the light screen is $z = 0.8\text{m}$. The parameter $N = 1000$ in the code indicates that the aperture screen and the light screen are divided into 1000×1000 pixels. The smaller the micro element, the higher the calculation accuracy and the clearer the diffraction image, but the longer the calculation time at the same time. The number of segmentation pixels selected in this code is moderate, and the accuracy has been high, so it can achieve the ideal effect. In the process of debugging, if the range of the displayed diffraction pattern is too small or too large, the reader can change the length and width of the light screen to obtain a more ideal image, or change the distance between the aperture screen and the light screen to observe the changes of diffraction images at different distances.

The description of aperture shape marked in the code is the mathematical description of 'U' shaped aperture. The reader's flexible replacement of this area is the key to the universal applicability of this code to holes of any shape. In addition to several letter screens provided in this article, readers can also write holes of other shapes by themselves, such as arbitrary polygon [2].

4. Simulation of circular aperture diffraction

For circular aperture diffraction, it has been observed experimentally that the centre of the diffraction image is a bright spot, which is called Airy disk. The outward diffraction image of the Airy disk forms a group of concentric rings with alternating bright and dark circles, and the brightness of each bright ring decreases sharply with the increase of the radius.

In the case of circular aperture diffraction, the expression of light intensity is:

$$I(P)=I_\theta=I_0\left[2J_1(u)/u\right]^2 \tag{10}$$

The half width of Airy disk is:

$$\theta=0.61\lambda/R=1.22\lambda/D \tag{11}$$

The accuracy of the program can be verified to a certain extent by comparing the Airy disk half-angle widths obtained by experimental measurement and program simulation under the same conditions. From the theoretical calculation, it is known that the half-angle width of the Airy disk under the circular aperture $D=200\mu\text{m}$ and the laser wavelength $\lambda=650\text{nm}$ is:

$$\theta=1.22\lambda/D=3.965\times10^{-3}\text{rad} \tag{12}$$

As shown in the left figure of Figure 3, we take the center of the two points A and B to be approximately the first-level dark pattern on the left, and the center of the two points C and D to be approximately the first-level dark pattern on the right. The coordinates of the four points of A, B, C, D are marked as x_A , x_B , x_C , x_D respectively. Then the first-order dark pattern diameter $D=\left[(x_C+x_D)-(x_A+x_B)\right]/2$ is the Airy disk diameter, as shown in Table 1. And then obtain the Airy disk half-angle width under this condition as $\theta_{exp}=\arctan\left(\overline{D}_{exp}/2L\right)=3.883\times10^{-3}\text{rad}$. We set the aperture diameter D to $200\mu\text{m}$, the laser wavelength λ to 650nm, and the distance L from the aperture screen to the light screen to 80cm.

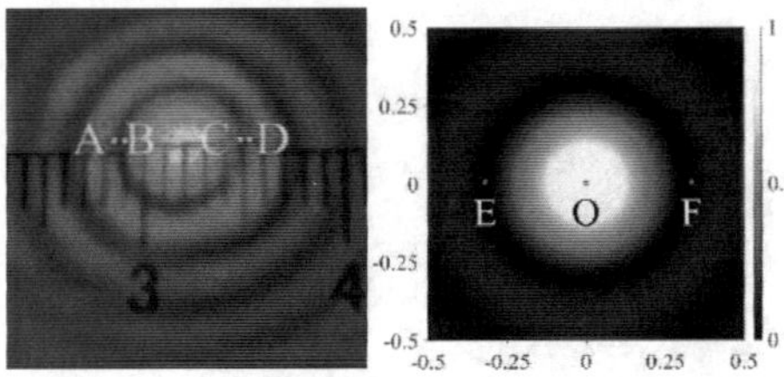

Figure 3. Experimental circular aperture diffraction image and program simulated circular aperture diffraction image. The scale units in both figures are cm.

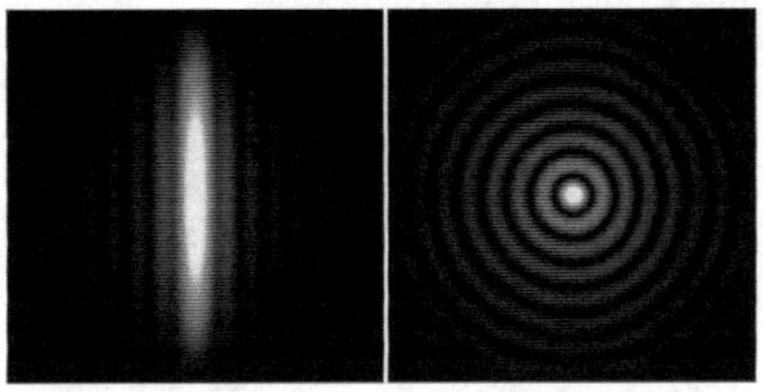

Figure 4. Light and dark stripes corresponding to slits and concentric rings corresponding to rings.

Table 1. Experimental data of circular aperture diffraction.

Coordinates of each point (cm)	1	2	3	4
x_A	3.22	3.25	2.87	2.86
x_B	3.30	3.30	2.80	2.79
x_C	3.86	3.87	3.41	3.42
x_D	3.91	3.92	3.49	3.48
$D_{exp}=\left[(x_C+x_D)-(x_A+x_B)\right]/2$	0.625	0.620	0.615	0.625
$\overline{D}_{exp}$	0.62125			

The resulting simulated diffraction pattern is shown on the right in Figure 3. Points E and F in the figure are the first-level darkest points on the left and right sides, both of which are 316 pixels away from the center point O. It can be calculated that the diameter of the first-level dark pattern is $D_{sim} = 0.632\text{cm}$, and the Airy disk half-angle width is $\theta_{sim} = \arctan(D_{sim}/2L) = 3.950\times10^{-3}\text{rad}$. We can get the experimental relative error as $|\theta_{exp} - \theta_{sim}|/\theta_{sim}\times100\% = 1.701\%$.

We analyze the simulated diffraction image and Airy disk half-angle width, which are basically consistent with the experimentally observed circular aperture diffraction image and have little difference in Airy disk half-angle width. Therefore, the diffraction image simulated by the principle and method used in this code is correct.

5. Further discussion on diffraction pattern

By observing the diffraction pattern, it is found that the slits on the aperture screen generally correspond to light and dark stripes perpendicular to the slits on the light screen, and the rings on the aperture screen correspond to concentric rings on the light screen, as shown in Figure 4. For concentric rings, the experimental verification has been partially discussed. Next, we focus on comparing the differences of diffraction patterns of letters 'J' and 'U'. The difference between 'J' and 'U' is mainly in the direction parallel to the axis, 'J' corresponds to single slit and 'U' corresponds to double slit. According to Kirchhoff diffraction formula, the light intensity distribution of single slit and double slit on the light screen is [7]:

$$I_S = I_0(\sin\alpha/\alpha)^2 \quad I_D = I_0[(\sin\alpha\cdot\sin2\gamma)/(\alpha\cdot\sin\gamma)]^2 \tag{13}$$

In the above formula, $\alpha = \pi a\sin\theta/\lambda, \gamma = \pi d\sin\theta/\lambda, \theta$ is the diffraction angle, a is the width of single slit, and $d = 2a$ is the center distance between two slits. For single-slit diffraction:

$$\sin\theta = \pm m\lambda/2a \ (m = 0,1,2\cdots) \tag{14}$$

Therefore, when $\sin\theta = 0$, it is the main maximum, when $\sin\theta = \pm\lambda/2a$, it is the submaximal, and so on. For double-slit diffraction:

$$\sin\theta = \pm m\lambda/2a \ (\text{when m} = 2,4,6\cdots \text{ the diffraction order is missing}) \tag{15}$$

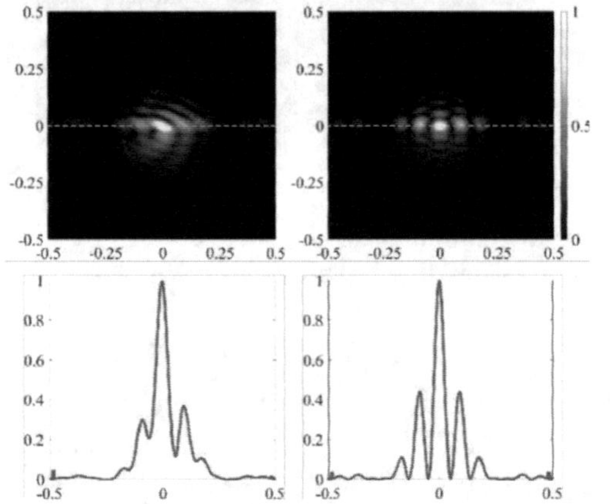

Figure 5. Diffraction patterns of the "J"-shaped hole and "U" shaped hole and their light intensity distribution along the dotted line.

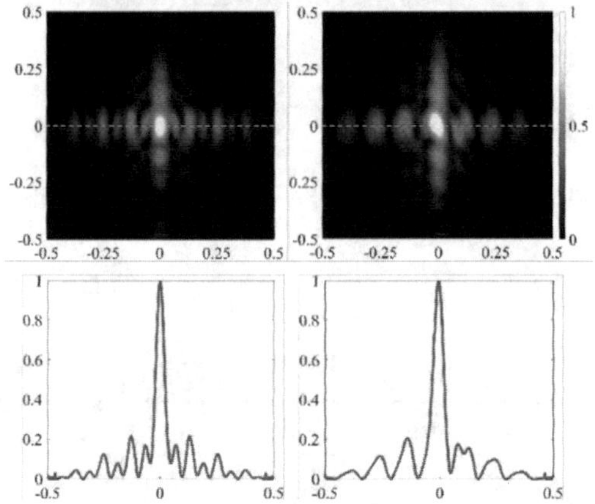

Figure 6. Diffraction patterns of "E"-shaped hole and "F"-shaped hole rotated 90° counterclockwise and their light intensity distribution along the dotted line.

As shown in the Figure 5, it can be seen that there are obvious differences in the light intensity distributions of the horizontal cross-sections of the diffraction patterns of these two holes: the 'J' -shaped hole is in good agreement with the single slit diffraction pattern, that is, it is consistent with the diffraction of 'I' aperture. The slight difference may be caused by the concentric ring in the lower half; the 'U' -shaped hole matches the double slit very well, the overall trend is consistent with the change caused by diffraction factor $(\sin\alpha/\alpha)^2$. And there is more detailed light intensity division in the envelope. For $d=3a$ in 'U' -shaped hole drawn by this program, it can be clearly found that there are missing diffraction orders at $m=\pm 3$, that is, the contribution of diffraction factor $(\sin 2\gamma/\gamma)^2$ to light intensity distribution. The previous literature [8] also discussed the difference of light intensity distribution between the 'E' -shaped and 'F' -shaped, and considered that there must be a sub extreme intensity between the two maxima of the diffraction pattern of the letter E; the diffraction pattern of the letter F is similar to that of a double slit. Based on the code operation provided in this article, the vertical section light intensity distribution of these two letters is shown in Figure 6.

It can be seen that the results of the code operation provided in this paper are highly consistent with the previous work results, which shows the correctness of the code operation results in this paper.

6. Summary and discussion

Based on Huygens principle and Kirchhoff diffraction theory, this paper provides a set of theoretical methods including computer codes that can be used to simulate Fraunhofer diffraction patterns of the holes of any shape. Taking some capital letters as an example, the relevant calculations are carried out and the result diffraction images are very clear. Theoretically, as long as the calculated micro-elements are made small enough, the diffraction patterns with high resolution can be obtained with the computer codes. Besides, to verify the accuracy of this method, we have operated a diffraction experiment on circular aperture and the theoretical simulation results are in good agreement with the experiment patterns.

Furthermore, the theoretical analysis method in this paper can be partly modified and extended to studying the Fresnel diffraction. As long as the operator masters a certain mathematical analysis software foundation, expresses the desired diffraction aperture mathematically, and flexibly debugs it into the code, the ideal simulated diffraction pattern can be obtained.

References

[1] Myung K K, Digital holographic microscopy: Principles, techniques, and applications (Springer, 2011).

[2] Song Y Z, 2017 Discussion on the Fraunhofer diffraction imaging of arbitrary regular polygon hole *J. Physical Experiment* **37 (11)** 48-51.

[3] Xia X, Dong Z C and Yang J H 2017 Computer simulation of Fraunhofer slit diffraction J. *Physics Bulletin* **(05)** 15-8.

[4] Li J F, Jiang Z F, Huang C J, He H Y and Zhu J F 2003 The general expressions of Fraunhofer diffraction formula J. *University Physics* **(11)** 9-14.

[5] Wang F, Fu H B and Zhao J G 2010 Computer simulation of Fraunhofer diffraction of arbitrary shape *J. Software Guide* **9 (12)** 190-1.

[6] Wang H T, Guo Z P and Xu M Y 2018 Fresnel diffraction based on the aperture of any regular polygon *J. Journal of Yanbian University (NATURAL SCIENCE EDITION)* **44 (04)** 344-7.

[7] Guo H S 2008 Electrodynamics third edition *M. Beijing: Higher Education Press* pp 176-80.

[8] Cai X H, Miao T F and Zhou J 2008 Analysis and computer simulation of the Fraunhofer diffraction of alphabet letters *J. Physical Experiment* **28(12)** 38-42.

Journal of Physics: Conference Series **2313** (2022) 012027 doi:10.1088/1742-6596/2313/1/012027

Upper critical field beyond Pauli limit in the exfoliated $RbCr_3As_3$ nanowires

Xiaomei Hu[1], Zhiyi Liu[1] and Xingyuan Hou[1, 2, 3]

[1]Information Materials and Intelligent Sensing Laboratory of Anhui Province, Institutes of Physical Science and Information Technology, Anhui University, Hefei 230601, China

[2]Key Laboratory of Structure and Functional Regulation of Hybrid Materials (Anhui University), Ministry of Education, Hefei, 230601, China

[3]Corresponding author's e-mail: xyhou@ahu.edu.cn

Abstract. Unconventional superconductivity in the quasi-two-dimensional cuprates and iron-based superconductors has attracted great attention in these years. Recently, the discovery of the Cr-based ternary compounds has aroused the research upsurge for the possible spin-triplet superconductivity revealed from the bulk samples. Here we carried out the electrical transport measurements of the air-stable $RbCr_3As_3$ after the dimensionality reduction. It is founded that the superconductivity is suppressed, while the upper critical field is still far larger than the Pauli paramagnetic limit, providing an ideal platform to shed light on the unconventional superconductivity.

1. Introduction

In the past few decades, cuprates and iron-based superconductors (IBS) have attracted the most attention due to the unconventional superconductivity and much higher superconducting transition temperature (T_c) [1-5]. Many novel physical properties and potential applications are uncovered for these systems. There are some common features between the two systems. On one hand, they exhibit a dome-shaped superconducting phase diagram concerning chemical doping or high pressure, which implies a complex interplay between superconductivity and other electronic states such as antiferromagnetism order, charge, or spin density waves [6-9]. On the other hand, both of them adopt a quasi-two-dimensional (Q2D) crystal structure and can be viewed as the carrier reservoir layers and the superconducting layers alternatively stacking along the c-direction [10,11]. Since then, enormous experimental and theoretical studies have been carried out focusing on the underlying mechanism of unconventional superconductivity, and lots of novel physical phenomena are reported. It is agreed that the pairing mechanism could not be understood in the framework of the well-established Bardeen-Cooper-Schrieffer (BCS) theory, in which the electrons with inverse spin momentum and orbital momentum combine into Cooper pairs based on the electron-phonon interactions. However, the underlying pairing mechanism is still a mystery in the field of condensed matter physics so far. Therefore, new platforms are always required to further understand the unconventional superconductivity.

Recently, novel superconductivity was discovered in Cr-based ternary compounds $A_2Cr_3As_3$ and ACr_3As_3 (A = alkali metals), which share the same $[Cr_3As_3]_\infty$ chains. They belong to the reduced dimensional crystal structure, namely quasi-one-dimensional (Q1D) structure [12-17]. Theoretically, one-dimensional materials are few to undergo a superconducting transition and support Tomonaga-Luttinger liquid (TLL) behavior in which spin and charge are separated [12,18,19]. However, the Q1D Cr-based superconductors exhibit relatively high T_c, and the T_c of $Na_2Cr_3As_3$ and $RbCr_3As_3$ is 8.6 K and 7.3 K, respectively [18,16,17]. Moreover, many exotic physical properties are reported. First, the upper critical field exceeds the Pauli limit, indicating the possible spin-triplet pairing component [12]. Second, complex magnetic orders were proposed in this system, including the intrachain and interchain magnetic phases [20]. Third, TLL behavior has been reported in the $K_2Cr_3As_3$ [14]. Hence, the newly discovered Cr-based superconductors provide an opportunity of studying the unconventional superconductivity and shed new light on the application in the future quantum computations for the possible topological states related to the spin-triplet pairing. However, the physical properties of most quantum materials are influenced by the size effect when they are fabricated into micro devices. For example, for the $NbSe_2$ thin flakes, the superconductivity is significantly suppressed, and dominated by a novel Ising superconductivity due to the inversion symmetry breaking in the monolayer limit [21,22]. Among the Cr-based superconductors, $A_2Cr_3As_3$ is extremely air sensitive. Fortunately, ACr_3As_3 is chemically stable in the atmosphere environments in contrast to $A_2Cr_3As_3$.

In this work, we obtained the nanowires of $RbCr_3As_3$ with different sizes by mechanical cleavage and performed electrical resistivity measurements to study the size that influenced superconductivity. It is found that the superconducting temperature T_c is suppressed by the reduced sample size. However, the upper critical field is still beyond the Pauli limiting, confirming the unconventional superconducting behavior.

2. Methods

The $Rb_2Cr_3As_3$ single crystals were acquired by deintercalating half of the Rb^+ ions using anhydrous ethanol from the $Rb_2Cr_3As_3$, which were grown by the high temperature solution growth method from the mixture of RbAs and CrAs [12]. Then the samples were annealed in a vacuum at 373K for 10 h to further improve the quality of the samples [18]. The $RbCr_3As_3$ nanowires are obtained by mechanical exfoliation method and transferred to the silicon oxide substrates. All the processes were carried out in an argon environment to avoid contamination. The appropriate $RbCr_3As_3$ flakes were selected by the scanning electron microscope (SEM), then the Pt electrodes for transport measurement were immediately deposited by the focused ion beam technique (FIB). The electrical transport measurements were carried out on the Quantum Design PPMS- DynaCool system.

3. Results

Figure 1(a) shows the schematic crystal of $RbCr_3As_3$ with a Q1D hexagonal structure ($P6_3/m$, No. 176), which consists of multiple parallel $[(Cr_3As_3)^-]_\infty$ chains [23]. It has been found that $RbCr_3As_3$ possesses very similar physical properties to the non-centrosymmetrical $Rb_2Cr_3As_3$, in which the missing inversion symmetry gives rise to mixed parity pairing in principle [16]. The SEM image shows the $RbCr_3As_3$ sample contains loose linear filaments with a diameter of microns and a length of millimeters as shown in Figure 1(b). The superconducting transition of bulk $RbCr_3As_3$ occurs around 7.25 K from the temperature dependent resistance curve in Figure 1(c), which is suppressed by the applied magnetic fields perpendicular to the c axis, consistent with previous reports [16,17]. The upper critical field could be derived from the fitting to the critical temperature under varied field (T_c vs $\mu_0 H_{c2}$) by the Ginzburg-Landau formula, $\mu_0 H_{c2}(T) = \mu_0 H_{c2}(0)(1 - t^2)/(1 + t^2)$, where t denotes a reduced temperature ($t = T/T_c$). The estimated upper critical field $H_{c2}(0)$ is about 72 T for the half-normal resistance criterion, which is very close to the estimation before [16,17]. This large value is much larger than the nominal Pauli paramagnetic limit of 13.4 T, which is roughly estimated by $\mu_0 H^P = 1.84T_c$. Since the singlet Cooper pair becomes energetically unstable for the Zeeman splitting energy exceeding the superconducting gap energy, it may imply the spin-triplet pairing in $RbCr_3As_3$.

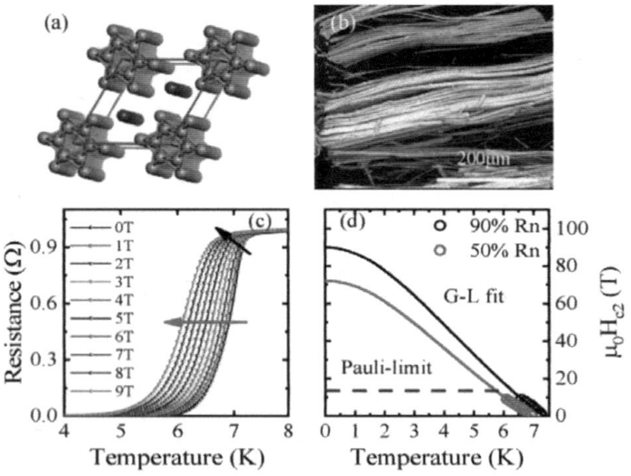

Figure 1. (a) Crystal structure of $RbCr_3As_3$ with the one-dimensional $[(Cr_3As_3)^-]_\infty$ chains. (b) SEM image of $RbCr_3As_3$ bulk sample. (c) The temperature dependent resistance curves under magnetic fields from 0 to 9 T. Each curve is normalized to the normal state resistance R_n. (d) Derived upper critical field, the blue dashed line denotes the Pauli limit.

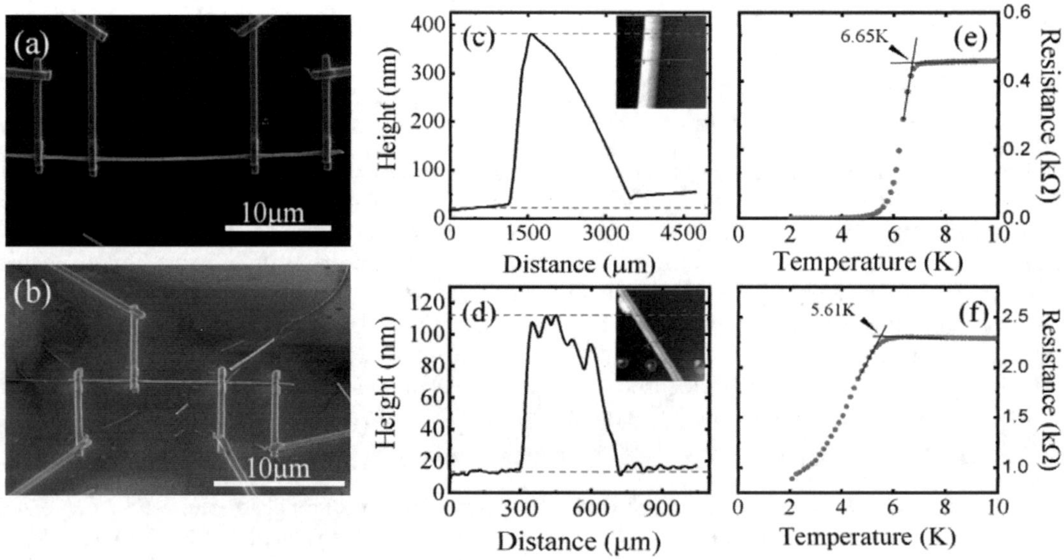

Figure 2. (a) and (b) are the SEM images of $RbCr_3As_3$ nanowires S01 and S02, respectively. (c) The profile of S01. The inset shows the AFM image of S01. (d) AFM image of S02. (e) Temperature dependence of the resistance for S01, showing a superconducting transition at about 6.65 K. (f) Temperature dependence of resistance for S02. The onset transition temperature is about 5.6 K.

It has been well noticed that topological edge states may exist on the boundaries of an intrinsic spin-triplet superconductor. However, the properties of most quantum materials will be modulated by dimensionality reduction when the micro devices are fabricated. Hence, the dimensionality dependence of $RbCr_3As_3$ should also be checked by the sophisticated microfabrication technique. The weak coupling between the loose filaments in the samples makes it easy to be exfoliated into nanowires by Scotch tape. Figure 2(a) and (b) show the SEM images of two $RbCr_3As_3$ nanowires (S01 and S02) with four standard electrodes fabricated by the FIB technique. The length and width of a nanowire could be measured by SEM directly, and its thickness could be measured by atomic force microscopy (AFM). The length of a nanowire is far larger than the width and thickness, and the width is close to each other. As illustrated in Figure 2(c) and (d), the length of S01 and S02 is about 40.5 μm and 16.9 μm, respectively, and the width is about 290 and 260 nm. While the thickness of S01 is about 362 nm, and that of S02 is about 98 nm. The two nanowires became superconducting at low temperature revealed from the resistive transport measurement in Figures 2(e) and (f). The resistance of sample S01 begins to drop at about 6.65 K and becomes zero at about 5 K. The onset Tc of S02 is only about 5.61 K, but no zero resistance occurs until the temperature limit of 2 K. It is clear that the T_c decreases gradually as the decreased thickness, showing the suppressed superconductivity and enhanced fluctuations by the dimensionality decrease of $RbCr_3As_3$.

The superconducting transition could be suppressed further by the magnetic fields. Figures 3(a) - (d) show the results of the temperature dependent resistance curves of the two samples in the fields from 0 to 14 T for $H\perp c$ and $H//c$, respectively. It has been found that, for the bulk $RbCr_3As_3$, the magnetic field suppresses the superconductivity a bit faster for $H\perp c$ than for $H//c$ near T_c. But the case is reversed at the lower temperature, giving a strange crossing behavior of $H_{c2}(T)$ [17,24]. We found that similar crossing behaviors are also observed in the two nanowires as shown in Figure 3(e). It should be pointed out that, the estimated $H_{c2}(0)$ are derived from the onset T_cs for the unavailable well-defined zero resistance transitions. However, the $H_{c2}(0)$ is still far larger than the Pauli paramagnetic limit.

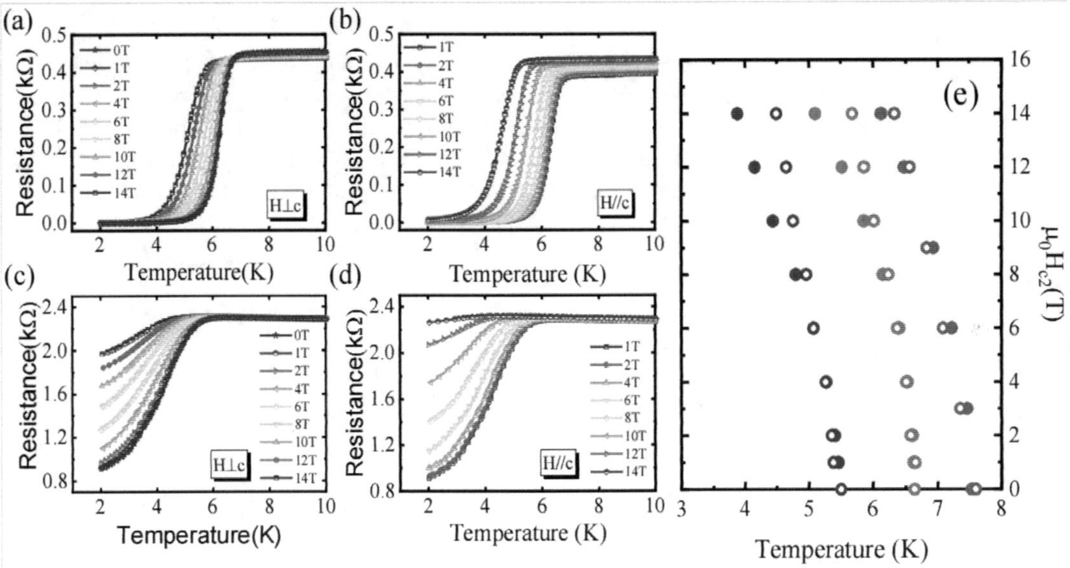

Figure 3. (a) The Field evolved superconducting transition for S01, the field is applied perpendicular to the c axis. (b) Parallel Field evolved superconducting transition for S01. (c) and (d) Field evolved superconducting transition for S02 with the vertical and parallel fields, respectively. (e) Derived temperature dependent critical fields with $H\perp c$ (hollow circle) and $H//c$ (filled circle) for the bulk (blue), S01 (red) and S01 nanowires (black).

It is unconventional that superconductivity survives under a magnetic field above the Pauli limit, which occurs in two general cases. The first one is that the Cooper pairs are formed by electrons with finite non-zero momentum, which is proposed in the so-called Fulde–Ferrell–Larkin–Ovchinnikov (FFLO) state, and boosts the upper critical field beyond the Pauli limit a little [25-27]. It is not the case in $RbCr_3As_3$, since the $H_{c2}(0)$ is much higher than the Pauli limit in both bulk and nanowire samples. The other case is the non-trivial pairing in presence of strong spin-orbit coupling (SOC), for instance, the Ising pairing or non-centrosymmetrical superconductors [12,28-30]. The strong SOC can boost the critical field well beyond the Pauli limit. Though $RbCr_3As_3$ possesses a centrosymmetrical crystal structure, different from the non-centrosymmetrical $Rb_2Cr_3As_3$, all of the results obtained above indicate that the superconducting properties are very similar in the two Cr-based compounds. As the results of the complex magnetic orders or spin fluctuations, various pairing symmetries have been proposed in the two systems, including the singlet s± wave pairing and triplet f wave or p_z wave pairing [31]. Therefore, it is widely believed unconventional spin-triplet pairing symmetry should be taken into account for $A_2Cr_3As_3$ superconductors [12]. In the $K_2Cr_3As_3$ system, the superconductivity is found to be suppressed dramatically by the high pressure. The upper critical field is also limited to below the Pauli paramagnetic limit μ_0H_p at ~2GPa [32]. The modifications of $[Cr_3As_3]_\infty$ chains and the interchain coupling may progressively change the electronic correlation and magnetic orders under the hydrostatic pressure, exhibiting a possible crossover of exotic spin triplet pairing to the spin singlet pairing [32]. By comparison, the exfoliation of $RbCr_3As_3$ should not significantly modify the intrachain couplings [33]. Though the superconductivity is also weakened dramatically, the large upper critical field related to the superconducting singularities is preserved. These results strongly refer to the magnetic orders in the $[Cr_3As_3]_\infty$ chains as the key factors for these novel Q1D superconductors, which deserve to be further studied. Thus, the air-stable ACr_3As_3 compound may provide an ideal platform to understand the superconducting mechanism in these Q1D Cr-based superconductors, and help to reveal the overall unconventional superconductivity.

4. Conclusions

In summary, the air-stable $RbCr_3As_3$ is exfoliated to nanowires with different thicknesses, whose resistive transport properties were measured. It is revealed that the superconductivity is suppressed by the reduced thicknesses, evidenced by the decreased transition temperatures and upper critical fields. However, the upper critical field is still far larger than the Pauli limit, indicating the unconventionality of the superconductivity in this system. These results help to understand the superconducting mechanism in the quasi-one-dimensional Cr-based superconductors.

Acknowledgments

We are grateful to Prof. Zhian Ren for the help on the samples and we thank Prof. Qingge Mu for fruitful discussions. This work was financially supported by the National Natural Science Foundation of China (Grant No. 11804379), the Major Basic Program of Natural Science Foundation of Shandong Province (Grant No. ZR2021ZD01).

References

[1] Lee P A, Nagaosa N, Wen X G. Doping a Mott insulator: Physics of high-temperature superconductivity[J]. Reviews of modern physics, 2006, 78(1): 17.
[2] Pickett W E. Electronic structure of the high-temperature oxide superconductors[J]. Reviews of Modern Physics, 1989, 61(2): 433.
[3] Chen X, Dai P, Feng D, et al. Iron-based high transition temperature superconductors[J]. National Science Review, 2014, 1(3): 371-395.
[4] Kamihara Y, Watanabe T, Hirano M, et al. Iron-based layered superconductor La $[O_{1-x} F_x]$ FeAs (x= 0.05− 0.12) with Tc= 26 K[J]. Journal of the American Chemical Society, 2008, 130(11): 3296-3297.
[5] Ren Z A, Yang J, Lu W, et al. Superconductivity in the iron-based F-doped layered quaternary compound Nd $[O_{1-x} F_x]$ FeAs[J]. EPL (Europhysics Letters), 2008, 82(5): 57002.
[6] Armitage N P, Fournier P, Greene R L. Progress and perspectives on electron-doped cuprates[J]. Reviews of Modern Physics, 2010, 82(3): 2421.
[7] Barišić N, Chan M K, Li Y, et al. Universal sheet resistance and revised phase diagram of the cuprate high-temperature superconductors[J]. Proceedings of the National Academy of Sciences, 2013, 110(30): 12235-12240.
[8] Chubukov A. Pairing mechanism in Fe-based superconductors[J]. Annu. Rev. Condens. Matter Phys., 2012, 3(1): 57-92.

[9] Luetkens H, Klauss H H, Kraken M, et al. The electronic phase diagram of the $LaO_{1-x}F_xFeAs$ superconductor[J]. Nature materials, 2009, 8(4): 305-309.
[10] Wu G, Liu R H, Chen H, et al. Transport properties and superconductivity in $Ba_{1-x}M_xFe_2As_2$ (M= La and K) with double FeAs layers[J]. EPL (Europhysics Letters), 2008, 84(2): 27010.
[11] Cava R J, Batlogg B, Van Dover R B, et al. Superconductivity at 60 K in $La_{2-x}Sr_xCaCu_2O_6$: the simplest double-layer cuprate[J]. Nature, 1990, 345(6276): 602-604.
[12] Bao J K, Liu J Y, Ma C W, et al. Superconductivity in quasi-one-dimensional $K_2Cr_3As_3$ with significant electron correlations[J]. Physical Review X, 2015, 5(1): 011013.
[13] Tang Z T, Bao J K, Wang Z, et al. Superconductivity in quasi-one-dimensional $Cs_2Cr_3As_3$ with large interchain distance[J]. Science China Materials, 2015, 58(1): 16-20.
[14] Watson M D, Feng Y, Nicholson C W, et al. Multiband one-dimensional electronic structure and spectroscopic signature of tomonaga-luttinger liquid behavior in $K_2Cr_3As_3$[J]. Physical review letters, 2017, 118(9): 097002.
[15] Zhang L D, Wu X, Fan H, et al. Revisitation of superconductivity in $K_2Cr_3As_3$ based on the six-band model[J]. EPL (Europhysics Letters), 2016, 113(3): 37003.
[16] Liu T, Mu Q G, Pan B J, et al. Superconductivity at 7.3 K in the 133-type Cr-based $RbCr_3As_3$ single crystals[J]. EPL (Europhysics Letters), 2018, 120(2): 27006.
[17] Liang Q, Liu T, Xi C, et al. Upper critical field and its anisotropy in $RbCr_3As_3$[J]. Physical Review B, 2019, 100(21): 214512.
[18] Mu Q G, Ruan B B, Pan B J, et al. Ion-exchange synthesis and superconductivity at 8.6 K of $Na_2Cr_3As_3$ with quasi-one-dimensional crystal structure[J]. Physical Review Materials, 2018, 2(3): 034803.
[19] Kong T, Bud'ko S L, Canfield P C. Anisotropic H_{c2}, thermodynamic and transport measurements, and pressure dependence of T_c in $K_2Cr_3As_3$ single crystals[J]. Physical Review B, 2015, 91(2): 020507.
[20] Cuono G, Forte F, Romano A, et al. Intrachain collinear magnetism and interchain magnetic phases in Cr_3As_3− K-based materials[J]. Physical Review B, 2021, 103(21): 214406.
[21] Xi X, Wang Z, Zhao W, et al. Ising pairing in superconducting $NbSe_2$ atomic layers[J]. Nature Physics, 2016, 12(2): 139-143.
[22] Baidya P, Sahani D, Kundu H K, et al. Transition from three-to two-dimensional Ising superconductivity in few-layer $NbSe_2$ by proximity effect from van der Waals heterostacking[J]. Physical Review B, 2021, 104(17): 174510.
[23] Bao J K, Li L, Tang Z T, et al. Cluster spin-glass ground state in quasi-one-dimensional KCr_3As_3[J]. Physical Review B, 2015, 91(18): 180404.
[24] Balakirev F F, Kong T, Jaime M, et al. Anisotropy reversal of the upper critical field at low temperatures and spin-locked superconductivity in $K_2Cr_3As_3$ [J]. Physical Review B, 2015, 91(22): 220505.
[25] Fulde P, Ferrell R A. Superconductivity in a strong spin-exchange field[J]. Physical Review, 1964, 135(3A): A550.
[26] Larkin A I, Ovchinnikov Y N. Zh. é ksp. teor. fiz. 47, 1136 1964 sov. phys[J]. JETP, 1965, 20: 762.
[27] Gruenberg L W, Gunther L. Fulde-Ferrell effect in type-II superconductors[J]. Physical Review Letters, 1966, 16(22): 996.
[28] Saito Y, Nakamura Y, Bahramy M S, et al. Superconductivity protected by spin–valley locking in ion-gated MoS2[J]. Nature Physics, 2016, 12(2): 144-149.
[29] Bauer E, Hilscher G, Michor H, et al. Heavy fermion superconductivity and magnetic order in noncentrosymmetric $CePt_3Si$[J]. Physical review letters, 2004, 92(2): 027003.
[30] Karki A B, Xiong Y M, Vekhter I, et al. Structure and physical properties of the noncentrosymmetric superconductor Mo_3Al_2C[J]. Physical Review B, 2010, 82(6): 064512.
[31] Zhang L D, Zhang X, Hao J J, et al. Singlet $s^{\pm}$-wave pairing in quasi-one-dimensional ACr_3As_3 (A= K, Rb, Cs) superconductors[J]. Physical Review B, 2019, 99(9): 094511.

[32] Sun J P, Jiao Y Y, Yang C L, et al. Effect of hydrostatic pressure on the superconducting properties of quasi-1D superconductor $K_2Cr_3As_3$[J]. Journal of Physics: Condensed Matter, 2017, 29(45): 455603.
[33] Cuono G, Forte F, Romano A, et al. Intrachain collinear magnetism and interchain magnetic phases in Cr_3As_3 − K-based materials[J]. Physical Review B, 2021, 103(21): 214406.

Exact calculation of the elastic antiferromagnetic Ising model under static external force on Husimi lattice

Xinyi Ning[1], Lisheng Tang[2], Xiaoyan He[2], Ran Huang[1, 2, 4] and Xin Ma[1, 3]

[1]Academy for Engineering and Applied Technology, Fudan University, Shanghai 200433, China
[2]Department of Materials Technology and Engineering, Research Institute of Zhejiang University-Taizhou, Taizhou, Zhejiang 318000, China
[3]Huashan Hospital, Fudan University, Shanghai 200433, China

[4]Corresponding author's e-mail: huangran@fudan.edu.cn

Abstract. We investigate an elastic Ising model with a modified Lennard-Jones potential on the Husimi recursive lattice, where the model can be solved exactly. An external force field to either stretch and press the lattice is applied to approach the elasticity in Ising model in a static way. As expected, both stretching and compressing induce less stability with lower T_c, nevertheless due to the asymmetric feature of the LJ well, trivial but realistic destabilization is observed in the stretch case implying the drawing crystallization, and interesting phenomenon of supercooled state and *Kauzmann Paradox* addressed by zero entropy at finite temperature is observed in the compression case on the left side of the LJ potential well.

1. Introduction

As one of the most classical and successful models to explore the fundamental thermodynamics in general particle systems, the study of Ising model with elasticity may date back half a century ago [1]. Not only the elastomer itself is of research interest, the elastic feature of interaction between spin pairs also expectedly acts as an effective tool in constructing more complicated models, such as compressible lattice [2]. In recent years, the studies have been extended into 3D and various lattices, e.g. triangular net, to observe multiple types of critical behavior(s) [3,4]. However, the stochastic feature of elasticity makes non-trivial difficulty in theoretical calculation, and simulation methods such as Monte Carlo are still the mainstream tools [5], nevertheless where the thermals are not directly observed, the nature of elastic Ising model remains an interesting subject to unfold [6,7].

In this work we applied the recursive lattice as an exact calculation technique to solve the 2D antiferromagnetic Ising model [8,9], with a modified Lennard-Jones potential embedded in the Hamiltonian. To avoid dealing with the stochasticity, we designed a static model with an external force field to approach the effects of elasticity. The model and method are described in Sec. II. Results of interesting observations are shown and discussed in Sec. III. And we conclude remarks in Sec. IV.

2. Model and methods

The original Husimi lattice is a fractal structure in which one square unit linking other four uniform squares on its four sites, it is a good approximation to a regular square lattice due the coordination numbers are as the same as four [8,9]. Now let us imagine a square unit is stretched along one edge to be a rectangular, and if there exist elastic interactions on the edge, a potential energy resisting the stretch shall raise to push the unit back to square shape. And if we define an external force field making the stretch all over the lattice, an elastic Ising model on Husimi lattice is established as shown in Figure 1. Note that Figure 1 only illustrates the external force to stretch, reversely the principle of compressing force field is the same.

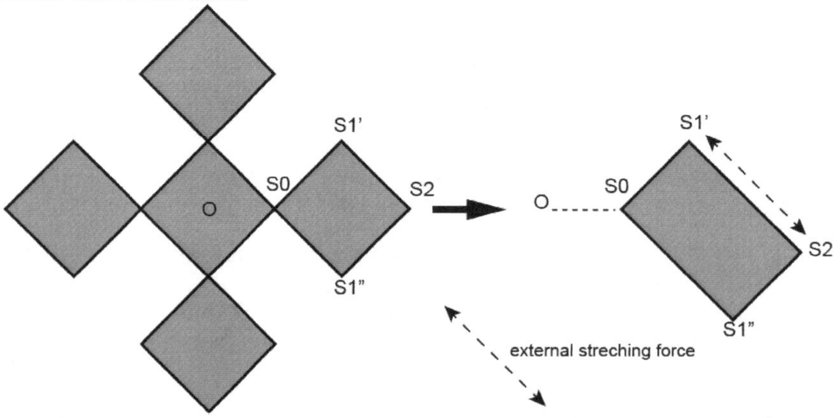

Figure 1. The local area of classical Husimi lattice around origin. The right part demonstrates the stretch between the parallel pairs of spins S1'-S2 and S0-S1" under a uniform external force.

There are two ways to add the elastic potential into the Hamiltonian of Ising model: define cohesive and magnetic interactions respectively between the neighboring spins [3], or make the basic spin-spin exchange J to be elastic, i.e. variable with distance [10]. We take the former method because the construction is more complicated and versatile to explore more situations, and it is always easy to be simplified to the latter case (e.g. the eqn. 3 in Ref. [10]). The Hamiltonian in this work is defined as:

$$\mathcal{H} = \sum_\alpha e_\alpha = -J\sum_{ij} S_i S_j - (S_i S_{i+1} + S_j S_{j+1})\left[\left(\frac{1}{d}\right)^{12} - 2\left(\frac{1}{d}\right)^{6} + 1\right], \quad (1)$$

where α denotes the index of recursive units and e_α is the local energy of the unit. The combination of spin pairs $(S_i S_{i+1} + S_j S_{j+1})$ in the second term is the stretched/compressed pairs as shown in Figure 1. Here we setup the LJ energy to be spin states dependent, that is, the potentials may act as attractive or repulsive in either way, otherwise the elasticity will become a trivial inert energy term of less thermodynamic interests. The correction +1 in LJ expression is to offset the lowest energy, i.e. the basin of the elastic potential to be 0, so that at equilibrium $d = 1$ we will simply have the regular Ising model. Note the distance d is defined with the unit length of a square edge. Overall the elasticity is a modified LJ-like potential, it can be understood as a special outside field similar to the magnetic field H, but only that it acts on pairs of spins along certain planary direction. For convenience, the magnetic field acting on single spin is ignored in this work.

The spin S is assigned the value of +1/−1 for up and down in the calculation. We select antiferromagnetic Ising model to investigate in this work ($J = -1$), where different neighboring spins states are preferred with lower energy (−1)·(+1)·(−1). The antiferro- setup makes the magnetic exchange J and elastic potential competing to each other, that is, the different neighboring spin states are preferred by anti-magnetization but with the highest potential (0), and the identical spins will be preferred in the LJ well while dislocation around the basin raises higher energy, which is more intuitive with our common sense. Please note this setup is optional, the model is in either way reasonable to describe various real systems, e.g. the ferromagnetic metallic glass [11] or antiferro-alloy [12], the methodology discussed in this work can be easily extended to other physical situations and more complicated case.

The calculation mainly follows the established technique we previously reported [13,14]. The exact calculation process will provide a set of solutions x to denote the probability of a site to be occupied by a +1 spin, based on which the partition function Z, free energy F and entropy S can be obtained, while in this work the temperature T is normalized with the Boltzmann constant $k = 1$ and has the unit of energy. Here we briefly review the key calculation process:

Consider that we are looking at the thermodynamic contributions on a random square unit at some level n, it has a "base site" S_{n0} close to the (imaginary) original site S_0 of the entire lattice on the zeroth level (Figure 1), then the sites nearest to S_{n0} are marked as S_{n+1} or S'_{n+1}, the site diagonal to S_{n0} is marked as S_{n+2}. In this way, on this unit there are three sub-trees $\mathcal{T}_{n+1}$, $\mathcal{T}'_{n+1}$ and $\mathcal{T}_{n+2}$ contributing onto the sites S_{n+1}, S'_{n+1} and S_{n+2}, and a larger sub-tree $\mathcal{T}_n$ is synthesized by adding the weights of upper sub-trees and the local unit. With the particular spin state of $S_{n0} = \pm 1$, we may define the partial partition function (PPF) $Z_n(S_{n0})$ on level n to count the contribution of sub-tree onto the site S_{n0} (not included), then the nth level PPF is the function of PPFs on the previous level n+1, n+2 and the weights of local unit $w(\gamma)$:

$$Z_n(+) = \sum_{\gamma=1 \text{ to } 8}^{8} Z_{n+1}(S_{n+1})Z_{n+1}(S'_{n+1})Z_{n+2}(S_{n+2})w(\gamma) \quad (2)$$

and

$$Z_n(-) = \sum_{\gamma=9 \text{ to } 16}^{8} Z_{n+1}(S_{n+1})Z_{n+1}(S'_{n+1})Z_{n+2}(S_{n+2})w(\gamma). \quad (3)$$

Remember the local unit has $\gamma = 16$ configurations of spins state, and the PPFs above are the summation over eight of them respecting to the +1/−1 state of base spin S_{n0}. Then the partition function (PF) of the whole sub-tree on nth level is given by

$$Z_n = Z_n(+)e^{\beta HS} + Z_n(-)e^{-\beta HS} \quad (4)$$

which sums all the weights at level n including the spin S_{n0}. Although in this work the we have ignored magnetic field $H = 0$ for convenience, the term $e^{\pm\beta HS}$ is still presented here to remind not missing the contribution of the base site S_{n0}.

With the PPFs we may introduce the ratios:

$$x_n = \frac{Z_n(+)}{Z_n(+)+Z_n(-)} \text{ and } y_n = \frac{Z_n(-)}{Z_n(+)+Z_n(-)}$$

to present the probability of one site to be occupied by a up or down spin. Since $Z_n(+/-)$ is the function of PPFs on higher levels n+1 and n+2 as shown in eqn. (2) and (3), for ratio x_n or y_n the recursion between successive levels must also hold, that is we can easily derive equation in such form:

$$x_n = f(x_{n+1}, x_{n+2}),$$

then starting with two initial seeds of xs we can recursively calculate the x on lower level.

In a homogeneous lattice the recursive calculation is done for many iterations until reaching a fixed cycling solution $x_k = x_{k+i}$ (the fixed-point solution). Usually we have 1-cycle (uniform) solution for ferro- case and 2-cycle (x_1 and x_2) for antiferro- case, the former means all the spins have the same probability to be up (or down), the latter indicates the neighboring spins have different probability, which is clearly the antiferro- case of spins' up-down alternation at $T = 0$.

3. Results and discussion

Remember in the LJ potential the dislocation in either way raises higher energy (the elasticity would either push or pull back the distance to be equilibrium), thus it is not surprising that the system under the external force to be either stretched or compressed has lower critical T_c corresponding to a less stable state. But the asymmetric feature of LJ potential well provides very distinguishable behaviors in different situations. In this section we will observe the stretch and compression along with the variation of d > 1 and < 1 and analyse the thermodynamics.

3.1. The Stretching Case

Figure 2 presents antiferromagnetic solution pairs under stretch with $d = 1.0$ to 2.8. Basically, the solutions are consistent to our expectation that the reference $d = 1.0$ gives highest T_c, and then with larger dislocation the system becomes less stable. The transition temperature tremendously decreases until $d = 1.8$, the system behaves amorphous at very low temperature, and further change of d affects so less that is hard to be distinguished at the resolution of $\Delta T = 0.1$.

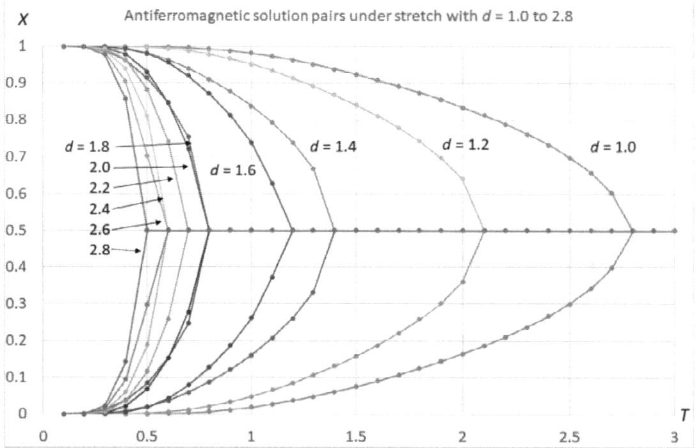

Figure 2. The antiferromagnetic solution pairs of x_1 and x_2 under stretch with d = 1.0 to 2.8: at high temperature the 0.5 solutions mean equal probability of up/down spins everywhere as amorphous state; with cooling, the solutions branch to 0 and 1, that is neighboring sites having opposite preference on spin state as antiferro-ordered crystal. The variation of d dramatically changes the position of Curie point, and all systems have perfect 1/0 ordering at near-zero T.

It can be seen that above d = 1.8 the trend is slow with less interest, which agrees with the long flat trail of the right side of LJ potential. Therefore, in Figure 3 we only present the free energy and entropy ranges from d = 1.0 to 1.8. Regardless of the dislocation, the critical behavior of Curie can be observed on all the curves, with the clear slope change on the entropy.

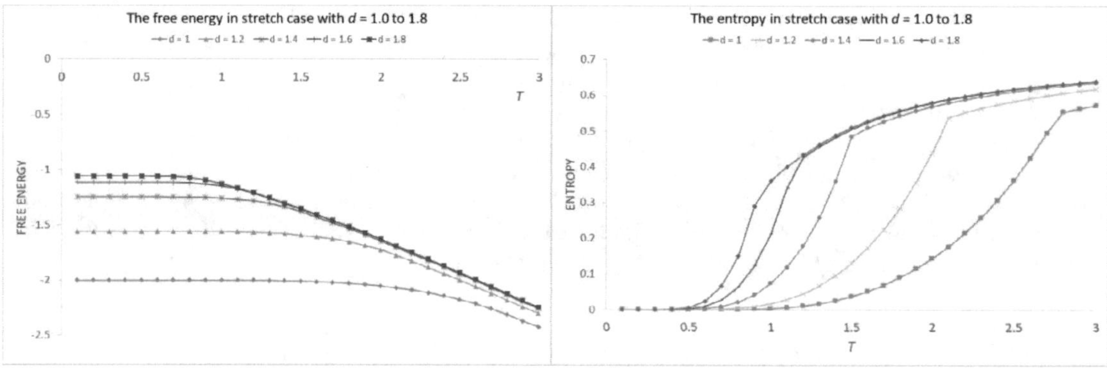

(a) (b)

Figure 3. The free energy (a) and entropy (b) in stretch case with d = 1.0 to 1.8.

3.2. The Compression Case: The Supercooled State with Kauzmann Paradox

Simple setup of $d < 1$ makes the model being able to describe the case of constant compression. Since the behavior in this part is so bizarre, that the data may better illustrate the sharp turning trend of solutions instead of figures, we present Table 1 to show solutions of compressed system with d ranged from 1.0 to 0.86. Remember the emergence of 0.5 solution indicates the Curie transition T_c, the system exhibits regular antiferromagnetic behavior as d decreases until 0.9, where the T_c is shrunk to 1.4~1.5 as expected since the elastic term competes with the neighbor exchange J and decomposes the stability of system. The data between d =1.0 and 0.9 are mediocre to the trend thus not presented in the table.

Table 1. The solutions of compressed system with d ranged from 1.0 to 0.86.

T	d = 1.0 (ref.)		0.9		0.89		0.88		0.87		0.86	
0.1	0	1	0	1	0.782811	0.782811	0.660651	0.869672	0.66	0.87	0.66	0.87
0.2	0	1	0.000005	0.999995	0.770612	0.770612	0.775625	0.787922	0.660245	0.869838	0.659997	0.869998
0.3	0.000002	0.999998	0.000304	0.999696	0.58046	0.58046	0.712778	0.712778	0.678003	0.847378	0.657571	0.868399
0.4	0.000045	0.999955	0.002475	0.997525	0.500348	0.500348	0.521203	0.521203	0.651562	0.676608	0.607346	0.826863
0.5	0.000336	0.999664	0.008964	0.991036	0.5	0.5	0.500017	0.500017	0.515115	0.515137	0.503105	0.658293
0.6	0.001274	0.998726	0.021543	0.978457	0.5	0.5	0.5	0.5	0.500026	0.500026	0.490912	0.517191
0.7	0.003309	0.996691	0.040824	0.959176	0.5	0.5	0.5	0.5	0.5	0.5	0.499831	0.500179
0.8	0.006784	0.993216	0.066615	0.933385	0.5	0.5	0.5	0.5	0.5	0.5	0.5	0.5
0.9	0.011887	0.988113	0.09847	0.90153	0.5	0.5	0.5	0.5	0.5	0.5	0.5	0.5
1	0.01867	0.98133	0.136153	0.863847	0.5	0.5	0.5	0.5	0.5	0.5	0.5	0.5
1.1	0.027101	0.972899	0.180052	0.819948	0.5	0.5	0.5	0.5	0.5	0.5	0.5	0.5
1.2	0.037096	0.962904	0.231784	0.768216	0.5	0.5	0.5	0.5	0.5	0.5	0.5	0.5
1.3	0.048557	0.951443	0.296031	0.703969	0.5	0.5	0.5	0.5	0.5	0.5	0.5	0.5
1.4	0.061392	0.938608	0.392029	0.607971	0.5	0.5	0.5	0.5	0.5	0.5	0.5	0.5
1.5	0.075524	0.924476	0.5	0.5	0.5	0.5	0.5	0.5	0.5	0.5	0.5	0.5

Immediately at $d = 0.89$, the 2-cycle solution of $x_1 \rightarrow 1$ and $x_2 \rightarrow 0$ is destroyed, the system becomes amorphous at very low temperature. In the near zero range, the spins exhibit uniformly with 1-cycle solution bias to fully occupying of +1 state in the system. (That is to say the numerical results shown here, for $H = 0$ the bias to either $x = 1$ and 0 makes no difference.) More interestingly, after this remarkable transition happens at $d = 0.89$, the emergence of 0.5 solution is pushed back to higher values with decreasing d, nevertheless, the 0.5 solution here is no longer the sign of Curie transition, since the systems exhibit as supercooled state, which will be detailed later. Descriptionally, the elasticity actually prefers a more stable state beyond its overwhelming the primary ordered state of antiferromagnetization.

The mysterious behaviors of solution is of course companied with the thermal functions. Figure 4 shows the free energy (a) and entropy (b) of the pressed system. The regular Curie transition of $d = 1.0$ and 0.9 are remarked by the red arrow in the Figure 4(b), after that, the transition disappears and the entropy of $d = 0.89$ gradually goes down to 0 at a non-zero T, which reminds us obviously the supercool state and *Kauzmann paradox*. Again, as the d decreases, the systems become more stable as the entropy curves go lower, and the zero-entropy temperature (boldly to call the ideal Glass Transition Temperature) increases. (The original idea of *Kauzmann paradox* considers the negative excess entropy, i.e. the difference between the supercooled liquid and crystal state, to determine the crisis. Since the extrapolation of the entropy of supercooled liquid goes down to zero immediate below the negative excess entropy, we simply take the zero entropy to distinguish the paradox.)

This observation is of quite interest but not strange to us, when we explored the exact calculation in the antiferromagnetic Ising system on recursive lattice [11,12], the destroy of 2-cycle ordered state and mandatory induction of 1-cycle solution can lead to the supercooled state without crystallization, and subsequent the *Kauzmann paradox* of zero entropy [15]. The difference is that in the previous reports these unphysical negatives are from calculation technique, and are controversial to be numerical but not really *Kauzmann*. Here in this work it is induced by an external force field and looks more realistic and convincing. We leave these complicated implications for future investigations.

Overall, Figure 5 presents the Curie temperatures varied with dislocation from $d = 1.8$ to 0.9, and the LJ potential is shown for reference. Clearly the effect of elastic potential drags down the stability with lower T_c in either direction. And on the left side of the potential well below $d = 0.9$, the potential increases so sharp that it destroys the cycling solution and Curie transitions.

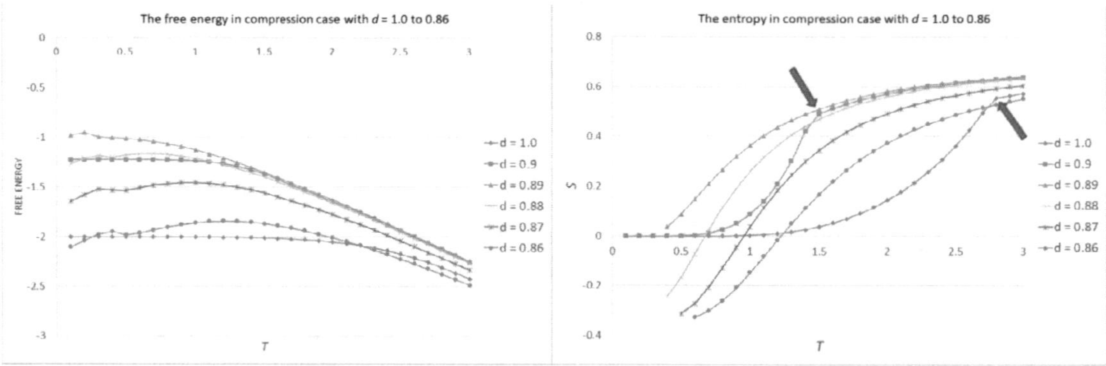

(a) (b)

Figure 4. The thermodynamic behaviors of system under compression with d = 1.0 to 0.86: (a) the free energy; (b) the entropy.

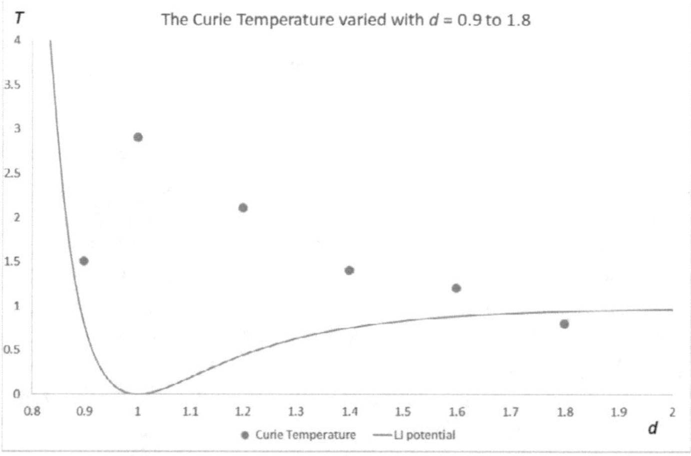

Figure 5. The Curie temperatures varied with dislocation from d = 1.8 to 0.9, and the LJ potential is presented for reference.

4. Conclusion

A modified LJ potential is added into antiferromagnetic Ising model on Husimi lattice to study the elasticity, the system is set to be under constant external force field to approach the static study, solutions of the spin density and thermodynamics are calculated by exact technique. We have observed that: 1) the system under the external force of either stretching or compressing has lower critical T_c corresponding to a less stable state; 2) the stretch cases show common behaviors agreeing to our empirical knowledge of drawing crystallization in materials science; 3) In the compression cases, an immediate change at d = 0.89 is observed with the disappearance of Curie critical behavior, and the results present the supercooled state followed by negative entropy which is believed to be the *Kauzmann Paradox* and the ideal glass transition. Further explorations are expected to uncover the underlying mechanism of a bunch of implications revealed in this primary research.

Acknowledgement

This work is financially supported by the National Key Research and Development Program China (2022YFC2009500), the Science and Technology Development Funds Shanghai (No. 21511102200), the Medical Engineering Fund of Fudan University (yg2021-005), and the Taizhou Municipal Science and Technology Program (1801gy16 and 21gya24).

References

[1] Renard, R., & Garland, C. W. (1966). Order-disorder phenomena. II. Elastic constants of a two-dimensional ising model. *The Journal of Chemical Physics*, 44(3), 1125–1129.

[2] Bergman, D. J., & Halperin, B. I. (1976). Critical behavior of an Ising model on a cubic compressible lattice. *Physical Review B*, 13(5), 2145–2175.

[3] Boubcheur, E. H., & Diep, H. T. (1999). Effect of elastic interaction on critical behavior of three-dimensional Ising model. *Journal of Applied Physics*, 85(8 II B), 6085–6087.

[4] Zhu, X., Landau, D. P., & Branco, N. S. (2006). Critical behavior of an elastic Ising model on a stacked triangular net at constant volume. *Physical Review B*, 73(6), 1–6.

[5] Yao, X., Lo, V. C., & Liu, J. M. (2009). Monte Carlo simulation on ferroelectric response to magnetic field in an elastic Ising spin chain. *Journal of Applied Physics*, 106(1), 013903.

[6] Najafi, M. N., Cheraghalizadeh, J., & Herrmann, H. J. (2019). Elastic backbone phase transition in the Ising model. *Physical Review E*, 100(4), 42132.

[7] Frechette, L. B., Dellago, C., & Geissler, P. L. (2020). Origin of mean-field behavior in an elastic Ising model. *Physical Review B*, 102(2), 1–9.

[8] Gujrati, P. D. (1995). Bethe or bethe-like lattice calculations are more reliable than conventional mean-field calculations. *Physical Review Letters*, 74(5), 809–812.

[9] Mézard, M., & Parisi, G. (2001). The Bethe lattice spin glass revisited. *European Physical Journal B*, 20(2), 217–233.

[10] Balcerzak, T., Szałowski, K., & Jaščur, M. (2020). Thermodynamic properties of the one-dimensional Ising model with magnetoelastic interac*tion. Journal of Magnetism and Magnetic Materials*, 507(March), 166825.

[11] Huang, R., Zhang, L., Chen, C., Wu, C., & Yan, L. (2015). Theoretical exploration on the magnetic properties of ferromagnetic metallic glass: An Ising model on random recursive lattice. *European Physical Journal Plus*, 130(7), 127.

[12] Nan, T., Quintela, C.X., Irwin, J. et al. (2020). Controlling spin current polarization through non-collinear antiferromagnetism. *Nature Communication* 11, 4671.

[13] Huang, R., & Chen, C. (2014). Thermodynamic transitions of antiferromagnetic ising model on the fractional multi-branched Husimi recursive lattice. *Communications in Theoretical Physics*, 62(5), 749-754.

[14] Huang, R. (2019). A Husimi rhombus lattice with random angles to present the configurational stochasticity in exact thermodynamic calculation. *Results in Physics*, 13, 102135.

[15] Huang, R., & Gujrati, P. D. (2019). Phase transitions of antiferromagnetic Ising spins on the zigzag surface of an asymmetrical Husimi lattice. *Royal Society Open Science*, 6(3), 181500.

Journal of Physics: Conference Series **2313** (2022) 012029 doi:10.1088/1742-6596/2313/1/012029

A novel dynamic mechanical analysis device to measure the in-vivo material properties of plantar soft tissue and primary finite elementary analysis results

Longyan Wu[1 #], Jun Zhu[1 #], Jun Zheng[2], Xiang Geng[3], Xiaoyan He[2], Lisheng Tang[2], Ran Huang[1, 2, 4] and Xin Ma[1, 3]

[1]Academy for Engineering and Applied Technology, Fudan University, Shanghai 200433, China
[2]Department of Materials Technology and Engineering; Department of Electrical and Control Engineering, Research Institute of Zhejiang University-Taizhou, Taizhou, Zhejiang 318000, China
[3]Huashan Hospital, Fudan University, Shanghai 200433, China

#equal contribution
[4]Corresponding author's e-mail: huangran@fudan.edu.cn

Abstract. We have designed a series of dynamic mechanical analysis (DMA)-like device to directly measure the material properties of living human plantar soft tissue. Various mechanical tests of plantar soft tissue such as vertical, horizontal shear and torsion can be carried out on the newly invented instruments, and periodic strain-stress outputs are obtained to analyse the viscoelasticity of the tissue. Pioneering finite element analysis has been done by coupling the machine and human foot FE model from different simulation environments, and the simulation tests show good engineering verification of the device design, and consistent theoretical results of the material properties of plantar soft tissue as expected.

1. Introduction

As a heavily-loaded organ under daily working conditions, plantar soft tissue has evolved to be with excellent characteristics of wear-resistant, pressure-resistant, interlayer displacement limitation in anatomical structure in order to bear the walking loading [1]. The characterizations and bionic research of plantar soft tissue have gained wide research interests in biomechanics, foot and ankle clinical medicine and biomedical engineering [2]. A number of clinical researches show that plantar soft tissue degeneration closely relates to some common foot and ankle diseases, such as plantar and heel pain (incidence rate 24.5%) [3], hallux valgus (incidence rate 23%) [4], and diabetes foot (incidence rate 5.4%) [5]. Especially, in the treatment of diabetes foot ulcer, based on the detection of the biomechanical and material properties of the diseased soft tissue, the early intervention of patients through customized foot pads, shoes and other professional aids is of important means [6].

However, the test of material properties of living soft tissue remains under-developed. Since living tissue is almost impossible to carry out the material tests based on standardized samples, nowadays the mechanical tests of soft tissue are still mainly carried out with cadaveric specimens [7], which is always controversial. In addition, the mechanical property of soft tissue is a large data set closely related to

Journal of Physics: Conference Series **2313** (2022) 012029 doi:10.1088/1742-6596/2313/1/012029

gender, age, occupation, health status, etc. [8]. Obviously, the diagnosis related to above variables can only be possible by obtaining the real-time data of living body.

Researches show that the plantar soft tissue has both viscosity and elasticity for energy absorption, slow rebound and creep characteristics [9]. Some studies show that the viscosity seems to be more relevant to diseases such as diabetes foot than hardness and elasticity, which are easier to observe [10]. In classical material mechanics, viscoelasticity is usually characterized by dynamic mechanical analysis (DMA) to obtain periodic "stress-strain" data, establish the constitutive equation of complex function, and calculate the elastic and viscous data. Therefore, long-term and periodic direct mechanical measurement should be inevitable in the study of soft tissue viscosity. In recent years, some reports have adopted in-vivo indentation tests of a single pulse to obtain force-displacement and stress-relaxation curves [11], which however, is still far from dynamic periodic stress-strain correspondence and viscoelastic analysis in DMA method.

On the other hand, finite element analysis (FEA) has also been widely used in various biomechanical problems in recent years, especially in the study of soft tissue materials. Based on the parameter data mainly obtained from in-vitro specimen, scholars have constructed a large number of foot FE models to analyze the mechanical behavior of plantar soft tissue [12]. However, the FEA method depends on the initial parameter input of the material, and how to accurately obtain these parameters is a fundamental question.

According to the above concerns, in this study we have explored to design a set of instruments and techniques for in-vivo detection inspired by DMA method from the perspective and principles of classical materials mechanics. Various mechanical tests of plantar soft tissue such as vertical, horizontal shear and torsion strain-stress can be carried out on the newly invented instruments, and periodic strain-stress output can be obtained to analyze in a DMA way. Under simulated working conditions, the dynamic simulation analysis of the device is carried out, combined with the finite element foot model for the engineering verification of device prototype. The FEA shows well consistence between the artificial material setup and the simulated testing results.

2. Device and characterization

We have designed three instruments to measure the periodic strain-stress of vertical motion, horizontal shear, and horizontal torsion respectively. In this section the structure and functional designs are described in details. The support and framework are universal in all the three devices. Therefore, these structure and design are only described in the Vertical Periodic Dynamic Test device in sec. 2.1, and only the key contact and test parts of horizontal shear and torsion periodic dynamic Test device are detailed in the sec. 2.2.

2.1. General design and vertical periodic dynamic test

As shown in Figure 1, the on-site DMA for soft tissues test device includes four main sections: support, test, drive and transmission. The support is to place the living body to be measured, i.e. the limb and foot in this case. The platform has a through-hole 111, where probe 21 can be positioned in and move up and down with the tissue fixed onto.

In the drive section, with controller 32 the motor 31 is to drive the probe 21 to reciprocate in straight at a predetermined frequency within a displacement range, thereby causing simultaneous deformation of the living soft tissues. In this work, the displacement range of the probe is ±20 mm with the periodic frequency <100 Hz.

The transmission section is to connect test and drive sections. It includes a crankshaft 41, a slider 42, a slide rail 43, a connecting rod 44, a drive coupling 45, a straight bearing 46, and a plum coupling 47. One end of the 41 is connected to 31 via 45. The 42 is arranged on the other end of 41 and moves in circle. Meanwhile, the 42 can be movably arranged on 43. The 43 is mounted on 13 by 46. One end of 44 is connected to probe 21 and the other end is connected to slide 43, which moves synchronously under the action of 42.

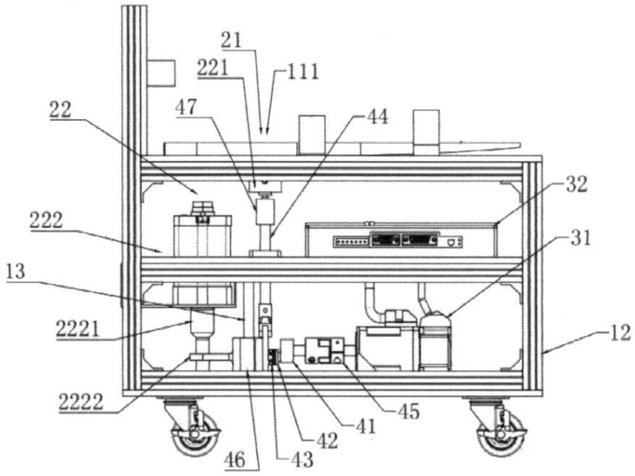

Figure 1. The horizontal view of vertical periodic dynamic test device. The detailed parts remarked and the working mechanism are described in the sec. 2.1.

The sensor assembly 22 includes a tension pressure sensor 221 and a displacement sensor 222. The 221 is mounted on 44 through 47 and connected to 21 to detect tension or pressure on living soft tissues. The 222 is mounted on 12 and connected to 43 to collect displacement data. The 222 employs a magnetic ring type high-precision displacement sensor, including a bar type sensor body 2221 and a magnetic ring 2222, which is fixed on 12 and moves synchronously.

2.2. Horizontal shear and torsion periodic dynamic test

The shear and torsion periodic dynamic test devices employ the similar support and framework designs as described in the sec. 2.1. We only present the key parts to illustrate the test of shear and torsion in Figure 2, it is clear that both devices employ the eccentric rotary to achieve the horizontal and torque motion. The torsion range is ±30° and the shear displacement is ±20 mm, both with periodic frequency up to 100 Hz.

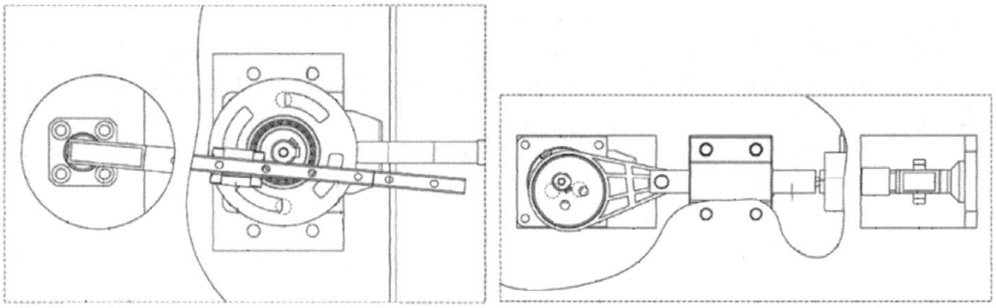

(a) (b)

Figure 2. The scheme of key functional parts in (a) torque and (b) shear test devices (vertical view).

In all the three test instruments, the "detect probe" is a blank contact point and not further detailed in the engineering drawing. Since the probe is required to have a firm joint with the surface of living tissue, i.e. the skin. We adopt a mini vacuum pump with a pipe-type probe to ensure the firm fixing between two sides, in this way the reversal "pull out" motion in vertical and horizontal tests are made possible. The overall design is less relevant to the main function and thus ignored in Figure 1 and 2.

2.3. Simulation test coupled with human foot finite elements model

Along with the manufacture of the devices illustrated above, the simulation test on plantar has been done to complete the engineering verification of the design, and pioneeringly test the plantar soft tissue by the

FEA. Figure 3 shows the coupling of a human foot FE model established in FEA software to the machine model. The testification is done in mechanic engineering software. As a primary exploration, we have adopted previously reported material parameters of bone, soft tissue and skin as shown in Table 1 to exercise the FE simulation [7,8].

Figure 3. A demonstration of the test device coupling with a foot FE model for a simulation test.

Table 1. Parameters in foot FE Model.

	Density (g/cm^3)	Young's modulus (MPa)	Poisson ratio (μ)
Bone	1.92	7300	0.3
Soft tissue	1.00	0.45	0.45
Skin	0.92	1	0.49

3. Results and discussion

Three types of simulation tests have been done to testify the design of the machines, and as primary research on the plantar properties with the novel technique. Figure 4a shows the cloud diagram of stress-strain of reciprocate motion. the plantar skin is pulled by the vacuum suction cup with the detection rod, and the corresponding deformation is in a radial shape, the maximum deformation is observed to be 4.54 mm. Figure 4b presents the simulation curve of periodical radial force and displacement, and the full travel length is 10 mm, which is approximately twice of the deformation observed, and the maximum tensile force in both directions is 30 N.

Figure 5a shows the cloud diagram of shearing stress-strain, and 5b shows the stress-strain behavior. The maximum displacement is 10 mm with the peak force of 15 N. Figure 6a shows the cloud diagram of torsion stress-strain, and 6b shows the curve. The maximum displacement angle is set to be 8°, with peak torque of 45 N·mm.

Basically, the simulation results provide consistent behavior with the initial setup of material properties in FE. The synchronous periodical variation of stress-strain in all three cases implicate the pure elastic property of the materials, which is as expected since the FE model adopts so-called "hyperelastic material" to build the block of soft tissue. Ironically, as introduced in section I, although nowadays viscoelasticity has been acknowledged as the basic feature of plantar soft tissue, the mainstream FE modelling of plantar still takes pure elasticity. The modelling of viscoelastic materials remains a challenge in FE method, one probable reason maybe that real living data are non-accessible, and the subsequent parameter assignment makes less sense in simulation if it cannot describe the objective reality. Again, the results here reveal such controversial situation and the further vision of our work is to approach a better material simulation.

Journal of Physics: Conference Series **2313** (2022) 012029 doi:10.1088/1742-6596/2313/1/012029

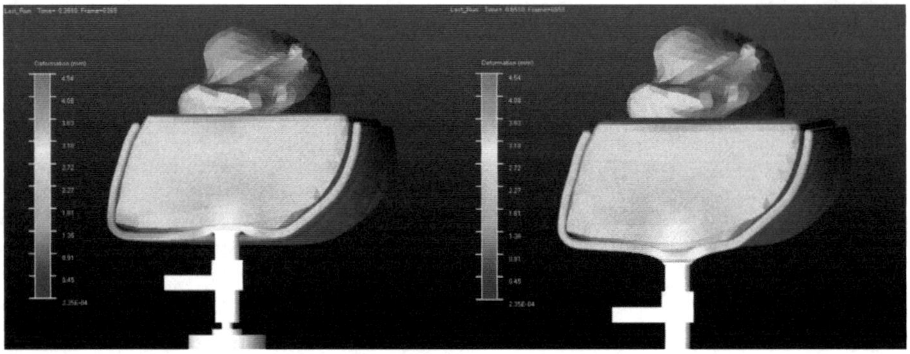

(a)

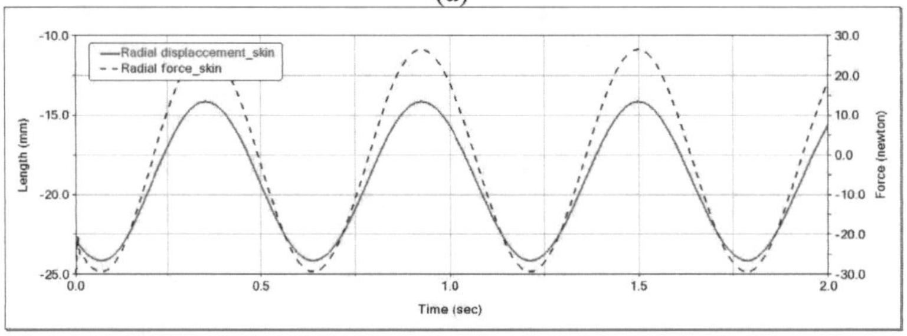

(b)

Figure 4. The simulation test of vertical reciprocate motion: (a) the cloud diagram of stress-strain during the test; (b) The periodical stress and strain curves.

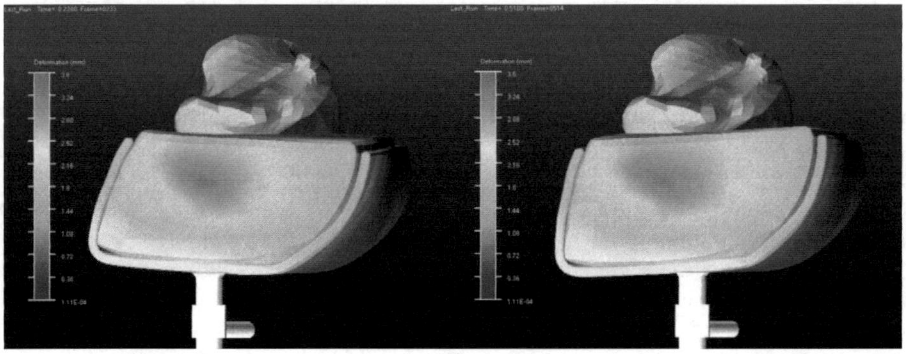

(a)

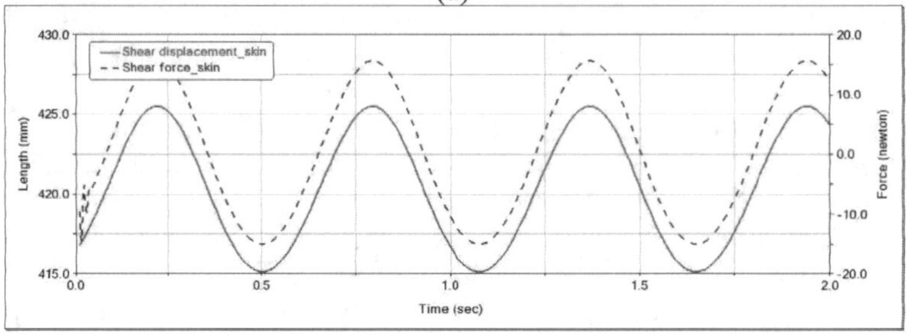

(b)

Figure 5. The simulation test of horizontal shearing motion: (a) the cloud diagram of stress-strain during the test; (b) The periodical stress and strain curves.

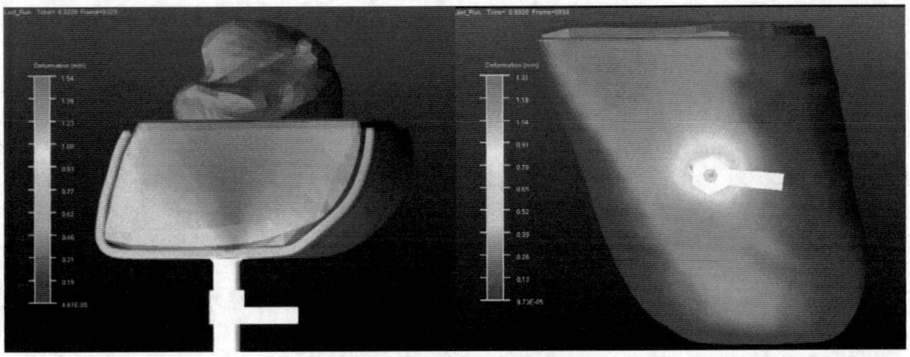

(a)

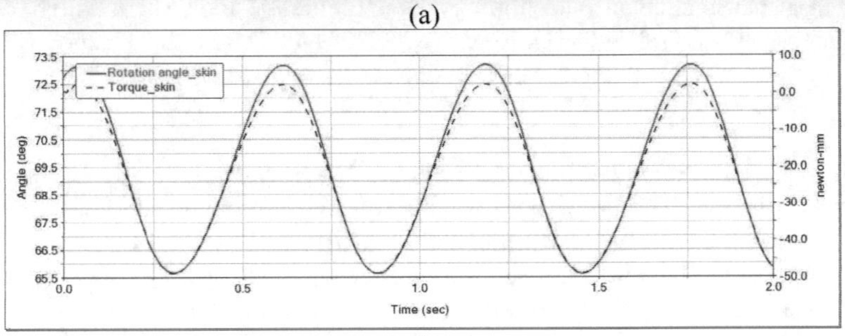

(b)

Figure 6. The simulation test of horizontal torsion: (a) the cloud diagram of stress-strain during the test; (b) The periodical stress and strain curves.

The other upcoming challenge is the estimation of modulus. Although the DMA-like test has been introduced, without standard sample of materials, the non-observable properties are baffled to be conducted: for example, the computation of Young's requires the original length and the change of length to achieve the relative strain (ΔL / L), but if we place an enough large piece of tissue onto the newly invented test machine, although obviously we shall only count in the tissue range involved in the deformation, but it is not a trivial question to determine the effective length of tissue, which is also one of the fundamental problem when testing the human body in-vivo, and a main reason lead us to the work presented in this paper.

Here we present a trial to estimate the modulus in the vertical test for a pioneering research. The contact point (vacuum pump) is fixed with foot tissue with a round area of D = 10 mm, so the effective area for the stress is 78.54 mm^2, then we have the stress 30 N / 78.54 mm^2 = 0.38 MPa, the entire thickness of the modelled plantar is 37.50mm, if it is taken as the initial length, the relative strain is 4.54 mm / 37.50 mm = 0.12, yielding the module to be 0.38 MPa / 0.12 = 3.17 MPa, which is somehow too large to the setup parameter (0.4 MPa for the soft tissue and 1 MPa for the skin), but still comparable. The modulus estimations in shearing and torsion are much more complicated, or even new definitions maybe required. Beside the fact that beyond the invention and manufacture of a new sets of devices, there are numerous works in both hardware and software to exercise to build a complete system to achieve the in-vivo human soft tissue characterizations, we will leave the rest for future investigations.

4. Conclusion

In this work, we pioneeringly apply the DMA-like characterizations onto the living plantar soft tissue, by design and testify a set of novel devices to make the periodic stress-strain tests possible, while meeting the harsh requirements for in-vivo. Three types of test devices for vertical reciprocation, horizontal shearing, and torsion are invented to gain a wide knowledge of the mechanical properties of plantar soft tissue. The machines are coupled with human foot finite elements model to run a series of simulation tests for the engineering verification, and primary explorations on these novel test techniques.

The results show good reliability of the design. The test processes correlate well with the foot FE model and present reasonable stress-strain behaviors, which is consistent to the initial setup parameters, and can serve as powerful tools to further research the viscoelasticity of plantar tissue, nevertheless, a number of serious fundamental questions and improvements are waiting to be investigated in the future researches.

This study is expected to establish a new standard method, which may have important value on biomechanical theory, clinical diagnosis and engineering application. The manufacture of machines is also in process and we expect a series of real tests being reported in the near future.

Acknowledgement

This work is financially supported by the National Key Research and Development Program China (2022YFC2009500), the Science and Technology Development Funds Shanghai (No. 21511102200), the Medical Engineering Fund of Fudan University (YG2021-005), and the Taizhou Municipal Science and Technology Program (1801GY16 and 21GYA24).

References

[1] Weissengruber, G. E., Egger, G. F., Hutchinson, J. R., Groenewald, H. B., Elsässer, L., Famini, D., & Forstenpointner, G. (2006). The structure of the cushions in the feet of African elephants (Loxodonta africana). *Journal of Anatomy*, 209(6), 781-792.
[2] Wang, Y. N., Lee, K., Shofer, J. B., & Ledoux, W. R. (2017). Histomorphological and biochemical properties of plantar soft tissue in diabetes. *Foot*, 33, 1-6.
[3] Wearing, S. C., Smeathers, J. E., Yates, B., Urry, S. R., & Dubois, P. (2009). Bulk compressive properties of the heel fat pad during walking: A pilot investigation in plantar heel pain. *Clinical Biomechanics*, 24(4), 397-402.
[4] Nix, S., Smith, M., & Vicenzino, B. (2010). Prevalence of hallux valgus in the general population: a systematic review and meta-analysis. *Journal of foot and ankle research*, 3(1), 1-9.
[5] Armstrong, D. G., Boulton, A. J., & Bus, S. A. (2017). Diabetic foot ulcers and their recurrence. *New England Journal of Medicine*, 376(24), 2367-2375.
[6] Westra, M., van Netten, J. J., Manning, H. A., van Baal, J. G., & Bus, S. A. (2018). Effect of different casting design characteristics on offloading the diabetic foot. *Gait & posture*, 64, 90-94.
[7] Grigoriadis, G., Newell, N., Carpanen, D., Christou, A., Bull, A. M., & Masouros, S. D. (2017). Material properties of the heel fat pad across strain rates. *Journal of the mechanical behavior of biomedical materials*, 65, 398-407.
[8] Taş, S. (2018). Effect of Gender on Mechanical Properties of the Plantar Fascia and Heel Fat Pad. *Foot and Ankle Specialist*, 11(5), 403-409.
[9] Jan, Y. K., Lung, C. W., Cuaderes, E., Rong, D., & Boyce, K. (2012). Effect of viscoelastic properties of plantar soft tissues on plantar pressures at the first metatarsal head in diabetics with peripheral neuropathy. *Physiological Measurement*, 34(1), 53.
[10] Teoh, J. C., & Lee, T. (2020). Identification of potential plantar ulceration among diabetes patients using plantar soft tissue stiffness. *Journal of the Mechanical Behavior of Biomedical Materials*, 103, 103567.
[11] Negishi, T., Ito, K., Kamono, A., Lee, T., & Ogihara, N. (2020). Strain-rate dependence of viscous properties of the plantar soft tissue identified by a spherical indentation test. *Journal of the Mechanical Behavior of Biomedical Materials*, 102, 103470.
[12] Ahanchian, N., Nester, C. J., Howard, D., Ren, L., & Parker, D. (2017). Estimating the material properties of heel pad sub-layers using inverse Finite Element Analysis. *Medical Engineering and Physics*, 40, 11-19.

Statistical evaluation of PMMA surface roughness

Zvikomborero Hweju, Fundiswa Kopi and Khaled Abou-El-Hossein[1]

Ultra-High Precision Engineering Research Unit, Department of Mechatronics, Nelson Mandela University, Port Elizabeth, South Africa.

[1]Corresponding author's e-mail: Khaled.Abou-El-Hossein@mandela.ac.za

Abstract. There is increased enthusiasm in polymer materials and yet limited research on single point diamond turning of Polymethyl methacrylate (PMMA) used to produce contact lenses. This study is a presentation of a statistical-based PMMA surface roughness prediction and parameter significance. The data utilized is obtained during dry single point diamond turning of PMMA. The experiment has been designed with the Central Composite Design (CCD) method and the Response Surface Methodology (RSM) has been used for surface roughness prediction and evaluation of cutting parameter importance. The surface roughness data used for regression model generation has been acquired manually using a profilometer. The resultant surface roughness (R_a) dataset has been split into training and testing datasets. The accuracy of the model has been determined based on the Mean Absolute Percentage Error (MAPE). Results have indicated that the generated regression model can predict surface roughness with 75.12 % accuracy. Furthermore, the order of parameter importance in decreasing order is as follows: feed rate, cutting speed, and depth of cut. The paired t-test results indicate that the difference between the averages of measured surface roughness and predicted surface roughness is not big enough to be statistically significant. Hence, the technique can be reliably utilized in predicting surface roughness during single point diamond turning of PMMA.

1. Introduction

Single point diamond turning is a widely used machining methods to produce smooth components with expected tolerances. Due to the high industrial usage of the method, a lot must be understood about the process with regards to surface quality [1, 2]. In the biomedical industry, SPDT has become popular in the production of contact lenses and intraocular lenses. The lenses are used for a variety of visual impairments, visual correction, cosmetic and therapeutic reasons, and their increased use has created a demand, increasing their manufacturing [3]. The increase in the demand has created a pressure in the manufacturing technique to increase the volume of production to meet the customer needs. Customer requirements include fitment, comfortability, and minimum susceptibility to bacterial adhesion and infections. Since SPDT requires no post-polishing, provides minimum manufacturing time and it can produce a product suited to customer needs that has created the popularity in utilizing it during polymer machining [4].

The quest for continuous release of defect-free machined components from the production line has resulted in the emergence of various surface quality prediction models. "Experimental investigation is time consuming and highly costly. Therefore, the use of prediction models and numerical methods [5] has gained popularity. Among this pool of prediction models is the Response Surface Methodology. This technique has proved its worth over the years. The RSM model has been utilized by Xiao et. al.

[6] to predict the surface roughness of turned stainless steel as well as optimize cutting parameters. The obtained results were satisfactory, and it has been concluded that the technique could be reliably used. Numerous other researchers have used the RSM for prediction and optimization in various studies [7-10].

Progress in the use of poly (methyl methacrylate) (PMMA) has unfolded extensive applications in the field of nanotechnology. PMMA has applications in biomedical, optical, solar, sensors, battery electrolytes, nanotechnology, viscosity, pneumatic actuation, molecular separations, and polymer conductivity fields [11]. This study utilizes statistical methods to evaluate the surface roughness of PMMA. The research paper is structured as follows: introduction, materials and methodology, results and analysis and finally the conclusion.

2. Materials and methodology

The dry single point diamond turning operation has been carried out on PMMA hydrophobic contact lens polymer with 12.7 mm diameter. Cutting speed, feed rate and depth of cut have been selected as the process control variables in the study. For each variable, three levels have been judiciously decided and the chosen limits are presented in Table 1. The experiment has been designed using the Central Composite Design method. The experiments were performed three times and each time a surface roughness measurement was taken using the Taylor Hobson PGI profilometer with a repeatability of < 100 nm. That was done for repeatability and the average surface roughness was computed and utilized for model generation. The experiments were carried out using the Precitech Nanoform ultra-grind 250 precision machine under dry cutting conditions. The machining setup is presented by Figure 1.

Table 1. Control process parameters.

Parameter	**Low**	**High**
Cutting Speed	200 rpm	4000 rpm
Feed Rate	2 mm/min	12 mm/min
Depth of cut	10 μm	40 μm

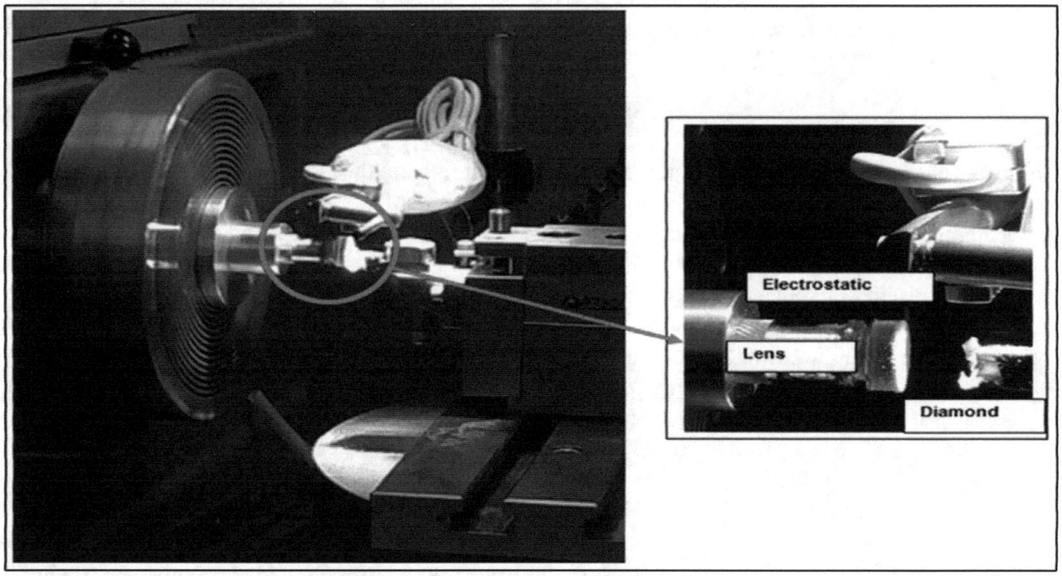

Figure 1. Machining process setup.

3. Results and analysis

The dry single point diamond turning of a 12.7 mm diameter PMMA specimen yielded the results presented in Table 2. The cutting variables used in the experiments are cutting speed, feed rate and depth of cut. Surface roughness average parameters (R_a) is the single response variable in the study. The test cases have been classified according increasing cutting speed, therefore the test cases with 200 rpm then for 2100 rpm and the 4000 rpm have been presented. The measured surface roughness varied from a minimum value of 4.8 nm to a maximum value of 203.2 nm over a range of 198.4 nm. That was an indication of the diversity in the experimental observations, over the various machining parameter combinations. Figure 2 displays a scatterplot of the experimental combinations. The scatter plot has been provided to show a visual trend in the data relationship.

Table 2. PMMA surface roughness presentation.

Run Order	Cutting Speed (rpm)	Feed rate (mm/min)	Depth of Cut (µm)	Surface Roughness (nm)
1	200	12	10	141.3
4	200	2	40	57.3
7	200	7	25	80.6
18	200	2	10	51.8
19	200	12	40	203.2
2	2100	7	25	24.9
6	2100	7	25	27.1
8	2100	7	25	23.2
9	2100	2	25	27.9
10	2100	7	10	21.1
11	2100	7	25	26.1
13	2100	12	25	31.4
14	2100	7	40	26.4
17	2100	7	25	24.0
20	2100	7	25	22.9
3	4000	2	10	4.8
5	4000	12	40	10.4
12	4000	7	25	8.8
15	4000	12	10	9.6
16	4000	2	40	6.7

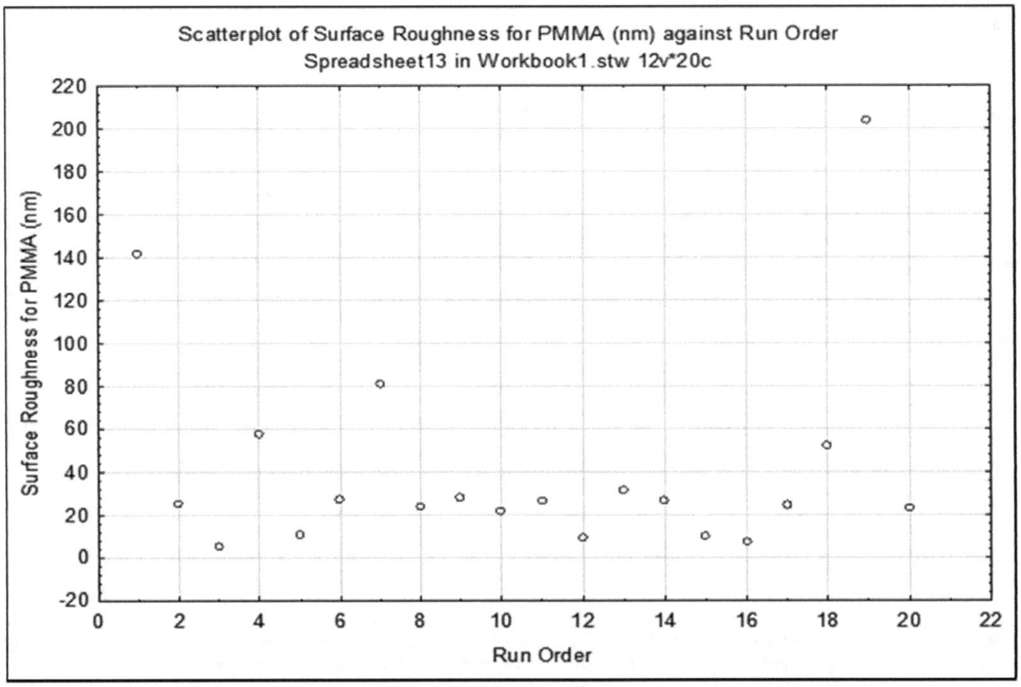

Figure 2. Scatter plot of surface roughness observations for PMMA.

To enhance data clarity, the obtained surface roughness values were clustered according to magnitude as follows: best, moderate, and worst. The generated clusters are presented in Table 3.

Table 3. Surface roughness groups.

Run Order	Range	Category
3, 5, 12, 15, 16	0 nm – 19.9 nm	Best
2, 6, 8, 9, 10, 11, 13, 14, 17, 20	20 nm – 39.9 nm	Moderate
1, 4, 7, 18, 19	40 nm – 204.9 nm	Worst

Best and moderate surface roughness observations were achieved using cutting speed between 2100 – 4000 rpm, feed rate between 2 – 7 mm/min and a depth of cut over the entire cutting range. The worst surface roughness values were achieved using a cutting speed of 200 rpm and a feed rate in the range 7 – 12 mm/min. Hence, a combination of low cutting speed and high feed rate results in poor PMMA surface quality.

Table 4 is a presentation of Analysis of Variance (ANOVA) synopsis, which has been utilized in examining data normality as well as assessment of the interrelationship between explanatory and response variables. The ANOVA summary shows that the utilized RSM model has a p-value of $0.00 < 0.05$. This value indicates that the model is significant for prediction purposes. Furthermore, the lack of fit p-value is $0.064 > 0.05$. This value is desirable since for the model to be adequate for prediction purposes, the lack of fit value is expected to be higher than 0.05. The total DF (bottom row) is 19. This is the total number of experimental runs (20) minus 1. It is the same regardless of any assumptions about repeated measures.

Journal of Physics: Conference Series **2313** (2022) 012030 doi:10.1088/1742-6596/2313/1/012030

Table 4. ANOVA summary for PMMA.

Source	**Degrees of freedom (DF)**	**Sum of Squares (SS)**	**Adjusted Mean Squares (MS)**	**F – value**	**P – value**
Model	4	41961	10490.1	37.32	0.000
Lack of fit	4	2221	555.1	3.06	0.064
Pure error	11	1996	181.5		
Total	19	46177			

Use is made of the least squares method to determine model parameters and coefficients, and the results are presented in Table 5. The cutting parameters are cutting speed (s), feed rate (f) and depth of cut (d). However, depth of cut has been observed to be insignificant and has thus been neglected in the prediction model.

Table 5. Regression model coefficients for PMMA.

Term	**Coefficient**	**Contribution**	**F – value**	**P – value**	**T – value**
Constant	38.05				
s	-0.0411	57.08 %	16.43	0.001	-4.053
f	11.43	15.12 %	45.99	0.000	6.782
s*s	0.00001	11.85 %	17.68	0.001	4.205
s*f	-0.003	15.95 %	23.81	0.000	-4.880

The global F-test results indicate that all the model's terms are significant since they all have p-values lower than 0.05. Additionally, the decreasing order of parameter importance is as follows: feed rate, cutting speed and depth of cut. A change in any of the parameters' constants results in a change in surface roughness value. The resultant prediction model is shown by Eq. (1). The model predictions based on Equation 1 are given in Table 6. Experimental runs 1-13 have been used to generate the model, while experimental runs 14-20 have been used for testing the model. The mean absolute percentage errors for both training and testing datasets have been provided in Table 6.

$$\widehat{SR_{PMMA}} = 38.05 - 0.0411s + 11.43f + 0.00001s^2 - 0.003sf \quad (1)$$

Table 6. Predicted surface roughness for PMMA.

Run Order	Measured R_a (nm)	Predicted R_a (nm)	Error (%)
1	141.3	159.9	13.16
2	24.9	25.5	2.40
3	4.8	11.7	143.75
4	57.3	53.4	6.80
5	10.4	4.4	57.69
6	27.1	25.5	5.90
7	80.6	106.8	32.50
8	23.2	25.5	9.91
9	27.9	0.7	97.49
10	21.1	25.5	20.85
11	26.1	25.4	2.68
12	8.8	8.1	7.95
13	31.4	50.2	59.87
14	26.4	25.5	3.40
15	9.6	4.4	54.16
16	6.7	11.7	74.62
17	24.0	25.5	6.25
18	51.8	53.4	3.08
19	203.2	159.9	21.30
20	22.9	25.5	11.35
Training Data Mean Absolute Percentage Error (MAPE)			**35.46**
Training Data Prediction Accuracy (%)			**64.54**
Testing Data Mean Absolute Percentage Error (MAPE)			**24.88**
Testing Data Prediction Accuracy (%)			**75.12**

The prediction accuracy of the model is 75.12 %, which is classified as high accuracy since it lies within the upper quartile region. Hence the RSM model can be reliably utilized for surface roughness prediction during dry single point diamond turning of PMMA. To assess the significance of the difference between the measured surface roughness and the predicted surface roughness, a paired t-test is performed. The null hypothesis postulates that the measured surface roughness average is equal to the predicted surface roughness average. The alternative hypothesis proposes that the average of the measured surface roughness is different from the predicted surface roughness average. The p-value equals 0.9881, ($P(x \leq -0.01505) = 0.4941$). It means that the chance of type I error, rejecting a correct null hypothesis, is too high: 0.9881 (98.81%). The larger the p-value the more it supports H0. Since the p-value > α (0.05), H0 cannot be rejected. In other words, the difference between the averages of measured surface roughness and predicted surface roughness is not big enough to be statistically significant.

4. Conclusion

The overall aim of this study was to investigate the surface roughness prediction accuracy of the RSM model during dry single point diamond turning of PMMA. Furthermore, the order of parameter importance has been assessed. A comparison of measured and predicted surface roughness values using MAPE has indicated that the RSM model's prediction accuracy is 75.12 %. The decreasing order of parameter importance is as follows: feed rate, cutting speed and depth of cut. Based on the paired t-test results, the difference between the averages of measured surface roughness and predicted surface roughness is not big enough to be statistically significant. The results are a testimony to the high effectiveness of the RSM technique in predicting surface roughness and assessing parameter importance during dry single point diamond turning of PMMA. The RSM model can therefore be reliably used to predict surface roughness of PMMA.

References

[1] Hatefi S and Abou-El-Hossein K 2020 *Int. J. Adv. Manuf.* **106** 2167
[2] Chen J and Zhao Q 2015 *Measurement* **69** 20
[3] OLUFAYO O A, ABOU-EL-HOSSEIN K and KADERNANI M M 2014 *Procedia Materials Science* **6**
[4] Hocheng H and Hsieh M L 2004 *Int. J. Mach. Tools Manuf.* **44**
[5] Kaselouris E, Papadoulis T, Variantza E, Baroutsos A and Dimitriou V 2017 *Solid State Phenomena* **261** 339
[6] Xiao M, Shen X, Ma Y, Yang F, Gao N and Wei W 2018 *Mathematical Problems in Engineering* **2018** 9051084
[7] Reddy B S, Kumar J S and Reddy K V K 2009 *Jordan Journal of Mechanical and Industrial Engineering* **3** 252
[8] Meddour I, Yallese M A, Bensouilah H, Khellaf A and Elbah M 2018 *Int. J. Adv. Manuf.* **97** 1931
[9] Mahesh G, Muthu S and Devadasan S R 2015 *Int. J. Adv. Manuf.* **77** 369
[10] Hweju Z and Abou-El-Hossein K 2020 *Journal of Physics: Conference Series*
[11] Ali U, Karim K J B A and Buang N A 2015 *Polymer Reviews* **55** 678

A reliability estimation method for spring group balanced system at different degradation temperatures

Haodi Ji[1,4], Xiaobing Ma[1] and Yanyan Wang[2,3]

[1]School of Reliability and Systems Engineering, Beihang university.
[2]Southwest Institute of Technology and Engineering.
[3]School of mechanical engineering, Nanjing University of Science and Technology.

[4]Corresponding author's e-mail: jhd2021@buaa.edu.cn

Abstract. The failure process of a balanced system consisting of several independent springs with the same structure is more complicated than that of a single spring. The balanced system of spring group should not only satisfy that the performance of all springs in the spring group is greater than the failure threshold, but also ensure that the performance difference between springs is not big. In this paper, the degradation parameters of spring compression ratio related to ambient temperature are estimated and the degradation parameters of other temperatures are extrapolated by Arrhenius model. Then the joint probability density function of the maximum and minimum order statistics of the spring compression ratio is used to estimate the reliability of the spring group balanced system at different degradation ambient temperatures. The results show that the Arrhenius model is effective in fitting the parameters of spring degenerate drift. The model established in this paper realizes the reliability evaluation of the spring balance system at different temperatures.

1. Introduction

With the rapid development of science and technology, new reliability systems such as balanced system appear in aerospace, wind power generation, ordnance equipment and other fields. Hua and Elsayed [1-2] first defined the equilibrium system, pointing out that there is at least a pair of vertical symmetry axes in the work unit to achieve the equilibrium of the system. Moreover, Cui et al. [3] evaluated the equilibrium degree of the system according to the state difference between the best component and the worst component. Gao [4] believes that the system is a state of equilibrium and that the number of units in each department should be the same at all times. Wang [5] considered two restricted rebalanced mechanisms and proposed random failures of unit forced-down function and resumed function respectively with the intent of modeling actual engineering situations. Zhao et al. [6] further defined the equilibrium degree of the system based on the state difference of two units.

Due to the difference between the traditional degradation process failure criteria and the degradation process failure criteria of the balanced system, the traditional degradation theory cannot effectively solve the problem of reliability assessment in the degradation process of the balanced system. In the degradation process of general systems, the Markov model is often used to describe the degradation path of components because of its feasibility and simplicity. [7-8] It is assumed that the residence time of the system obeys exponential distribution under discrete operating conditions, and the Markov model is established to deduce the system performance index. However, during the creep process, the internal structure of the spring gradually changes. Due to the randomness of spring degradation process and the influence of working conditions, the performance degradation of spring at a certain moment is continuous and random. Therefore, the Wiener process can be considered to describe the performance degradation of the spring. Wang et al. [9] extended the drift coefficient in the Wiener process to state and estimated the state with Kalman filter to obtain the remaining life distribution with improved accuracy. Si et al. [10] considered the uncertainty of the drift state and proposed an iterative degradation model based on the Wiener process, which reduced the uncertainty of residual life distribution. Jin et al. [11] used normal distribution to describe the drift parameters and diffusion parameters of the degradation model of the Wiener process of the unit and predicted the overall life. Peng et al. [12-13] used Wiener process to evaluate the reliability of metallized film capacitor. Zhang [14] used Wiener process to describe the individual difference of leakage current degradation of different optocouples, and improved the linear Wiener process to establish random variable Arrhenius equation. The above research proves that the Wiener process can be used to reasonably and accurately describe the degradation state of units.

However, due to the difference in the degradation environment of each component in a balanced system [15], the degradation mechanism is the same but the degradation path is different. Therefore, this paper proposes a reliability evaluation method for a balanced system with n identical components degraded under different environmental conditions based on order statistics and accelerated degradation tests. The drift coefficient and diffusion coefficient in the Wiener process are unknown parameters, which need to be estimated. In this paper, the drift coefficient of spring degradation process is considered as a function of ambient temperature T and the relationship between them is established according to the available data of spring compression ratio degradation. According to this function, the drift coefficient can be reasonably extrapolated to describe the degradation process of the spring at any temperature, and the reliability of the spring balanced system at different ambient temperatures is estimated by using the joint probability density function of the maximum and minimum order statistics.

2. Model description

2.1. System description

The spring group discussed in this paper is a balanced system composed of n independent springs with the same distribution. The spring compression ratio $C(t)$ refers to the ratio of the compressed length of the spring to the free shape length. The compression ratio $C(t)$ of each spring gradually degrades with the time t, and the degradation process is random. As shown in Figure 1, if a balanced system consists of two springs, assume that their initial compression ratios are C_{01} and C_{02}. The spring group has two failure modes. Once the difference between the states of the two springs exceeds the critical threshold d, the entire spring group is out of balance. In addition, when the compression ratio of any spring falls below the failure threshold C_{fail}, the spring group will fail due to insufficient resilience. Therefore, the reliability of the balanced system for the spring group discussed in this paper depends on the minimum compression ratio state difference and maximum compression ratio loss of the spring group.

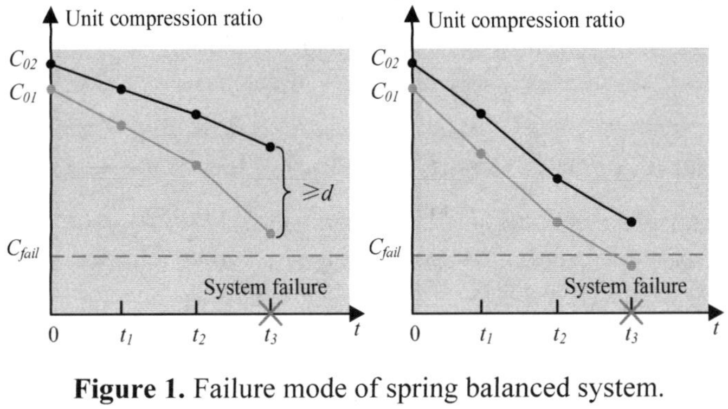

Figure 1. Failure mode of spring balanced system.

2.2. Description of degradation process for spring

The health status of springs essentially reflects the degradation level of spring compression ratio $C(t)$, so the research on the health status estimation of springs is the compression ratio degradation process of springs. Wiener process can not only describe the degradation process with increasing or decreasing tendency, but also has good mathematical properties, so it can be used to describe the compression ratio degradation process of spring. Based on the basic Wiener process and considering the spring compression ratio degradation process under different ambient temperatures, the drift coefficient representing the degradation rate of spring compression ratio is regarded as a function related to temperature T, denoted as $v(T)$. Therefore, the basic nonlinear Wiener process expression can be written as:[14]

$$X(\Lambda(t)) = X_0 + v(T)\cdot\Lambda(t) + \sigma_B B(\Lambda(t)) \tag{1}$$

where $\Lambda(t)$ is a nonlinear function of time t; X_0 is the initial state, the initial compression ratio of the spring is 1; $v(T)$ is drift coefficient, which can represent velocity change; σ_B is diffusion coefficient; $B(\cdot)$ is the standard Brownian motion.

It is assumed that the failure mechanism of the spring is fixed under different temperature stress levels and the diffusion coefficient σ_B in Wiener process is also fixed. The relationship function between drift coefficient v and ambient temperature T can be established by the compression ratio degradation test data of springs at different temperatures.

In addition, the Wiener process has stable independent increment, Δt time interval generated compression ratio loss $\Delta C \sim N(v(T)\cdot\Lambda(\Delta t), \sigma_B^2\cdot\Lambda(\Delta t))$. Therefore, the drift coefficient $v(T)$ and diffusion coefficient σ_B under various temperature stresses can be estimated by the maximum likelihood method. The probability density function $f_{C(t|T)}$ of compression ratio degradation of spring at different temperatures is:

$$f_{C(t|T)}(x) = \frac{1}{\sqrt{2\pi\left(\sigma_B^2\cdot\Lambda(t)\right)}}\exp\left(-\frac{(x - x_0 - v(T)\cdot\Lambda(t))^2}{2\left(\sigma_B^2\cdot\Lambda(t)\right)}\right) \tag{2}$$

2.3. System reliability assessment method

The reliability of the balanced system of the spring group is determined by the maximum compression ratio state difference and maximum compression ratio loss of the spring group. Therefore, determining the probability density function of maximum compression ratio state difference and maximum compression ratio loss is the key to reliability evaluation. According to Eq (2), the probability density function of compression ratio degradation of each spring can be obtained. The probability density function of maximum degradation compression ratio can be obtained through order statistics. The probability density function of maximum compression ratio state difference of spring group can be calculated according to the joint probability density function of maximum and minimum order statistics.

Assuming that $X_1, X_2, \ldots, X_n$ are samples from different populations with observation values $x_1, x_2, \ldots, x_n$. Rearrange the observations from small to large:

$$x_{(1)} \le x_{(2)} \le \ldots \le x_{(n)}$$

When $(X_1, X_2, \ldots, X_n)$ is identical to $(x_1, x_2, \ldots, x_n)$, $X_{(1)}$ is equal to $x_{(1)}$. The resulting random variable $(X_{(1)}, X_{(2)}, \ldots, X_{(n)})$ is called the order statistic of $(X_1, X_2, \ldots, X_n)$. Obviously, $X_1 = \min_{1\le i\le n} X_i$, $X_n = \max_{1\le i\le n} X_i$. Where $X_{(1)}$ is the minimum order statistic and $X_{(n)}$ is the maximum order statistic. Assuming that the probability density function of the population distribution of each subsample $X_1, X_2, \ldots, X_n$ is $f_{X_1}, f_{X_2}, \ldots f_{X_n}$. And their cumulative probability functions are $F_{X_1}, F_{X_2}, \ldots, F_{X_n}$ respectively. Then the probability density function of the maximum order statistic is:[16]

$$f_{X_{(n)}}(x)=\sum_{i=1}^{n} f_{X_i}(x) \prod_{j=1\&j\neq i}^{n} F_{X_j}(x) \tag{3}$$

The probability density function of the joint distribution of the maximum and minimum order statistics is:

$$f_{X_{(1)},X_{(n)}}(x_{(1)},x_{(n)})=\sum_{i=1}^{n}\sum_{j=1\&j\neq i}^{n} f_{X_i}(x_{(1)})f_{X_j}(x_{(n)}) \prod_{k=1\&k\neq i\&k\neq j}^{n} [F_{X_k}(x_{(n)})-F_{X_k}(x_{(1)})] \tag{4}$$

The reliability of the system represents the probability that the system can complete the specified function in the specified time and under the specified conditions. Therefore, the reliability of the balanced system described in this paper is the probability that the two failure behaviors in Figure 1 will not occur. According to the maximum compression ratio state difference d and maximum compression ratio loss C_{fail} of the spring, the reliability $R(t)$ of spring pack can be given as follows:

$$R(t)=P\{X_{(n)}-X_{(1)}<d \,\&\, X_{(1)}>C_{fail}\} \tag{5}$$

3. Method application case

3.1. Experimental data source

The experimental data of spring compression ratio degradation is from reference [17]. The free length of a certain missile spring is 59 mm, and the spring is compressed to 22 mm in the storage process of the missile, and the storage temperature is 25℃. After a period of storage, the free length decreases due to the creep of the spring. As long as the spring free length is greater than 51 mm, the missile can be ensured to push normally. Therefore, the failure compression ratio threshold C_{fail} of the spring is 0.86. In addition, the critical threshold d is conservatively assumed to be 0.05. Spring compression ratio degradation data at different temperatures are shown in Figure 2. The test was carried out at 140℃, 110℃, 80℃ and 50℃, and samples were taken for 13 times at each test temperature level.

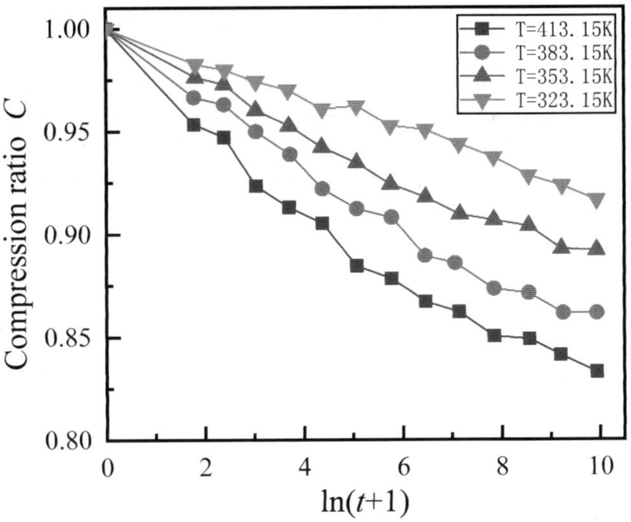

Figure 2. Spring compression ratio degradation data.

3.2. Parameter estimation

The maximum likelihood method is used to estimate the drift coefficient and diffusion coefficient in Eq (1) at different ambient temperatures. As Feng found in his experiments, the observed spring creep over logarithmic time intervals was nearly equal. [18] Therefore, the spring degradation in each logarithmic time interval fluctuates around a certain value, so it can be assumed that the spring degradation in logarithmic time interval follows a normal distribution. And it can be assumed that $\Lambda(t) = \ln(t+1)$. Then, there are five parameters $v(T_1)$, $v(T_2)$, $v(T_3)$, $v(T_4)$, and σ_B^2 need to be estimated. Suppose there are K_l intervals per sample at each stress, $\Delta C_{l,m}$ is the compression ratio loss in the mth interval under lth temperature stress. It is known that $\Delta C \sim N(v(T)\cdot\Lambda(\Delta t), \sigma_B^2\cdot\Lambda(\Delta t))$, so it can be obtained from revised maximum likelihood estimation:

$$\hat{v}(T_l) = \frac{\sum_{m=1}^{K_l} \Delta C_{l,m}}{\sum_{m=1}^{K_l} \Lambda(\Delta t_{l,m})} \tag{6}$$

$$\hat{\sigma}_B^2 = \sum_{l=1}^{4}\sum_{m=1}^{K_l} \frac{(\Delta C_{l,m} - \hat{v}(T_l)\cdot\Lambda(\Delta t_{l,m}))^2}{(4K_l - 1)\cdot\Lambda(\Delta t_{l,m})} \tag{7}$$

The parameter estimation results using spring compression ratio degradation data are shown in Table 1.

Table 1. Parameter estimation results of compression ratio degradation.

Parameter	Value
$\hat{v}(T_1)$	-0.0168
$\hat{v}(T_2)$	-0.0140
$\hat{v}(T_3)$	-0.0108
$\hat{v}(T_4)$	-0.0084
$\hat{\sigma}_B^2$	5.63E-5

3.3. System reliability assessment

Suppose a spring balanced system consists of four Springs. Because the distance between each spring and the engine is different, the ambient temperature is also different. Assume that the four batteries are exposed to ambient temperatures of 293.15 K, 298.15 K, 303.15 K, and 308.15 K near room temperature. The degradation parameters of springs at these four temperatures are calculated by fitting or interpolation according to the parameters in Table 1.

The acceleration stress used in the spring degradation test is temperature, so the Arrhenius acceleration equation is chosen to fit the relationship between the mean initial compression ratio and temperature. According to the fitting results of test parameters, it is found that the mean value of initial compression ratio decreases with the decrease of temperature, so the Arrhenius model can be rewritten as:[14]

$$\ln(-v(T)) = a + \frac{b}{T} \tag{8}$$

where a and b are undetermined parameters that are determined by fitting with test data. The above equation shows that the logarithm of the initial compression ratio is a linear function of the reciprocal of temperature. According to the parameters in Table 1, the Arrhenius model is fitted by least squares, and the estimated results are as follows: a=-1.563, b=-1042, R^2=0.9983. The estimated parameters are substituted into Eq(8), and the spring degradation parameters at four temperatures in the model are extrapolated. The extrapolation results of the final parameters are shown in Table 2.

Table 2. The extrapolation results.

Temperature T	Drift coefficient $\hat{\nu}(T)$
293.15 K	-0.0060
298.15 K	-0.0064
303.15 K	-0.0067
308.15 K	-0.0071

The probability density function of spring compression ratio and the joint probability density function of maximum & minimum order statistics at different times can be obtained by Eq (2) and (4). Finally, the time-varying curve of the spring compression ratio degradation and system reliability obtained according to Eq (5) are shown in Figure 3. As can be seen from Figure 3, the initial performance of springs at different temperatures is the same, but gradually produces differences with time. The higher the ambient temperature, the faster the spring degrades. Moreover, the fluctuation of degradation increases with time. Therefore, in the early stage of degradation, the spring difference is small and the performance is good, so the reliability of the balanced system is high. In the middle stage of degradation, the main reason for the decrease of the reliability of the balanced system is the difference between the springs. While in the later stage of degradation, the main reason why the reliability of the balanced system decreases is that the spring performance is generally low. The reliability R of the system is the probability that the system works properly at this time and condition. In this case, reliability decreases with time and decreases more slowly in the late degradation period than in the early degradation period. This may be due to the tendency of later spring compression ratio to be similar. Compared with other reliability assessment methods using Markov processes, this method can take continuous degradation processes into account. Therefore, this method is more convenient and fits the continuous degradation system in practice.

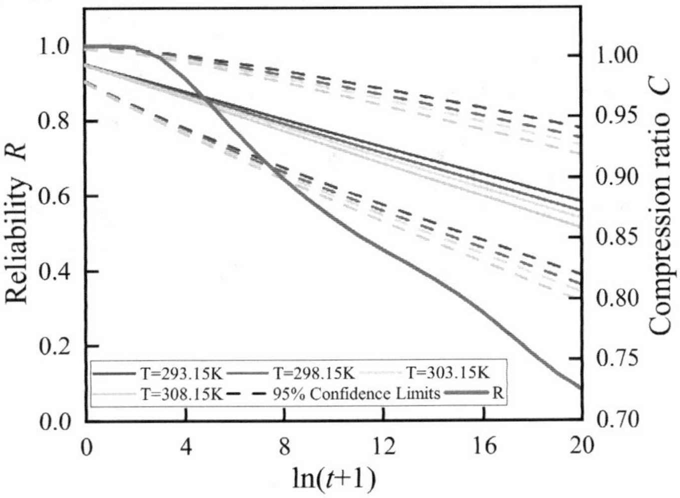

Figure 3. System reliability and compression ratio time-varying curves.

4. Conclusions

In this paper, a reliability evaluation method for spring group under different degradation temperatures is proposed. The maximum likelihood method is used to estimate the degradation parameters of spring compression ratio, and the relationship model between ambient temperature and initial compression ratio and degradation rate of spring is established and extrapolated. By extrapolating the model, the compression ratio degradation parameters can be obtained at any temperature. According to the joint probability density function of the maximum and minimum order statistics, the reliability of the spring group balanced system is calculated. It has certain reference significance and value for spring group reliability prediction with a small sample size.

Acknowledgment

This work was supported in part by National Natural Science Foundation of China under 52075020 and Southwest Institute of Technology and Engineering Cooperation Fund under WZDW592019080301.

References

[1] Hua D and Elsayed E A 2016 Degradation Analysis of k-out-of-n Pairs:G Balanced System With Spatially Distributed Units *IEEE Transactions on Reliability.* **65** 941-56.

[2] Hua D and Elsayed E A 2016 Reliability Estimation of k-out-of-n Pairs:G Balanced Systems With Spatially Distributed Units *IEEE Transactions on Reliability.* **65** 886-900.

[3] Cui L R, Chen J H and Li X C 2019 Balanced reliability systems under Markov processes *IISE Transactions.* **51** 1025-35.

[4] Gao H D, Cui L R and Yi H 2019 Availability analysis of k-out-of-n: F repairable balanced systems with m sectors *Reliab eng syst safe.* **191** 106572.

[5] Wang X Y, Zhao X, Wu C S and Lin C 2020 Reliability assessment for balanced systems with restricted rebalanced mechanisms *Computers & Industrial Engineering.* **149** 106801.

[6] Zhao X, Wang S Q, Wang X Y and Fan Y 2020 Multi-state balanced systems in a shock environment *Reliab eng syst safe.* **193** 106592.

[7] Wang J J and Miao Y H 2021 Optimal preventive maintenance policy of the balanced system under the semi-Markov model *Reliab eng syst safe.* **213** 107690.

[8] Chen Z, Li Y P and Xia T B 2019 Hidden Markov model with auto-correlated observations for remaining useful life prediction and optimal maintenance policy *Reliab eng syst safe.* **184** 123–36.

[9] Wang W B, Carr M, Xu W J and Kobbacy K 2011 A model for residual life prediction based on Brownian motion with an adaptive drift *Microelectronics Reliability.* **51** 285-93.

[10] Si X S, Wang W B, Hu C H, Chen M Y and Zhou D H 2013 A Wiener-process-based degradation model with a recursive filter algorithm for remaining useful life estimation *Mechanical Systems and Signal Processing.* **35** 219-37.

[11] Jin G, Matthews D E and Zhou Z 2013 A Bayesian framework for on-line degradation assessment and residual life prediction of secondary batteries inspacecraft *Reliab eng syst safe.* **113** 7-20.

[12] Peng B X, Zhou J L, Pan Z Q 2010 Bayesian method for reliability assessment ofproducts with Wiener process degradation *System Eng. Theor. Prac.* **30** 543-9.

[13] Peng B X, Jin G, Pan Z Q 2009 Reliability assessment of metallized film capacitor using multiple reliability information sources *High Power Laser and Particle Beams* **21** 1271-5.

[14] Zhang X G, Mu X H, 2019 Evaluation of Optocoupler Storage Life of Terminal Guided Projectile Based on Nonlinear Wiener Process *Journal of Ordnance Equipment Engineering* **40** 75-8.

[15] Zhong L, Zhang C B, He Y and Chen Z H 2014 A method for the estimation of the battery pack state of charge based on in-pack cells uniformity analysis *Appl. Energy* **113** 558-64.

[16] Mao S S, Wang J L, Pu X L 1989 Advanced Mathematical Statistics(Beijing: China Higher Education Press) pp 53-60.

[17] Feng X H 2012 Creep dynamic model and its application in creep of spring *Chin. Sci. Bull.* **57** 2354-8.

[18] Feng X H 2015 Basement and Application of Dynamics of Environmental Effect Equipment Environmental Engineering 12 15-8.

Journal of Physics: Conference Series **2313** (2022) 012032 doi:10.1088/1742-6596/2313/1/012032

Radiation evaluation of household appliances and stone materials from residences in Hangzhou

Xiangyu Zhang[1], Fang Zheng[2], Xiaodong Wu[3, 5], Qiuxiang Liu[1], Jun Jiang[4] and Zhangsen Chen[1]

[1]Zhejiang Light Industrial Products Inspection and Research Institute, Hangzhou 310018, China;
[2]LanRun Angel Foreign Language Experimental School, Huzhou 313300, China;
[3]Eco-environmental Science Research & Design Institute of Zhejiang, Hangzhou 310000, China;
[4]Zhejiang Tmall Technology Co., Ltd, Hangzhou 310023, China.

[5]Corresponding author's e-mail: 57285478@qq.com

Abstract. The electromagnetic radiation of household appliances and radioactivity of stone materials from residences were determined using integrated magnetometer and γ-ray spectrometer techniques. The radiation levels of desktop computers were greater than that of notebook computers, while microwave ovens and induction cookers produced low-frequency electromagnetic fields during the operation. The safety distances for hair drier, desk lamp, and patch board were *ca.* 50 cm, whereas those of the rest appliances were *ca.* 30 cm. Radioactive tests of stones collected from certain regions were conducted. The specific activities of ^{226}Ra, ^{232}Th, and ^{40}K were used to calculate internal exposure indexes (I_{Ra}) and external exposure indexes (I_γ). Moreover, both the I_{Ra} and I_γ for wall coatings were putty < tile adhesive, thereby the I_{Ra} and I_γ for bathroom furniture were ceramic facet < pottery basin < flush toilet. Natural radionuclides widely exist in interior decorations, indicating that humans in open radioactive environments receive background radiation all the time.

1. Introduction

The application of electromagnetic technology in China has reached an international advanced level due to the progress of modern science [1−3]. With huge benefits brought by electromagnetic developments, the radiations of indoor household appliances and stone materials have also severely disrupted our ecological environments [4,5]. The electromagnetic radiations of mobile phones, computers, televisions, induction cookers, and printers are familiar as well as their problems, which has become a priority of humanity [6−9]. Long term exposure to high electromagnetic radiations may cause various health concerns. Therefore, it is urgent to find useful anti-radiation methods for reducing the electromagnetic pollution.

In terms of real estate and infrastructure construction, new types of building materials are emerging one after another, including stone, ceramic tile, cement, brick and industrial residue [10−13]. More attention should be paid to the radioactivity of interior decoration, and these building materials are harmful to the environment and cause various radioactive hazards [14,15]. Relevant investigations, such as the impact of appliances [16,17] and stone materials [18,19] on daily life are conducted to reduce and monitor radioactive hazards in the construction industry. Thus, radioactive detection has a quite significance for correctly selecting raw materials for decorations.

External exposure mainly refers to a biological effect produced by direct irradiation of radioactive elements on humans, causing damage to haematopoietic, nervous, digestive, and reproductive systems. Internal irradiation usually refers to products originated from the decay of natural radionuclides, such as Ra and Th. After inhalation of Ra by children, it will continue to deposit on the surface of the respiratory tract and gradually accumulate in the body. However, the incubation period is long and sometimes the harm appears after decades. To determine the stone material radioactivity, we used the mean value of specific activities.

This study aimed to evaluate the radiation levels of household appliances and stone materials. The electromagnetic radiation of different sources and the indoor environment of buildings where these materials existed were investigated. Typical magnetic field strength and electric field intensity were assessed via measurements of frequented resident rooms. Here, we focused on the characterisation of electromagnetic field exposure and effective safety distances for protection from radiations of household appliances and stone materials.

2. Methods

2.1. Instruments and test sites

Instruments, including a NARDA ELT-400 integrated magnetometer (Germany), a FYFS 400X multi-channel γ-ray spectrometer (China), a MKCA6-5J ultrafine pulverizer (Japan), and an EHP-50B electromagnetic field probe (Italy) were used.

Test sites were certain residences in Hangzhou, Zhejiang Province, China. To avoid the influence of other community facilities on the experiments, indoor household appliances were measured in triplicate. Other equipments that might make sounds were turned off, and only the test object remained operational. The environmental background radiation was low, and the results were completely based on the objects being measured [20].

2.2. Electromagnetic radiation measurement

The electromagnetic radiation was zero when the electric appliance was unopened based on Chinese mandatory standard GB 8702-2014 'Controlling limits for electromagnetic environment'. We used the magnetometer to measure radiation at different distances. According to the regulations on electromagnetic radiation protection and relevant national standards, the derived limit of public exposure in the frequency band of 0.1–3 MHz was 0.1 A m^{-1}. Considering there was a rare magnetic medium in air, the magnetic field strength (0.13 μT) and electric field intensity (40 V m^{-1}) were selected as the average dose limit for household appliances. Each datum is the average of three parallel experiments.

2.3. Stone radioactivity analysis

Two stone samples (≥2 kg) were obtained, one chosen as the inspection sample and the other as a standby. The stone sample was smashed to a particle size ≤0.16 mm, and it was put into a box consistent with the geometric shape. They were then weighed (0.1 g) and sealed for radioactivity analysis. When the radioactive decay in the specimen reached equilibrium [21], a low background multi-channel γ was accessed as well as the standard under the same conditions. The specific activities of ^{226}Ra, ^{232}Th, and ^{40}K were determined using an energy spectrometer, and the internal and external exposure indices were calculated based on GB 6566-2010 'Limits of radionuclides in building materials'.

3. Results and discussion

3.1. Characterisation of electromagnetic radiation levels

Television (TV) sets included traditional picture tube, plasma, rear-projection, and liquid crystal TVs. For the traditional picture tube, the front radiation generated after start-up was 0.17 μT at a distance of

Journal of Physics: Conference Series **2313** (2022) 012032 doi:10.1088/1742-6596/2313/1/012032

0.5 m. Its magnetic field strength was 0.35 μT (normal viewing), 0.29 μT (channel changing) and 0.16 μT (standby mode). In addition, the magnetic field strength reached 5.1 μT, and it did not decrease across the glass and wood. However, it was still 1.0 μT when measured behind a 10 cm thick wall, showing that the TV back had greater radiation. In terms of plasma, rear-projection, and liquid crystal TV, they were measured at the front and back, whose strengths were ≤0.3 μT. Hence, an appropriate position is desirable when using a TV, resulting in reasonable magnetic radiations. Figure 1 shows a photograph of the household appliances. The host, display, mouse, and keyboard around computers produced radiation. For desktop computers, the magnetic field strengths on the front and side of the host were *ca.* 0.3 μT. Those of the chassis centre and back were 0.23 and 0.49 μT at a distance of 30 cm, while it was 0.18 μT at the moment of start-up. For notebook computers, the magnetic field strength was 0.13 μT (front of the display screen), 0.16 μT (above the keyboard) and 0.2 μT (side of the power adapter). It can be inferred that the radiation level of a desktop computer is greater than that of a notebook computer, especially behind the host.

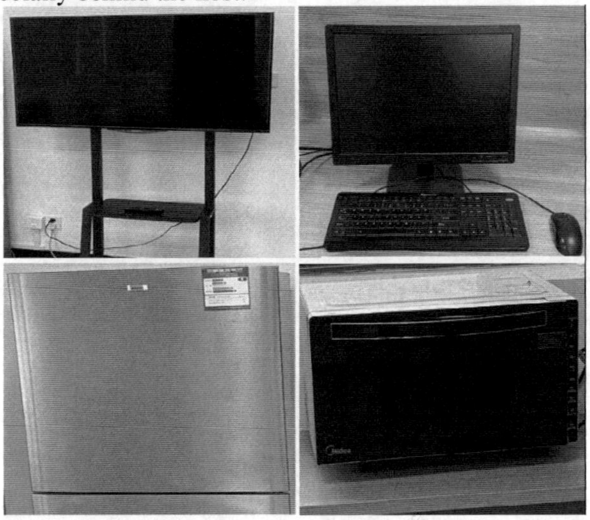

Figure 1. Schematic graph of common household appliances.

It had been established that both the electric field intensities of microwave oven and induction cooker were <2 V m^{-1} within 100 cm. The magnetic field strength of microwave oven was 41 μT (directly above), 64 μT (straight ahead), and 52 μT (side). In terms of induction cooker, the magnetic field strength was 8.7 μT (100 cm away from cooker) and 4.3 μT (20 cm away from cooker). There was no safe area within 30 cm around it. The microwave oven and induction cooker demonstrated low-frequency electromagnetic field during operation, which was easily induced by magnetism.

The magnetic field strength was 0.018−0.09 μT, and the electric field intensity ranged from 0.7−88.5 V m^{-1} for other household appliances within 100 cm. When the distance was increased to 200 cm, both the magnetic field strength (0.015−0.068 μT except for mobile phone) and the electric field intensity (0.3−14.9 V m^{-1}) decreased. Followed by the increase in distance, the radiation level decayed rapidly, and the attenuation speed was closely regarded as inversely proportional to the squared distance. The safety distances for hair drier, desk lamp, and patch board were 50 cm, whereas those of the rest appliances were 30 cm. The relevant parameters are shown in Table 1, and the safety distances when using household appliances should be noted.

Table 1. Electromagnetic radiation levels at different distances for other household appliances.

Household appliance	Distance (cm)							
	30 cm	50 cm	100 cm	200 cm	30 cm	50 cm	100 cm	200 cm
	Magnetic field strength (μT)				Electric field intensity (V m^{-1})			
Air conditioner	0.022	0.019	0.018	0.016	19.3	9.6	1.8	0.4
Desk lamp	0.047	0.042	0.035	0.028	79.2	25.1	20.7	14.9
Hair drier	0.078	0.038	0.025	0.023	88.5	39.8	25.0	14.7
Mobile phone	-	-	-	-	2.7	1.6	0.7	0.3
Patch board	0.034	0.026	0.024	0.021	38.1	35.6	16.8	4.2
Electric kettle	0.090	0.035	0.029	0.026	17.1	10.2	8.4	6.5
Rice cooker	0.051	0.024	0.019	0.015	6.7	4.9	2.8	0.9
Refrigerator	0.074	0.073	0.070	0.068	14.3	6.4	2.3	1.9

3.2. Characterisation of stone material radioactivity

Radioactive detections of building materials (such as cement, wall coatings, ceramic tiles, bathroom furniture, tourmaline powder, and brick, etc.) were performed, and the radioactivity data for different stone materials (Figure 2) were obtained.

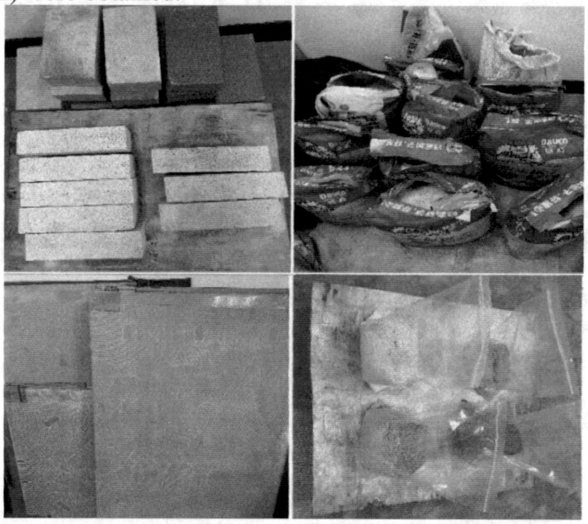

Figure 2. Schematic graph of common stone materials.

The internal exposure index (I_{Ra}) is calculated using the following equation:

$$I_{Ra} = \frac{C_{Ra}}{200} \quad (1)$$

where C_{Ra} (bq kg^{-1}) is the specific activity of natural radionuclide ^{226}Ra, which is used to express the relative degree of internal exposure dose caused by stone materials. The external exposure index (I_γ) can be calculated by Equation (2):

$$I_\gamma = \frac{C_{Ra}}{370} + \frac{C_{Th}}{260} + \frac{C_K}{4200} \quad (2)$$

where C_{Th} and C_K (bq kg^{-1}) is the specific activity of ^{232}Th and ^{40}K, respectively. The sum of the quotient is obtained by dividing the specific activity of each standard.

Table 2 displays the evaluating indicators of the main building materials, of which the I_{Ra} was ordinary silicate cement < slag silicate cement < composite silicate cement < granite < cement mortar < sand < stone powder < artificial ceramic tile < ash silicate cement < slag cement; the I_γ was ordinary silicate cement < slag silicate cement < stone powder < cement mortar < granite < composite silicate cement < ash silicate cement < sand < artificial ceramic tile < slag cement.

Table 2. Evaluating index of the main building materials.

Specimen	I_{Ra}	I_γ
Artificial ceramic tile	0.6	0.9
Granite	0.2	0.6
Cement mortar	0.3	0.5
Slag cement	0.8	0.9
Sand	0.4	0.9
Stone powder	0.5	0.5
Ordinary silicate cement	0.1	0.5
Composite silicate cement	0.2	0.6
Slag silicate cement	0.1	0.5
Ash silicate cement	0.6	0.8

Table 3. Statistics of wall coatings and bathroom furniture.

Specimen	I_{Ra}	I_γ
Ceramic tile adhesive	0.2	0.2
Vitrified tile adhesive	0.2	0.3
Stone tile adhesive	0.1	0.4
Bottom putty (standard type)	<0.1	0.1
Bottom putty (gypsum base)	0.1	0.1
Mould proof putty	<0.1	0.2
Water proof putty	0.1	0.2
Integrated toilet	0.9	1.2
Split toilet	1.0	1.1
Ceramic facet	0.3	0.7
Quartz stone cabinet	0.4	0.6
Column basin	0.8	1.2
Cabinet basin	0.3	0.8
Art basin	0.5	1.3

Subsequently, we reported the competitive capacity of I_{Ra} and I_γ versus various wall coatings and bathroom furniture used in interior decoration (Table 3). For wall coatings, both the I_{Ra} and I_γ were putty < tile adhesive, while the I_{Ra} and I_γ for bathroom furniture were ceramic facet < pottery basin < flush toilet. Table 4 presents the summary of the mean values of radioactivity measurements, of which the I_{Ra} was tourmaline powder < natural marble < fireproof brick < wall coating < floor tile < cement < bathroom furniture; and the I_γ was natural marble < fireproof brick < wall coating < floor tile < cement < bathroom furniture < tourmaline powder. People do not have to fear radiation level of stone materials, and radioactive problems can be approached from a scientific point of view. To reduce the radiation enhancement effect of subsequent processing steps for stones, we should explore low-level radioactive materials or alternative raw materials.

Table 4. Summary of radioactivity measurement data.

Factor	I_{Ra}	I_γ	I_{Ra}[22,23]	I_γ[22,23]
Cement	0.1−0.6	0.2−0.8	0−0.53	0.27−0.69
Wall coating	0−0.4	0−0.5	-	-
Floor tile	0−0.7	0.1−1.2	0.18−0.5	0.22−1.1
Bathroom furniture	0.1−1.1	0.3−1.4	-	-
Tourmaline powder	<0.2	22.7−305	-	-
Natural marble	0.2−0.4	0.2−0.6	0.21−0.4	0.25−0.67
Fireproof brick	0−0.3	0−0.4	0−0.33	0−0.42

4. Conclusion

We investigated the electromagnetic radiation of household appliances and the radioactivity of stone materials used in decorations of Hangzhou residences. The electromagnetic measurements of microwave oven and induction cooker indicated that electric field intensities were <2 V m^{-1} within 100 cm, and the radiation effects were fairly low. Meanwhile, the magnetic field strength was 0.018−0.09 μT, and the electric field intensity was 0.7−88.5 V m^{-1} for other household appliances within 100 cm. The corresponding data for stones were presented, and radiation level of stones was conducted according to the specific activities of ^{226}Ra, ^{232}Th, and ^{40}K, since the I_{Ra} and I_γ were natural marble < fireproof brick < wall coating < floor tile < cement. This study provides reliable distances for external and internal hazard indices, extending knowledge of typical electromagnetic radiation and radioactivity pollution.

References

[1] Yang P, Wang C, Cheng Y, Li D and Liu J 2021 Analysis of indoor radon concentrations in Changchun *Chinese Journal of Public Health Engineering* **20** 190
[2] Xu N, Wei X, Kuang F, Zhang L and Liu H 2018 Study on the natural radioactivity level of stone coal-bearing strata in East China *Environmental Geology* **77** 726
[3] Ma L, Lu Y, Zou P and Zhou X 2012 Research on indoor electromagnetic radiation field of multiple antenna system *Journal of Electronics (China)* **29** 400
[4] Fu T T, Chen Y, Han L L and Qin Q Z 2012 Preliminary report on the indoor electromagnetic radiation in a municipality of Western P.R. China: Up-to-now still within the range *Journal of Electromagnetic Analysis and Applications* **4** 199
[5] Zheng J 2013 Electromagnetic fields and waves-second edition. Tsinghua University Press. ISBN: 9787302302674
[6] Mostafa R M, Elmoemen E A, Fawzy M S and Hagras A M 2012 Possible impact(s) of cell phone electromagnetic radiation on human sperm parameters *Human Andrology* **2** 49
[7] Przystupa K, Pohrebennyk V, Mitryasova O and Kochan O 2020 Assessment of the electromagnetic radiation level of vehicles, computers, household appliances, personal care products, mobile phones and smartphones *Przeglad Elektrotechniczny* **1** 190
[8] Bastide M, Youbicier-Simoa B J, Lebecq J C and Giaimis J 2001 Toxicologic study of electromagnetic radiation emitted by television and video display screens and cellular

telephones on chickens and mice *Indoor & Built Environment* **10** 291

[9] Wang D, Guo J F, Shi J S and Liu B H 2013 Electromagnetic radiation intensity survey of induction cooker *Environmental Science Survey* **32** 109

[10] Yan X Y, Zhang G B, Yuan R H, Ye X Y, Chen X X and He H H 2022 Size-controlled C/N-doped TiO_2 nanoparticles for visible-light-driven photocatalysis of formaldehyde from plywood *Solid State Phenomena* **330** 169

[11] Rahman S, Matiullah, Mujahid S A and Hussain S 2008 Assessment of radiological hazards due to the presence of natural radionuclides in samples of building materials collected from the northwestern areas of Pakistan *Journal of Radiological Protection* **28** 205

[12] Li C Y 2005 Analysis of radioactivity level of building materials *Chinese Journal of Radiological Health* **14** 281

[13] Brígido Flores O, Montalván Estrada A, Rosa Suárez R, Tomás Zerquera J and Hernández Pérez A 2008 Natural radionuclide content in building materials and gamma dose rate in dwellings in Cuba *Journal of Environmental Radioactivity* **99** 1834

[14] Kuzmanović P, Todorović N, Petrović L F, Mrđa D and Knežević J 2020 Radioactivity of building materials in Serbia and assessment of radiological hazard of gamma radiation and radon exhalation *Journal of Radioanalytical and Nuclear Chemistry* **324** 1077

[15] Zheng Y X, Yang X L and Yu R G 2004 Level of γ radiation in decoration materials used in Mt.Yimeng area and its dose effect on residents *Industrial Health and Occupational Diseases* **30** 166

[16] Yu H F 2010 The reasons for potential safety hazards of household appliances and its preventive measures *Electronics and Packaging* **10** 42

[17] Aerts S, Calderon C, Vali B, Maslanyj M and Joseph W 2017 Measurements of intermediate-frequency electric and magnetic fields in households *Environmental Research* **154** 160

[18] Turhan S 2009 Radiological impacts of the usability of clay and kaolinas raw material in manufacturing of structuralbuilding materials in Turkey *Journal of Radiological Protection* **29** 75

[19] El-Arabi A M 2007 ^{226}Ra, ^{232}Th and ^{40}K concentrations in igneous rocks from eastern desert, Egypt and its radiological implications *Radiation Measurements* **42** 94

[20] Che X H, Yuan R H, Tang L C, Xiong Y D, Wu M F and Zhu M H 2021 Study of catecholate-functionalized zinc complex on carbon steel and its application to corrosion resistance *Journal of Physics: Conference Series* **1978** 012009

[21] Saleh H, Hamideen M, Al-Hwaiti M and Al-Kharoof S 2018 Radiological risk measurements due to natural radioactivity of building stones used in Jordanian houses *Jordan Journal of Physics* **11** 193

[22] Xing H U and Hou X 2011 Detection of specific activity of building materials based on low-background Ge(Li)γ-spectroscopy *Guangdong Trace Elements Science* **18** 57

[23] Wang X, Zhang L, Luo Z and Yang J 2017 Effect of red mud on properties and mineral composition of Portland cement for road *Journal of Building Materials* **20** 774

Journal of Physics: Conference Series **2313** (2022) 012033 doi:10.1088/1742-6596/2313/1/012033

Reliability evaluation of balanced systems with common bus performance sharing considering balance degree base on battery storage systems

TianZi Tian[1, 2], and Jun Yang[1]

[1]School of Reliability and Systems Engineering, Beihang University, China

[2]Corresponding author's e-mail: tiantianzi2000@163.com

Abstract. To solve the problem of inconsistency of batteries, many people have studied the the battery system with the automatic balancing mechanism which can be regarded as a balanced system. In this paper, a reliability evaluation model for balanced systems with common bus performance sharing considering the degree of balance is proposed. The balance degree of the system is represented by the difference in performance between the components. If the performance difference between components exceeds the balance degree threshold, the system is unbalanced and performance sharing is required to redistribute performance through a common bus. This paper also takes into account transfer capacity limits. When the output performance exceeds the maximum transmission capacity, the performance transmission is performed according to the maximum limit to achieve the best balance. The criterion for system failure is that the balance is still out of balance after performance sharing or the total system performance is lower than the specified requirements. Firstly, by solving the Markov equation, the steady-state probability of component performance levels under the long-term operation of the system is obtained, and then the reliability of the system is evaluated by UGF (Universal generating function) method. Finally, a numerical example is carried out to illustrate the effectiveness of the method.

1. Introduction

With the promotion and popularization of electric vehicles, reliability evaluation of battery packs has attracted much attention. For a large-scale battery pack, it is difficult to ensure the consistency of the single cells in the battery pack. When the unbalanced phenomenon occurs for a long time, some single cells with relatively high or low energy will be damaged due to long-term overcharge or over-discharge. It not only reduces the maximum energy value of the battery but even causes safety problems in severe cases [1]. Therefore, in order to use batteries safely and effectively, many people have researched the auto-balancing circuit matched with the batteries, so that the performance of the batteries tends to be consistent [2]. The studied auto-balancing mechanism for battery systems is achieved through performance sharing. So, it is valuable to evaluate the reliability of the battery balanced system with common bus performance sharing.

In recent years, performance sharing systems have been extensively studied. In a system with common bus performance sharing, each unit has random performance and should satisfy its individual random requirements. This means that the component has multiple states, and the state space in a system composed of multiple batteries will be large. The Universal Generation Function (UGF) method can be used as an effective tool to solve the above problems. Yu et al. [3] proposed a method for evaluating the

Journal of Physics: Conference Series **2313** (2022) 012033 doi:10.1088/1742-6596/2313/1/012033

reliability of multi-stage bus performance sharing systems based on UGF. Wang et al. [4] build a reliability model for evaluating tandem multi-state systems with performance sharing between adjacent cells based on the UGF.

The research on the balanced system started in 2016, and the relevant content still needs to be improved. The current research on balanced systems is mainly divided into two categories. The first category can be explained as the forced shutdown or restoration of the unit to rebalance the system when the unit failure causes the system balance to be disrupted, such as the port system [5], the drum input belt system [6], and multi-stage unmanned aerial system [7] and so on. Wang et al. [7] used a Markov process imbedding approach to analyze the reliability of balanced systems with restricted rebalanced mechanisms. This paper belongs to the second category, which is to introduce a new balance concept based on engineering applications, and believes that once the balance requirements are not met, the system will fail. Zhao et al. [8] proposed a balance function which the multiple states of the system are determined according to different system balance degrees based on the operating states of all components. Wang et al. [9] also considered the effect of balance degree threshold in balanced system. Wu et al. [10] studied a balanced system with performance sharing but only considered that the performance of all components remains the same at all times. The content of reliability assessment of balanced system with performance sharing considering the balance threshold has not been studied so far. In addition, the transmission limit is also an important characteristic of the performance sharing system, which is also considered in this paper.

The remainder of this paper is organized as follows. Section 2 of this paper presents the related concepts of the battery balanced systems and describes the studied system through a mathematical model. Besides, a method combining Markov stochastic process and UGF technology is proposed to solve the system reliability. Section 3 gives a numerical example to illustrate the effectiveness of the proposed method. Finally, the conclusion and future outlook are given in Section 5.

2. System model

2.1. Model descriptions

The battery performance sharing system consists of *n* batteries and a common bus connecting all batteries. The system structure is shown in Figure 1. The essence of the transmission performance of batteries is to perform switching operations through the controller system. All individual cell equalizers (ICEs) are connected. Ignoring the balancing time caused by the control switch and transmission process, the system can be regarded as a balancing system with the common bus performance sharing [10]. The balancing process can be described as: turn on the switch SCS1 of the output performance, transfer the performance that needs to be transmitted to the inductance equalizer L, and then turn off the switch SCS1. Then transfer the performance to the corresponding components by turning on the switch SCS2. However, inductive equalization cannot withstand large equalization energy, so it can be considered that the transmission capacity of the common bus is limited [2]. Since automatic balancing consumes energy, when the cells' performance difference is within a tolerant range [11], we can also call the system balanced that doesn't need balancing operation.

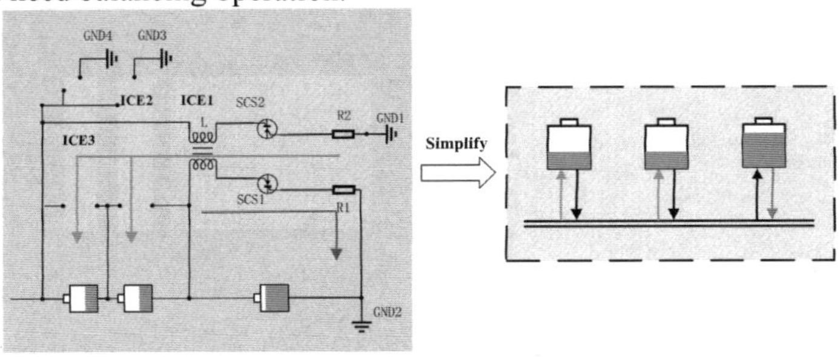

Figure 1. System structure of battery balanced system with performance sharing.

Journal of Physics: Conference Series **2313** (2022) 012033 doi:10.1088/1742-6596/2313/1/012033

For such a system, the concepts of system balance, failure modes, and balancing operation are as follows.

System Balance: A system is balanced if the difference between the performance of each component does not exceed the balance degree threshold (ε).

Failure mode: If the difference between the performance of each component still exceeds the balance degree threshold after automatic balancing, or the performance of the overall system is below the system performance requirements (σ), the system fails.

Balance operation: When cell equalizers detect the unbalanced phenomenon, the system performs balanced operation that perfectly balances the performance of the components within the system, i.e., the performance of each component is completely consistent (*perfect balance*); If the transmission performance exceeds the capacity limit of the common bus during the rebalanced process, the maximum performance is transferred according to the maximum capacity. At this point, the system performs a rebalance operation that is as balanced as possible (*imperfect balance*).

Subsequently, according to the above concepts, we can obtain the system model as follow. The random performance of the component *i* over time *t* is denoted by $g_i(t)$, and the total performance of all components can be derived,

$$S(t) = \sum_{i=1}^{n} g_i(t) \tag{1}$$

We define $\boldsymbol{G}$ as the set of states of all components at time *t*.

$$\boldsymbol{G} = \{g_1(t), g_2(t), \ldots, g_n(t)\} \tag{2}$$

The balance degree of the system can be calculated as,

$$\delta(t) = \max \boldsymbol{G} - \min \boldsymbol{G} \tag{3}$$

when $\delta(t) > \varepsilon$, that is, if the performance difference between components exceeds the balance degree threshold, the system needs to perform balanced operation.

$$\Delta_i(t) = g_i(t) - S(t)/n \tag{4}$$

Considering the performance sharing through the common bus, we give the performance that the component needs to pass out or pass in Equation (4).

C indicates the maximum capacity of the common bus. When $\Delta_i(t) \le C$, if the transmission performance does not exceed the capacity limit, then the perfect balance can be performed; if $S(t) \ge \sigma$ at this time, the system is reliable. When $\Delta_i(t) > C$, imperfect balance operation is required. After performing the above operations, we recalculate the balance degree of the system. If $\delta(t) \le \varepsilon$, the system is reliable. Assuming that the rebalancing operation does not take time, which means that if the common bus has no capacity constraints, the system is always in perfect balance.

Therefore, the system reliability can be derived,

$$R(t) = \Pr\{S(t) \ge \sigma, \delta(t) \le \varepsilon\} \tag{5}$$

In order to better understand the system model, we give the flow chart to describe the process of balance operation in Figure 2.

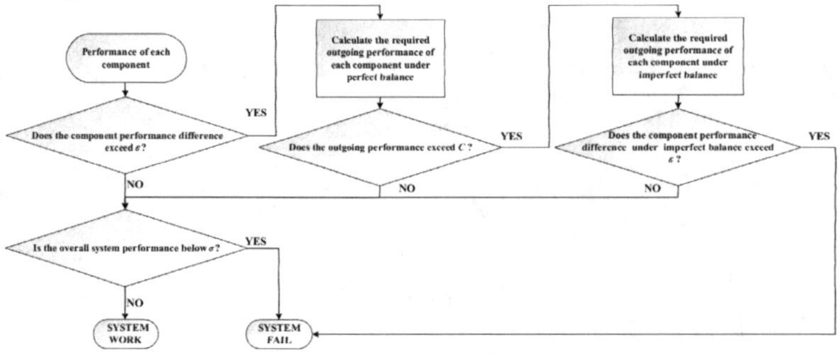

Figure 2. The process of balance operation.

2.2. *Markov model for multiple performance levels components*

In this subsection, we build a Markov model to address the problem of describing changes in component performance levels. We assume that the performance level of each component follows a continuous-time discrete-state Markov stochastic process. We only research the stationary performance when the system runs for a long time, so the stationary distribution of the involved Markov random processes was focused. Taking the component i as an example, the component i has L_i performance levels. $\lambda_{l_i,m}$ and u_{m,l_i} $(m > l_i)$ is the transition intensity between level j and level l_i. Its Markov model is shown in Figure 3.

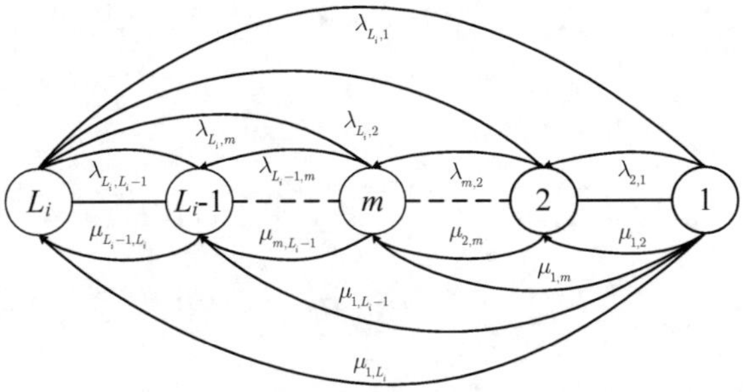

Figure 3. Multiple performance levels components Markov model.

The performance of the i-th component at any instant t obeys a Markov process, which obtains values from the set $\boldsymbol{x}_i = \{x_{i,1}, x_{i,2}, ..., x_{i,L_i}\}$, and the corresponding probability set is denoted as $p_{i,l_i}(t) = \{p_{i,1}(t), p_{i,2}(t), ..., p_{i,L_i}(t)\}$. Its Markov state transition process can be expressed by the following differential equation.

$$\begin{cases} \dfrac{dp_{i,1}(t)}{dt} = \sum_{l_i=2}^{L_i} p_{i,l_i}(t)\lambda_{j,1} - p_{i,1}(t)\sum_{l_i=2}^{L_i} \mu_{1,l_i} \\ \dfrac{dp_{i,m}(t)}{dt} = \sum_{l_i=1}^{m-1} p_{i,l_i}(t)\mu_{l_i,m} + \sum_{l_i=m+1}^{L_i} p_{i,l_i}(t)\lambda_{l_i,m} - p_{i,m}(t)\left(\sum_{l_i=1}^{m-1}\lambda_{m,l_i} + \sum_{j=m+1}^{L_i}\mu_{m,l_i}\right) \\ \dfrac{dp_{i,L_i}(t)}{dt} = \sum_{l_i=1}^{L_i-1} p_{i,l_i}(t)\mu_{l_i,L_i} - p_{i,L_i}(t)\sum_{l_i=1}^{L_i-1}\lambda_{L_i,l_i} \end{cases} \tag{6}$$

According to the equation for calculating the stationary distribution, let the formula on the left equal 0. Combining with the equation $\sum_{l_i=1}^{L_i} p_{i,l_i}(t) = 1$, we can obtain the set of steady-state probabilities of component performance levels under the stable operating state of the system, $p_{i,l_i} = \{p_{i,1}, p_{i,2}, ..., p_{i,L_i}\}$.

2.3. *The calculation of the system model based on UGF*

This part uses the UGF method to solve the system reliability of the battery performance sharing system considering the balance degree. The UGF expression of the performance level of component i is,

$$u_i(z,t) = \sum_{l_i=1}^{L_i} p_{i,l_i}(t) z^{x_{i,l_i}} \tag{7}$$

where $p_{i,l_i}(t)(l_i = 1,2,...,L_i)$ represents the probability that component i is in level l_i . x_{i,l_i} represents the performance value of component i at level l_i .

$$U_n(z,t) = \sum_{s_n=1}^{S_n} \alpha_{n,s_n}(t) z^{x_{n,s_n}, l_{n,s_n}} \tag{8}$$

Equation (8) is the overall performance of all components. S_n represents the number of all possible realizations of the vector, which contains the performance levels of n components in the system. α_{n,s_n} is the corresponding probability. $\boldsymbol{x}_{n,s_n} = (x_1, x_2, ..., x_n)_{s_n}$, which is one of S_n realizations, and a vector that contains the performance levels of n components in the system. t_{n,s_n} is one of S_n realizations of performance for all components. The recursive algorithm is used to calculate the overall performance of the system as follows:

1 Assign $U_0(z,t) = z^{[\,],0}$

2 Main loop. For i=1, 2, …., n, repeat: $U_i(z,t) = U_{i-1}(z,t) \underset{+}{\otimes} u_i(z,t)$. The detailed expansion process is given in Equation (9).

$$\begin{aligned} U_i(z,t) &= U_{i-1}(z,t) \underset{\mid}{\otimes} u_i(z,t) \\ &= \sum_{s_{i-1}=1}^{S_{i-1}} \alpha_{i-1,s_{i-1}}(t) z^{\boldsymbol{x}_{i-1,s_{i-1}}, t_{i-1,s_{i-1}}} \underset{+}{\otimes} \sum_{l_i=1}^{L_i} p_{i,l_i}(t) z^{x_{i,l_i}} \\ &= \sum_{s_{i-1}=1}^{S_{i-1}} \sum_{l_i=1}^{L_i} \alpha_{i-1,s_{i-1}}(t) p_{i,l_i}(t) z^{[\boldsymbol{x}_{i-1,s_{i-1}}, x_{i,l_i}], t_{i-1,s_{i-1}} + x_{i,l_i}} \\ &= \sum_{s_i=1}^{S_i} \alpha_{i,s_i}(t) z^{\boldsymbol{x}_{i,s_i}, t_{i,s_i}} \end{aligned} \tag{9}$$

3 Obtain the overall performance of all components $U_n(z,t)$.

where $\underset{+}{\otimes}$ is the operator for recursively computing the first i element and the i-th component. When the new component i enters the loop, the new vector formed, that is, $[\boldsymbol{x}_{i-1,s_{i-1}}, x_{i,l_i}]$. To simplify the expression, a realization of vectors consisting of the performance level of each component belonging to the first i components is defined as $\boldsymbol{x}_{i,s_i}$. Subsequently, S_i is used to represent the number of all possible realizations of $\boldsymbol{x}_{i,s_i}$. The value of t_{i,s_i} is a realization of the total performance of $\boldsymbol{x}_{i,s_i}$. α_{i,s_i} is the corresponding probability.

To judge whether the different realizations of the vector composed of the performance levels of all components exceed the threshold of balance degree ε, function F_1 is proposed.

$$F_1\left(\boldsymbol{x}_{n,s_n}\right) = 1(\max\{\boldsymbol{x}_{n,s_n}\} - \min\{\boldsymbol{x}_{n,s_n}\} < \varepsilon) \tag{10}$$

Equation (10) represents that if $\max\{\boldsymbol{x}_{n,s_n}\} - \min\{\boldsymbol{x}_{n,s_n}\} < \varepsilon$, then $F_1 = 1$, otherwise, $F_1 = 0$ .and its function is to compare the maximum and minimum performance levels of all components to determine whether the system is out of balance. $F_1 = 1$ indicates that the system is balanced, otherwise, it is out of balance. The operator θ_{F_1} is applied to $U_n(z,t)$. θ_{F_1} is a function operation on the exponential term in the UGF expression. All of the terms with the exponents producing the zero acceptability function ($F_1 = 0$) from $U_n(z,t)$ constitute $U_n^0(z,t)$, and collecting terms with $F_1 = 1$ in $U_n(z,t)$ constitute $U_n^1(z,t)$. For the situation where the system is still balanced, it is necessary to judge whether the overall performance of the system is lower than the minimum performance requirements σ. Here we propose F_2.

$$F_2(t_{n,s_n}) = 1(t_{n,s_n} > \sigma) \tag{11}$$

$F_2 = 1$ indicates that the system performance is not lower than σ, and $F_2 = 0$ means lower than. By applying the operator θ_{F_2} over $U_n^1(z,t)$ can obtain the probability that the group consists of meeting the system performance requestions. $\theta_{F_2}(U_n^1(z,t))$ is the coefficients of terms meeting $F_2 = 1$. Add value $\theta_{F_2}(U_n^1(z,t))$ to y. y is a value that is initially 0.

In order to calculate whether the output performance of the components would exceed C if perfect balance was performed, and obtain the performance of the components after rebalancing operation, one has to apply to newly obtained u-function $U_n^0(z,t)$ the following operation, which:

- performs F_3 on the elements of the vector $\boldsymbol{x}_{n,s_n} = (x_1, x_2, ..., x_n)_{s_n}$ in $U_n^0(z,t)$;
- obtains a new vector $\boldsymbol{g}_{n,s_n} = (g_1, g_2, ..., g_n)_{s_n}$ by replacing all of the elements of $\boldsymbol{x}_{n,s_n}$ with the value of F_3, then performs F_4 on vector $\boldsymbol{g}_{n,s_n}$;
- collects terms with $F_4 = 0$ from $U_n^0(z,t)$ constitute $V_n^0(z,t)$, and terms with $F_4 = 1$ from $U_n^0(z,t)$ constitute $V_n^1(z,t)$;
- rewrites the $V_n^0(z,t)$ after imperfect balance operation, and updates the performance of the components. Functions F_3 and F_4 are given as follows.

$$F_3(x_i) = 1(x_i - t_{n,a_n} / n \le C) \tag{12}$$

$$F_4(\boldsymbol{g}_{n,s_n}) = \prod_{i=1}^{n} g_i \tag{13}$$

F_3 is used to determine whether the output performance of component i exceeds C. As for F_4, it is used to determine whether there is a situation where the output performance of components exceeds C. For example, $\boldsymbol{g}_{n,s_n} = (0,1,1...,1)_{s_n}$ indicates that the output performance of component 1 exceeds C, and $\prod_{i=1}^{n} g_i = 0$ at this point. Then the operator θ_{F_1} is applied to $V_n^0(z,t)$. Removing all of the terms with $F_1 = 0$ from $V_n^0(z,t)$ constitute $W_n^1(z,t)$. Next, the operator θ_{F_2} is applied to $V_n^0(z,t)$ and $W_n^1(z,t)$, and value $\theta_{F_1}(V_n^1(z,t))$ and $\theta_{F_1}(W_n^1(z,t))$ are added to y. Ultimately, the reliability of the system is $R(t) = y$.

3. Illustrative examples

Table 1. Performance transitions intensity, performance levels set, and steady-state probabilities.

Component	$\{\lambda_{12}, \lambda_{13}, \lambda_{23}\}$	$\{\mu_{32}, \mu_{31}, \mu_{21}\}$	Performance level	Steady-state probabilities
Battery 1	{0.001, 0.001, 0.001}	{0.001, 0.001, 0.001}	{1, 3, 8}	$\{\frac{1}{3}, \frac{1}{3}, \frac{1}{3}\}$
Battery 2	{0.001, 0.002, 0.002}	{0.001, 0.001, 0.001}	{1, 3, 8}	$\{\frac{1}{4}, \frac{1}{4}, \frac{1}{2}\}$
Battery 3	{0.001, 0.001, 0.001}	{0.001, 0.002, 0.001}	{1, 3, 8}	$\{\frac{5}{12}, \frac{1}{3}, \frac{1}{4}\}$

Table 2. System reliability under varying parameters.

$\varepsilon\ (\sigma = 6\ \ C = 3)$	R	$\sigma\ (\varepsilon = 3\ \ C = 3)$	R	$C\ (\varepsilon = 3\ \ \sigma = 6)$	R
0.1	0.40972	5	0.96528	0.6	0.15972
1	0.51389	6	0.86806	1	0.27778
2	0.74306	7	0.86806	2	0.74306
3	0.86806	8	0.77778	3	0.86806
4	0.86806	15	0.29168	4	0.86806

In this case, we consider a performance-sharing balanced system with three batteries (n = 3). In this system, the change in the performance level of each component follows a Markov process. The performance transitions intensity of batteries is given in Table 1. Table 1 shows the set of performance levels and the set of steady-state probabilities for each battery. We set the original parameters as $\varepsilon = 3$, $\sigma = 6$, $C = 3$, and change the values of the balance degree thresholds ε, the minimum system performance requirements σ, and the maximum transmission capacity C, respectively, to give the corresponding system reliability. The results are shown in Table 2.

Considering the balance degree threshold, the system reliability increases as the balance threshold increases. When ε is greater than a certain value, the system reliability is not affected by ε. The reason why the system reliability cannot be improved at this time is that the system performance is insufficient. Similarly, system reliability decreases as σ increases and increases as C increases. When C is gradually increased, more situations will be able to achieve the balance allowed by the system through performance sharing. When C increases to a certain value, the system reliability will not change. Even with unlimited performance sharing, system reliability will not improve when the overall performance is insufficient.

4. Conclusion

In this work, an evaluation method for the reliability of a performance sharing balanced system considering the balance degree threshold is studied, and a battery balanced system is given as a practical engineering application background to illustrate the effectiveness of the proposed method. By taking different values of balance degree thresholds, the minimum system performance required, and the maximum transmission capacity, the changes in the system reliability are given. Then we can obtain the influence of different factors on the reliability of the lithium battery balanced system. In the future, combined with the cost of the corresponding parameters, the proposed method can be studied to optimize the appropriate parameters to minimize the economic cost of the system and maintain its stability of the system's reliability.

Acknowledgment

This work was supported by the scientific research project on Technical Basis of China with No. JSZL2017601B006.

References

[1] Liu X T, Wan Z H, He Y, Zheng X X, Zeng G J and Zhang J F 2018 A Unified Control Strategy for Inductor-Based Active Battery Equalisation Schemes *Energies.* **11** 405.

[2] Zhang Z L, Gui H D, Gu D J, Yang Y and Ren X Y 2017 A Hierarchical Active Balancing Architecture for Lithium-Ion Batteries *IEEE T Power Electr.* **32** 2757-2768.

[3] Yu H, Yang J, Lin J and Zhao Y 2017 Reliability evaluation of non-repairable phased-mission common bus systems with common cause failures *Computers & Industrial Engineering.* **111** 445-457 445.

[4] Wang G J, Duan F J and Zhou Y F 2018 Reliability evaluation of multi-state series systems with performance sharing *Reliab eng syst safe.* **173** 58-63 58.

[5] Fang C and Cui L R 2020 Reliability analysis for balanced engine systems with m sectors by considering start-up probability *Reliab eng syst safe.* **197** 106829.

[6] Fang C and Cui L 2021 Reliability evaluation of consecutive k-out-of-m: F balanced systems with a symmetry line Proc. Inst. *Mech. Eng., Part O: J. Risk Reliab.* (2021)12.

[7] Wang X Y, Zhao X, Wu C S and Lin C 2020 Reliability assessment for balanced systems with restricted rebalanced mechanisms *Comput. Ind Eng.* **149** 106801.

[8] Zhao X, Wang S Q, Wang X Y and Fan Y 2020 Multi-state balanced systems in a shock environment *Reliab eng syst safe.* **193** 106592.

[9] Wu C S, Zhao X, Wang X Y and Wang S Q 2021 Reliability analysis of performance-based balanced systems with common bus performance sharing *Reliab eng syst safe.* **215** 107865.

[10] Wang S, Zhao X, Tian Z and Zuo M J 2021 Optimal mission abort policy with multiple abort criteria for a balanced system with multi-state components *Comput. Ind Eng.* **160** 107544.

[11] Dong G Z, Yang F F, Tsui K L and Zou C F 2021 Active Balancing of Lithium-Ion Batteries Using Graph Theory and A-Star Search Algorithm *IEEE Trans. Ind. Inf.* **17** 2587-2599.

AUTHOR INDEX